Debashish Bhattacharya (ed.)

Origins of Algae and their Plastids

SpringerWienNewYork

Dr Debashish Bhattacharya
University of Iowa, Department of Biological Sciences, Iowa City, United States

Printed in Austria

Typesetting: Thomson Press (India) Ltd., New Delhi

Printing: Adolf Holzhausens Nfg. GesmbH, A-1070 Wien

Graphic design: Ecke Bonk

Printed on acid-free and chlorine-free bleached paper

SPIN: 10636447

With 55 Figures

CIP data applied for

ISBN 3-211-83036-7 (hard cover) Springer-Verlag Wien New York
ISBN 3-211-83035-9 Plant Systematics and Evolution [Suppl 11] (soft cover)
Springer-Verlag Wien New York

Preface

The algae are photosynthetic eukaryotes that include the dominant primary producers on our planet. Research in the past ten years using molecular evolutionary methods has led to many important insights that have fundamentally changed our view of algal origin and phylogeny. This book reports on this dynamic period of research by dealing in separate chapters with the significant advances in each of the major algal lineages. It is my hope that these data will convince readers not familiar with algal evolution that the origin of this group constitutes one of the most interesting aspects of biology. It is also my hope that the phylogenies shown in the following chapters will not only be further tested, added to, and refined by the respective authors and others interested in those groups but that the trees will be used to formulate deeper questions about cellular and genetic evolution. The most important role of molecular phylogenetics is not to underpin certain hypotheses that arise from tree topologies but to present trees as experimental tools for a better understanding of the evolution of organisms.

In organizing this special supplement, I have asked, as much as possible, young researchers to provide their data and perspectives on algal phylogeny and evolution. I am grateful to those authors and to their mentors for making the book possible. In addition I thank Prof. EHRENDORFER who gave me the chance to organize this volume as well as SILVIA SCHILGERIUS at Springer-Wien for her help and especially Dr IRMGARD KRISAI-GREILHUBER (University of Wien) who acted as the second reviewer of the manuscripts.

September 29, 1997 *Debashish Bhattacharya*

Contents

Listed in Current Contents

Contents

List of Contributors

ROBERT A. ANDERSEN, Bigelow Laboratory for Ocean Sciences, West Boothbay Harbor, Maine 04575, USA

DEBASHISH BHATTACHARYA, University of Iowa, Department of Biological Science, Biology Building, Iowa City 52242-1324, Iowa, USA (e-mail: dbhattac@blue.weeg.uiowa.edu)

H. J. BOHNERT, Department of Biochemistry, University of Arizona, Tucson, AZ 85721, USA

D. A. BRYANT, Department of Biochemistry and Molecular Biology, The Pennsylvania State University, University Park, PA 16802, USA

CHARLES F. DELWICHE, Department of Plant Biology, University of Maryland, College Park, MD 20742, USA

L. D. DRUEHL, Bamfield Marine Station, Bamfield, B. C. Canada V0H 1B0

MARTIN FRAUNHOFER, Universität Marburg, Zellbiologie, Karl-von-Frisch-Strasse, D-35032 Marburg, Federal Republic of Germany

THOMAS FRIEDL, Fachbereich Biologie, Allgemeine Botanik, Universität Kaiserslautern, Postfach 3049, D-67653 Kaiserslautern, Federal Republic of Germany, (fax: +49 631 2052998; e-mail: friedl@rhrk.uni-kl.de)

PAUL R. GILSON, Plant Cell Biology Research Centre, School of Botany, University of Melbourne, Parkville 3052, Australia

D. R. A. HILL, School of Botany, University of Melbourne, Parkville, Victoria 3052, Australia

CLAUDIA J. B. HOFMANN, Lehrstuhl für Pflanzensystematik, Universität Bayreuth, D-95440 Bayreuth, Federal Republic of Germany

VOLKER A. R. HUSS, Institut für Botanik und Pharmazeutische Biologie der Universität, Staudtstrasse 5, D-91058 Erlangen, Federal Republic of Germany

WIEBE H. C. F. KOOISTRA, NAOS Smithsonian Tropical Research Institute, APO AA 34002 0948, USA

G. T. KRAFT, School of Botany, University of Melbourne, Parkville, Victoria 3052, Australia

HARALD D. KRANZ, Max-Planck-Institut für Züchtungsforschung, Abteilung Biochemie, Carl-von-Linne-Weg 10, D-50829 Köln, Federal Republic of Germany

NAOMI LANG-UNNASCH, Division of Geographic Medicine, Department of Medicine, University of Alabama at Birmingham, Birmingham, Alabama 35294–2170, USA

W. Löffelhardt, Institut für Biochemie und Molekulare Zellbiologie der Universität Wien und Ludwig-Boltzmann-Forschungsstelle für Biochemie. Dr. Bohrgasse, 9, A-1030 Vienna, Austria

Uwe-G. Maier, Universität Marburg, Zellbiologie, Karl-von-Frisch-Strasse, D-35032 Marburg, Federal Republic of Germany

C. Mayes, Biological Sciences, Simon Fraser University, Burnaby, B. C., Canada V5A 1S6

Geoffrey I. McFadden, Plant Cell Biology Research Centre, School of Botany, University of Melbourne, Parkville 3052, Australia

Linda K. Medlin, Alfred-Wegener-Institute, Am Handelshafen 12, D-27570 Bremerhaven, Germany

Jeffrey D. Palmer, Department of Biology, Indiana University, Bloomington, IN 47405, USA

Daniel Potter, Department of Pomology, University of California, Davis 95616, USA

Michael E. Reith, Institute for Marine Biosciences, National Research Council, 1411 Oxford St, Halifax B3H 3ZI, NS, Canada

Stefan A. Rensing, Universität Freiburg, Lehrstuhl für Zellbiologie, Schänzlestrasse I, D-79104 Freiburg, Federal Republic of Germany

G. W. Saunders, Department of Biology, University of New Brunswick, Fredericton, N. B., Canada, E3B 6E1

Margitta M. Scherzinger, Universität Freiburg, Lehrstuhl für Zellbiologie, Schänzlestrasse I, D-79104 Freiburg, Federal Republic of Germany

Heiko A. Schmidt, Deutsches Krebsforschungszentrum, Theoretische Bioinformatik (815), Im Neuenheimer Feld 280, D-69120 Heidelberg, Federal Republic of Germany

J. P. Sexton, Bigelow Laboratory for Ocean Sciences, West Boothbay Harbor, ME 04575, USA

I. H. Tan, Biological Sciences, Simon Fraser University, Burnaby, B. C., Canada V5A 1S6

Seán Turner, Department of Biology, Jordan Hall 142, Indiana University, Bloomington, Indiana 47405, USA (e-mail: sturner@bio.indiana.edu)

Ross F. Waller, Plant Cell Biology Research Centre, School of Botany, University of Melbourne, VIC 3052, Australia

Jürgen Wastl, Universität Marburg, Zellbiologie, Karl-von-Frisch-Strasse, D-35032 Marburg, Federal Republic of Germany

Stefan Zauner, Universität Marburg, Zellbiologie, Karl-von-Frisch-Strasse, D-35032 Marburg, Federal Republic of Germany

An introduction to algal phylogeny and phylogenetic methods

Debashish Bhattacharya

Key words: Algae. – Endosymbiosis, evolution, phylogeny, plastids, systematics, taxonomy.

Abstract: The algae are an assemblage of morphologically diverse, photosynthetic protists that are ubiquitous in distribution. The understanding of algal phylogeny has been revolutionized with molecular evolutionary methods. These analyses have shown that the algae are of a polyphyletic origin and diverge nearly simultaneously from each other in the crown group radiation (except euglenophytes). An interesting perspective on algal origin is gained by the analysis of plastid origin because many plastids (e.g., in cryptophytes, dinoflagellates, chlorarachniophytes, and euglenophytes) have arisen multiple, independent times through secondary endosymbioses. This chapter is an introduction to algal phylogeny and to the most common methods used in the construction of evolutionary trees. The subsequent 13 chapters in this book deal with specific aspects of the phylogeny of cyanobacteria (the primary source of the algal plastid), plastids and all the major algal lineages.

The striking morphology of many micro- and macroalgae (e.g., desmids, kelps) has often been sufficient reason for many scientists to engage in phycological research. Algae have played central roles in ecological studies of aquatic and terrestrial ecosystems, been used as "model" protists in physiological and biochemical studies and in reconstructing and predicting climate change, and have posed many fundamental questions in biology due to their diverse and complex life histories (Bold & Wynne 1985). Algae have also had a long history in the food (e.g., nori, wakame) and pharmaceutical (e.g., agar-agar, carrageenan, alginic acid) industries. Studies of algal systematics have led to the description of thousands of taxa (5000–5500 species in the *Rhodophyta* alone, Van den Hoek & al. 1993) to aid in their taxonomic classification and further the understanding of their origins and evolutionary relationships.

Many of these studies, using "traditional" markers of species relationships, such as gross morphology or vegetative and reproductive cell ultrastructure (Kies 1974, Stewart & Mattox 1975, Taylor 1976, Moestrup 1978, Kraft 1981, Cavalier-Smith 1982, O'Kelly 1992), have allowed a deeper understanding of the uniqueness and complexity of algal cells and lineages and the phylogeny of members within these lineages. Traditional phylogenetics has not, however, proven to be very powerful in uncovering the relationships of algae with other eukaryotes (glaucocystophytes, see chapter 7 of this volume; chlorarachniophytes, chapter 10; dinoflagellates, chapter 13). One problem arises from the difficulties in assigning phylogenetic value to certain key characters (e.g., ultrastructure of flagellar hairs) in distantly related groups that either vary substantially in morphology and are of questionable homology (found in the common ancestor of a group) or seem to be homologous but may have arisen from convergence (independently evolved in different groups). Another potential problem comes from the use of plastid characters to define host cell lineages and to infer relationships among them (see below). A great advantage of modern phylogenetic methods using protein or nucleotide sequence comparisons is that they offer the possibility to arrange species in natural groups independent of their morphology. Such analyses have revolutionized the field of algal evolution by providing insights into the phylogenetic position of algae in the larger eukaryotic phylogeny and insights into the origin of the key morphological characters that define these clades (CAVALIER-SMITH 1993, FRESHWATER & al. 1994, SCHLEGEL 1994, BHATTACHARYA & MEDLIN 1995, PALMER & DELWICHE 1996).

Goals of this book

The main goal of this book is to provide the current results and hypotheses regarding the origins of the algae and their plastids (photosynthetic organelles) using gene/protein sequence comparisons. The sequence analyses should be interpreted in terms of phylogeny ("The evolutionary history of a group of taxa or genes and their ancestors", LI & GRAUR 1991: 244) and not necessarily taxonomy ("The principles and procedures according to which species are named and assigned to taxonomic groups", LI & GRAUR 1991: 250). Though molecular sequence data may be interpreted to create new or confirm existing taxonomic schemes there is no a priori reason why this has to be done. The connection between phylogeny and taxonomy remains a highly controversial area in biology and does not form the main focus of this book.

The wealth of phylogenetic data regarding the evolutionary relationships of algal groups that have been published in the past 5–6 years show that the algae are of polyphyletic origins (reviewed in BHATTACHARYA & MEDLIN 1995). The algae comprise multiple, often distantly related (e.g., euglenophytes, heterokonts, dinoflagellates) lineages that are synthetically grouped on the basis of photosynthetic capacity. Many algal lineages, for example the euglenophytes and chlorarachniophytes, share a closer phylogenetic relationship with non-algal taxa (kinetoplastids and filose amoebae, respectively) than with other photosynthetic eukaryotes.

The understanding of plastid endosymbiosis (the accepted theory for the origin of the plastid; MERESCHKOWSKI 1905, 1910; MARGULIS 1981) is critical for the understanding of algal origin (s) since plastids have been laterally transferred multiple times in algal evolution. Resolution of plastid origins has been greatly advanced by the recent availability of plastid sequence data (or complete genome sequences, REITH & MUNHOLLAND 1995, KOWALLIK & al. 1995; HALLICK & al. 1993, LÖFFELHARDT & al. 1996) from all the major algal groups for comparative analyses. These data are, in general, consistent with the hypothesis that the origin of all algal plastids can be traced to a single **primary** endosymbiosis (CAVALIER-SMITH 1982, BHATTACHARYA & MEDLIN 1995, PALMER & DELWICHE 1996) involving a coccoid cyanobacterium and a nonphotosynthetic free-living unicellular eukaryote. This "proto-alga" had two plastid bounding membranes. The origins of all algae cannot, however, be directly traced to a unique common ancestor. Many algae that contain plastids with more than two bounding-membranes have derived their photosynthetic capacity from the engulfment of an existing alga (e.g., heterokonts, chlorarachniophytes, euglenophytes-**secondary** endosymbiosis; GIBBS 1993; see also chapters 9, 10, 11). In some cases there is evidence that a plastid of a secondary endosymbiotic origin replaced an existing plastid of a primary endosymbiotic origin (see chapter 9). Algal and plastid origins may therefore be dealt with as interdependent topics because plastid lateral transfer via secondary endosymbiosis has in many cases transformed a previously nonphotosynthetic protist into an "alga".

For these reasons resolution of the processes of primary and secondary endosymbioses that have resulted in the origin of algal plastids continues to be one of the most interesting (and disputed) problems in biology (see DOUGLAS & TURNER 1991, MARTIN & al. 1992, MORDEN & al. 1992, LOISEAUX-DE-GÖER 1994, BHATTACHARYA & MEDLIN 1995) and forms a major focus of this book (e.g., chapters 2, 3, 7, 8). The most common "for" and "against" theories regarding plastid origins involve those who support fewer endosymbiotic events (both primary and secondary) as explanation for the origin of all plastids (e.g., CAVALIER-SMITH 1982) and those who favor a greater number of events (e.g., RAVEN 1970, MARGULIS 1981, WHATLEY 1993). Support for either theory is often tied with assumptions regarding secondary plastid loss (or plastid replacement), a phenomenon that is not yet well understood in algal evolution.

Molecular markers

The small subunit coding region of ribosomal DNA (SSU rDNA) is the most common molecular marker currently used in phylogenetic reconstruction. The eukaryotic nuclear encoded rDNA genes are encoded in an operon (SSU-5.8S-LSU) that normally exists as multicopy tandem repeats. Strong concerted evolution resulting from unequal crossing-over and/or gene conversion (see DOVER 1982) results in the homogenization of all gene copies so that the entire gene family is normally represented by a unique copy of each coding region (for exceptions, see MCCUTCHAN & al. 1988, SCHOLIN & al. 1993). The small and large subunit coding region of the rDNAs exist within all living organisms (excluding phages/viruses) and these molecules share enough sequence conservation that a satisfactory alignment for phylogenetic analyses can be created that includes eukaryotes, eubacteria and archaebacteria. There are, however, also less well conserved and highly variable regions in rDNA that can be used to resolve the phylogeny of more closely related taxa.

That rRNA secondary structure is also highly conserved facilitates greatly the alignment of homologous regions from distantly related taxa (SOGIN & al. 1989, MAIDAK & al. 1994, VAN DE PEER & al. 1996a). Further characteristics and advantages offered by SSU rDNA as a molecular marker include a relatively large size (approx. 1800 base pairs in eukaryotes), and ease of isolation using PCR methods because these coding regions exist in multiple copies and contain highly conserved sequence islands at the 5′ and 3′ termini that can be used as targets for amplification (MEDLIN & al. 1988). Finally, evidence for the utility of SSU rDNA sequences as a evolutionary marker comes ultimately from the general consistency that is found between phylogenetic relationships resolved in rDNA trees and hypotheses based on morphological and life history traits. For these reasons there now exist large databases of eukaryotic and bacterial SSU rDNAs (e.g., MAIDAK & al. 1994, VAN DE PEER & al. 1996a) for sequence comparisons that allow a realistic attempt at a global phylogeny of living organisms.

Other nuclear-encoded molecular markers that offer an alternative to rDNA include coding regions of actin (BHATTACHARYA & EHLTING 1995), elongation factors Tu/1α and G/2 (IWABE & al. 1989), heat-shock proteins (GUPTA & GOLDING 1993), tubulins (KEELING & DOOLITTLE 1996) and glyceraldehyde-3-phosphate-dehydrogenase (GAPDH, LIAUD & al. 1994) among a myriad of other gene/protein sequences. DOOLITTLE & al. (1996), for example, compared the sequences from 57 different enzymes to study the divergence times of the major kingdoms of living things. Protein sequences all have associated advantages and disadvantages with regard to phylogeny reconstruction that often are due to paralogous evolution (i.e., multi-copy gene families within organisms but all members are not identical) and the uncertain timing of the origin/loss of gene family members. Importantly, these markers provide perspectives on gene/species evolution that is independent of rDNA. The phylogeny of plastids has been extensively studied using plastid-encoded SSU rDNA and the coding regions and/or protein sequences of components of the photosynthetic apparatus (see chapter 3; DOUGLAS & TURNER 1991, MARTIN & al. 1992, MORDEN & al. 1992). The sequence of the large subunit of ribulose-1,5-bisphosphate carboxylase/oxygenase (rubisco), for example, has been most extensively applied to the study of land plant phylogeny (WOLF & al. 1994).

Phylogenetic methods

The most commonly used molecular phylogenetic methods are distance, maximum parsimony and maximum likelihood methods. There are also a number of other methods that are less frequently used but nevertheless offer important insights into phylogeny reconstruction (e.g., LogDet, LOCKHART & al. 1994; split decomposition, BANDELT & DRESS 1994; quartet puzzling, STRIMMER & VON HAESELER 1996; maximum posterior probability, RANNALA & YANG 1996). Detailed discussions of many of these methods are available in a number of research articles that regularly appear in journals such as, "Molecular Biology and Evolution" and "The Journal of Molecular Evolution" (e.g., BERRY & GASCUEL 1996, KUMAR & RZHETSKY 1996, VAN DE PEER & al. 1996b, WILKINSON 1996). The books, "Fundamentals of molecular evolution" written by LI & GRAUR (1991) and "Molecular systematics" edited by HILLIS & al. (1996) are also highly recommended as sources of theoretical and practical discussions of problems in molecular evolution. Here, the basic characteristics of distance, maximum parsimony and maximum likelihood methods will be described so that the reader who is unfamiliar with phylogenetic analysis can better interpret the trees shown in this book. Phylogenetic methods generally involve two steps, (1) the creation of a tree that best describes the relationships of the studied sequences under a given evolutionary model, and (2) the testing of the "fitness" of this best tree against many or all other possible species groupings for the data set. The second problem is nontrivial when one considers that there are:

$$N_u = \frac{(2n-5)!}{2^{n-3}(n-3)!}$$

possible unrooted, bifurcating trees (N_u) for $n \geq 3$ species in the analysis (Li & Graur 1991). This equation shows, for example, that with 10 species there are 2,027,025 possible unrooted trees to explain their interrelationships. The creation of large phylogenies with 15 or more sequences often demands the use of procedures that search only a fraction of all possible trees (i.e., heuristic searches such as stepwise insertion or clustering methods). This characteristic of phylogenetic methods suggests that they should not be expected to find the best solution each time.

Distance methods. The distance methods involve two steps for the inference of a phylogeny, (1) the computation of a distance matrix that describes the pairwise relationships of all studied sequences, and (2) the inference of a tree using that matrix and some functional relationship between the distance values. Distance estimates are computed from the raw identity values between any two sequences using an evolutionary model that attempts to correct for various characteristics of sequence evolution (e.g., superimposed mutations, transition/transversion mutation bias, nucleotide composition bias). The better the evolutionary model that describes sequence evolution, the more accurate are the inferred distance values and the more realistic (presumably) is the phylogeny. Widely used evolutionary models in order of increasing complexity include the Jukes-Cantor single-parameter model (Jukes & Cantor 1969) that assumes random substitutions of the four types of nucleotides, the Kimura two-parameter model (Kimura 1980) that allows for different rates of transitions and transversions and the Jin-Nei model (Jin & Nei 1990) that as well allows for rate variation across sites. The two most commonly used tree building methods with distance data are the neighbor-joining method (Saitou & Nei 1987) and the Fitch-Margoliash method (Fitch & Margoliash 1967); these trees normally show evolutionary distances as the sum of the horizontal branch lengths separating taxa within the phylogeny. Distance analyses are available in several computer packages such as the PHYLIP package (Felsenstein 1993), the MEGA package (Kumar & al. 1994), the Treecon packagae (Van de Peer & de Wachter 1993) and the SPLITS TREE package (Bandelt & Dress 1994). Distance analysis programs have also incorporated weighting or correction schemes to allow a more refined understanding of sequence evolution in single-stranded (e.g., loops) and double-stranded (e.g., stems-compensating base changes) regions of rRNA (Rzhetsky 1995, Kumar & Rzhetsky 1996) and to reduce or remove the influence of sequence positions that have extreme sequence divergence rates (Van de Peer & al. 1996b).

Maximum parsimony method. The most popular character-based phylogenetic method currently used is the maximum parsimony (MP) method (reviewed in Li & Graur 1991, Swofford & Olsen 1990, Swofford & al. 1996). The MP method uses character states at each sequence position and builds the tree that minimizes the pathway to these characters among all sequences. This in essence is the principle of parsimony, that the best tree should be the one that requires the smallest number of substitutions to explain the differences among the sequences being studied (i.e., has the shortest tree length). This method often results in the identification of multiple equally parsimonious solutions for a given data set. This should not be interpreted as a weakness of the MP method but as evidence that the data cannot resolve completely all nodes within the phylogeny. Often, a consensus phylogeny is constructed from the equally parsimonious trees that shows, as collapsed polytomies, the unresolved nodes. An important difference between the MP and distance methods is that the former does not use the complete data set but only the informative sites (i.e., those that are represented by at least two different character states found in at least two different sequences). The uninformative sites (e.g., constant) do not enter the analysis because these do not permit the discrimination between possible phylogenies for the data set. MP searches can be exhaustive (all possible trees are studied) or heuristic (a subset of all possible trees are investigated). Most analyses that contain greater than 7–10 taxa normally employ a heuristic search method to limit the computation time (see SWOFFORD 1993 for details). MP methods are also often used with weighting schemes for the columns of sequence data based on different calculations (e.g., consistency index, retention index, number of steps) to reduce the influence of those sites that have extreme rates of sequence divergence (MADDISON & MADDISON 1992; SWOFFORD 1993; BHATTACHARYA 1996). The most popular computer programs for doing maximum parsimony analyses are PAUP (SWOFFORD 1993) and DNAPARS (in the PHYLIP package, FELSENSTEIN 1993).

Maximum likelihood method. The maximum-likelihood (ML) method has often been heralded as the most robust of all phylogenetic reconstruction methods (FELSENSTEIN 1988). The ML method uses all sequence positions and a model of sequence evolution to find the phylogeny with corresponding branch lengths that has "the highest probability of evolving the observed data" (FELSENSTEIN 1981: 369); the inferred phylogenies are those with the highest likelihoods (reviewed in SWOFFORD & OLSEN 1990). The generalized 2-parameter model of KISHINO & HASEGAWA (1989), for example, is often used to calculate the site-by-site probabilities of one nucleotide changing to another (e.g., DNAML in the PHYLIP package, FELSENSTEIN 1993; fastDNAml, OLSEN & al. 1994). These probabilities follow a Markov process; i.e., the probability of a base changing may depend on its current identity but not on its past history (FELSENSTEIN 1981). The probability is calculated for any given tree and is used to infer the tree likelihood. The KISHINO-HASEGAWA model takes into account the different rates of change for transitions versus transversions as well as the nucleotide contents within the studied sequences. Two important characteristics of the ML method are the possibility of comparing statistically (likelihood ratio test, KISHINO & HASEGAWA 1989) differing topologies for any data set (i.e., hypothesis-testing) and the availability of variance estimates for all branch lengths in a tree allowing one to gain some idea about the significance of these branches.

In spite of its power, the ML method is plagued by the highest computation time requirements of any of the major phylogenetic methods and has therefore not been widely used in analyses containing greater than 20–25 sequences (even more rarely in bootstrap analyses, for exceptions see WAINRIGHT & al. 1993, DELWICHE & al. 1995). The high computation time requirement of the ML method also necessitates the use of heuristic methods for constructing trees (for details, see FELSENSTEIN 1988, 1993; OLSEN & al. 1994). A possibility for decreasing computation times is via parallelization of the problem. The concept of parallel computing is to divide a problem into a number of smaller partial problems which can each be solved simultaneously by independent processors (OTTMANN & WIDMAYER 1993). Parallel computing, due to its capacity for rapid analysis of complex problems, has seen wide application in molecular biology (e.g., RNA secondary structure prediction, NAKAYA & al. 1995). The availability of parallel maximum likelihood programs (fastDNAml, MATSUDA & al., unpubl.; pfastDNAml V2.0, SCHMIDT & BHATTACHARYA, unpubl.) is anticipated to bring this important technique into more common use in phylogeny reconstruction.

The bootstrap method. Once a "best" tree has been determined with a phylogeny program, the most popular method for testing the stability of inferred sequence groupings within this tree is the bootstrap method (FELSENSTEIN 1985). Other methods include the decay index (DONOGHUE & al. 1992) and the likelihood ratio test (KISHINO & HASEGAWA 1989, see below). The bootstrap is a statistical method for calculating a nonparametric estimate of the monophyly of groups in a phylogeny. Pseudosamples (normally 100–2000 data sets) are created from the original data set by repeated, random resampling of columns of data within the alignment with replacement. The number of columns of data that are drawn is the same as the total number of positions in the original data set. These pseudosamples are used as input for phylogenetic analyses and a consensus tree is calculated in which the bootstrap values are shown at the branches. These bootstrap values are the percentages that particular groupings were recovered in the resampled trees. Though simple to implement, results of bootstrap analyses with real phylogenetic data are often difficult to interpret (HILLIS & BULL 1993, WAINRIGHT & al. 1993). The apparent instability of bootstrap values with different species inputs (see BALDAUF & PALMER 1993; BHATTACHARYA & WEBER 1997), outgroups, and weighting schemes has led to a flood of theoretical papers to better understand these phenomena (e.g., FELSENSTEIN & KISHINO 1993, HILLIS & BULL 1993, ZHARKIKH & LI 1995, BHATTACHARYA 1996, EFRON & al. 1996, SITNIKOVA 1996, WILKINSON 1996). These analyses suggest that the bootstrap method, though useful in phylogeny reconstruction, must in practice be interpreted with caution.

Concluding remarks

The algae are both an interesting and a confounding group of organisms. They vary in size from the world's largest protists (kelps, see chapter 12) to minute, non-flagellated unicells (trebouxiophytes, members of the genus *Chlorella*; see chapter 4) to thecate amoebae (*Paulinella chromatophora*). Algae can be both toxic to humans (some dinoflagellates, see Chapter 13) and important food sources (*Porphyra* [nori]) as well as model organisms for cell/molecular biology (*Chlamydomonas reinhardtii*). Studying algal phylogeny and evolution provides insights not only into the diversity of living things but also into how complex organisms may be built from simpler ones (land plants and green algae, respectively; see Chapter 5). Understanding the origins of the algae and their plastids will offer basic insights into the development of the chimaeric eukaryotic cell and often provide surprising new insights, such as the finding of plastids in apicomplexan parasites (see Chapter 14). With this perspective, one of the great benefits offered by the recent advances in the molecular phylogenetic analyses of algae has been to provide an objective framework (i.e., sequence-based phylogenies) for asking deeper questions about algal/protist evolution. This book is meant to provide an overview of the important advances in this field.

I thank LINDA MEDLIN (Bremerhaven) and HEIKO A. SCHMIDT (Heidelberg) for their critical reading of this manuscript and KLAUS WEBER (Göttingen) for his support. This research was financed in part by a grant from the Deutsche Forschungsgemeinschaft (Bh 4/1-2).

References

BALDAUF, S. L., PALMER, J. D., 1993: Animals and fungi are each other's closest relatives: congruent evidence from multiple proteins. – Proc. Natl. Acad. Sci. USA **90**: 11558–11562.

BANDELT, H. J., DRESS, A. W. M., 1994: Splits Tree V1.0 – Department of Mathematics, University of Bielefeld.

BERRY, V., GASCUEL, O., 1996: On the interpretation of bootstrap trees: appropriate threshold of clade selection and induced gain. – Molec. Biol. Evol. **13**: 999–1011.

BHATTACHARYA, D., 1996: Analysis of the distribution of bootstrap tree lengths using the maximum parsimony method. – Molec. Phyl. Evol. **6**: 339–350.

– MEDLIN, L., 1995: The phylogeny of plastids: a review based on comparisons of small subunit ribosomal RNA coding regions. – J. Phycol. **31**: 489–498.

– EHLTING, J., 1995: Actin coding regions: gene family evolution and use as a phylogenetic marker. – Arch. Protistenk. **145**: 155–164.

– WEBER, K., 1997: The actin gene of the glaucocystophyte *Cyanophora paradoxa*: analysis of the coding region and introns and an actin phylogeny of eukaryotes. – Curr. Genet.

BOLD, H. C., WYNNE, M. J., 1985: Introduction to the algae, structure and reproduction. 2nd edn. – Englewood Cliffs: Prentice Hall.

CAVALIER-SMITH, T., 1982: The origins of plastids. – Biol. J. Linn. Soc. **17**: 289–306.

– 1993: Kingdom *Protozoa* and its 18 phyla. – Microbiol. Rev. **57**: 953–994.

DELWICHE, C. F., KUHSEL, M., PALMER, J. D., 1995: Phylogenetic analysis of *tuf*A sequences indicates a cyanobacterial origin of all plastids. – Molec. Phyl. Evol. **4**: 110–128.

DONOGHUE, M. J., OLMSTEAD, R. G., SMITH, J. F., PALMER, J. D., 1992: Phylogenetic relationships of *Dipsacales* based on *rbc*L sequences. – Ann. Missouri Bot. Gard. **79**: 333–345.

DOOLITTLE, R. F., FENG, D.-F., TSANG, S., CHO, G., LITTLE, E., 1996: Determining divergence times of the major kingdoms of living organisms with a protein clock. – Science **271**: 470–477.

DOUGLAS, S. E., TURNER, S., 1991: Molecular evidence for the origin of plastids from a cyanobacterium-like ancestor. – J. Molec. Evol. **33**: 267–73.

DOVER, G. A., 1982: Molecular drive: a cohesive mode of species evolution. – Nature **299**: 111–117.

EFRON, B., HALLORAN, E., HOLMES, S., 1996: Bootstrap confidence levels for phylogenetic trees. – Proc. Natl. Acad. Sci. USA **93**: 7085–7090.

FELSENSTEIN, J., 1981: Evolutionary trees from DNA sequences: a maximum likelihood approach. – J. Molec. Evol. **17**: 368–376.
– 1985: Confidence limits on phylogenies: an approach using the bootstrap. – Evolution **39**: 783–91.
– 1988: Phylogenies from molecular sequences: inference and reliability. – Annu. Rev. Genet. **22**: 521–565.
– 1993: PHYLIP manual, version 3.5c. – University of Washington: Department of Genetics.
– KISHINO, H., 1993: Is there something wrong with the bootstrap on phylogenies? A reply to HILLIS and BULL. – Syst. Biol. **42**: 193–200.
FITCH, W. M., MARGOLIASH, E., 1967: Construction of phylogenetic trees. A method based on mutation distances as estimated from cytochrome c sequences is of general applicability. – Science **155**: 279–284.
FRESHWATER, D. W., FREDERIQ, S., BUTLER, B. S., HOMMERSAND, M. H., CHASE, M. W., 1994: A gene phylogeny of the red algae (*Rhodophyta*) based on plastid *rbc*L. – Proc. Natl. Acad. Sci. USA **91**: 7281–7285.
GIBBS, S., 1993: The evolution of algal chloroplasts. In LEWIN, R. A., (Ed.): Origins of plastids, pp. 107–21. – New York: Chapman & Hall.
GUPTA, R. S., GOLDING, G. B., 1993: Evolution of HSP70 gene and its implications regarding relationships between archaebacteria, eubacteria and eukaryotes. – J. Molec. Evol. **37**: 573–582.
HALLICK, R. B., HONG, L., DRAGER, R. G., FAVREAU, M. R., MONFORT, A., ORSAT, B., SPIELMANN, A., STUTZ, E., 1993: Complete sequence of *Euglena gracilis* chloroplast DNA. – Nucl. Acids Res. **21**: 3537–3544.
HILLIS, D. M., BULL, J. J., 1993: An empirical test of bootstrapping as a method for assessing confidence in phylogenetic analysis. – Syst. Biol **42**: 182–192.
– MORITZ, C., MABLE, B. K., 1996: Molecular systematics. 2nd edn. – Sunderland: Sinauer.
IWABE, N., KUMA, K.-I-., HASEGAWA, M., OSAWA, S., MIYATA, T., 1989: Evolutionary relationships of archaebacteria, eubacteria, and eukaryotes inferred from phylogenetic trees of duplicated genes. – Proc. Natl. Acad. Sci. USA **86**: 9355–9359.
JIN, L., NEI, M., 1990: Limitations of the evolutionary parsimony method of phylogenetic analysis. – Molec. Biol. Evol. **7**: 82–102.
JUKES, T. H., CANTOR, C. R., 1969: Evolution of protein molecules. – In MUNRO, H. N., (Ed.): Mammalian protein metabolism, pp. 21–132. – New York: Academic Press.
KEELING, P. J., DOOLITTLE, W. F., 1996: Alpha-tubulin from early-diverging eukaryotic lineages and the evolution of the tubulin family. – Molec. Biol. Evol. **13**: 1297–1305.
KIES, 1974: Elektronenmikroskopische Untersuchungen an *Paulinella chromatophora* LAUTERBORN, einer Thekamöbe mit blaugrünen Endosymbionten Cyanellen. – Protoplasma **80**: 69–89.
KIMURA, M., 1980: A simple method for estimating evolutionary rates of base substitution through comparative studies of sequence evolution. – J. Molec. Evol. **16**: 111–20.
KISHINO, H., HASEGAWA, M., 1989: Evaluation of the maximum likelihood estimate of the evolutionary tree topologies from DNA sequence data, and the branching order of the *Hominoidea*. – J. Molec. Evol. **29**: 170–179.
KOWALLIK, K. V., STOEBE, B., SCHAFFRAN, I., FREIER, U., 1995: The chloroplast genome of a chlorophyll $a + c$ containing alga, *Odontella sinensis*. – Pl. Molec. Reporter **13**: 336–342.
KRAFT, G. T., 1981: *Rhodophyta*: morphology and classification. – In LOBBAN, C. S., WYNNE, M. J., (Eds): The biology of seaweeds, pp. 6–51. – Oxford: Blackwell.

KUMAR, S., RZHETSKY, A., 1996: Evolutionary relationships of eukaryotic kingdoms. – J. Molec. Evol. **42**: 183–193.
– TAMURA, K., NEI, M., 1994: MEGA: molecular evolutionary genetics analysis software for microcomputers. CABIOS **10**: 189–191.
LI, W.-H., GRAUR, D., 1991: Fundamentals of molecular evolution. – Sunderland: Sinauer.
LIAUD, M.-F., VALENTIN, C., MARTIN, W., BOUGET, F.-Y., KLOAREG, B., CERFF, R., 1994: The evolutionary origin of red algae as deduced from the nuclear genes encoding cytosolic and chloroplast glyceraldehyde-3-phosphate dehydrogenases from *Chondrus crispus*. – J. Molec. Evol. **38**: 319–327.
LOCKHART, P. J., STEEL, M. A., HENDY, M. D., PENNY, D., 1994: Recovering evolutionary trees under a more realistic model of sequence evolution. – Molec. Biol. Evol. **11**: 605–612.
LÖFFELHARDT, W., STIREWALT, V. L., MICHALOWSKI, C. B., ANNARELLA, M., FARLEY, J.Y., SCHLUCHTER, W. M., CHUNG, S., NEUMANN-SPALLART, C., STEINER, J. M., JAKOWITSCH, J., BOHNERT, H. J., BRYANT, D.A., 1996: The complete sequence of the cyanelle genome from *Cyanophora paradoxa:* the genetic complexity of a primitive plastid. In SCHENK, H. E. A., (Ed.): Endocytobiology VI – Heidelberg: Springer (in press).
LOISEAUX-DE-GÖER, S., 1994: Plastid lineages. – In ROUND, F. E., CHAPMAN, D. J., (Eds): Progress in phycological research **10**, pp. 137–177. – Bristol: Biopress.
MADDISON, W. P., MADDISON, D. R., 1992: "MacClade: analysis of phylogeny and character evolution, v. 3, Sunderland: Sinauer.
MAIDAK, B. L., LARSEN, N., MCCAUGHEY, M. J., OVERBEEK, R., OLSEN, G. J., FOGEL, K., BLANDY, J., WOESE, C. R., 1994: The ribosomal database project. – Nucl. Acids Res. **22**: 3485–3487.
MARGULIS, L., 1981: Symbiosis in cell evolution. – Chicago: Freeman.
MARTIN, W., SOMERVILLE, C. C., LOISEAUX-DE GOER, S., 1992: Molecular phylogenies of plastid origins and algal evolution. – J. Molec. Evol. **35**: 385–404.
MCCUTCHAN, T. F., DELA CRUZ, V. F., LAL, A. A., GUNDERSON, J. H., ELWOOD, H. J., SOGIN, M. L., 1988: Primary sequences of two small subunit ribosomal RNA genes from *Plasmodium falciparum*. – Molec. Biochem. Parasitol. **28**: 63–68.
MEDLIN, L., ELWOOD, H. J., STICKEL, S., SOGIN, M. L., 1988: The characterization of enzymatically amplified eukaryotic 16S-like rRNA coding regions. – Gene **71**: 491–499.
MERESCHKOWSKY, C., 1905: Über Natur und Ursprung der Chromatophoren im Pflanzenreiche. – Biol. Zentralbl. **25**: 593–604.
– 1910: Theorien der zwei Plasmaarten als Grundlage der Symbiogenesis, einer neuen Lehre der Entstehung der Organismen. – Biol. Zentralbl. **30**: 278–303.
MOESTRUP, Ø., 1978: On the phylogenetic validity of the flagellar apparatus in green algae and other chlorophyll a and b containing plants. – BioSystems **10**: 117–144.
MORDEN, C. W., DELWICHE, C. F., KUHSEL, M., PALMER, J. D., 1992: Gene phylogenies and the endosymbiotic origin of plastids. – BioSystems **28**: 75–90.
NAKAYA, A., YAMAMOTO, K., YONEZAWA, A., 1995: RNA secondary structure prediction using highly parallel computers. – CABIOS **11**: 685–692.
O'KELLY, C. J., 1992: Flagellar apparatus Architecture and the phylogeny of "green" algae: chlorophytes, euglenoids, glaucophytes. – In MENZEL, D., (Ed.): The cytoskeleton of the algae, pp. 315–345. – Boca Raton: CRC Press.
OLSEN, G. J., MATSUDA, H., HAGSTROM, R., OVERBEEK, R., 1994: fastDNAml: a tool for construction of phylogenetic trees of DNA sequences using maximum likelihood. CABIOS **10**: 41–48.
OTTMANN, T., WIDMAYER, P., 1993: Algorithmen und Datenstrukturen. – Mannheim: BI Wissenschaftsverlag.

PALMER, J. D., DELWICHE, C. F., 1996: Second-hand chloroplasts and the case of the disappearing nucleus. – Proc. Natl. Acad. Sci. USA **93**: 7432–7435.

RANNALA, B., YANG, Z., 1996: Probability distribution of molecular evolutionary trees: a new method of phylogenetic inference. – J. Molec. Evol. **43**: 304–311.

RAVEN, P. H., 1970: A multiple origin of plastids and mitochondria. – Science **169**: 641–649.

REITH, M., MUNHOLLAND, J., 1995: Complete nucleotide sequence of the *Porphyra purpurea* chloroplast genome. – Pl. Molec. Biol. Reporter **13**: 332–335.

RZHETSKY, A., 1995: Estimating substitution rates in ribosomal RNA genes. – Genetics **141**: 771–783.

SAITOU, N., NEI, M., 1987: The neighbor-joining method: a new method for reconstructing phylogenetic trees. – Molec. Biol. Evol. **4**: 406–425.

SCHLEGEL, M., 1994: Molecular phylogeny of eukaryotes. – TREE **9**: 330–335.

SCHOLIN, C. A., ANDERSON, D. M., SOGIN, M. L., 1993: Two distinct small-subunit ribosomal RNA genes in the North American toxic dinoflagellate *Alexandrium fundyense Dinophyceae.* – J. Phycol. **29**: 209–216.

SITNIKOVA, T., 1996: Bootstrap method of interior-branch test for phylogenetic trees. – Molec. Biol. Evol. **13**: 605–611.

SOGIN, M. L., GUNDERSON, G. J., ELWOOD, H. J., ALONSO, R. A., PEATTIE, D. A., 1989: Phylogenetic meaning of the kingom concept: an unusual ribosomal RNA from *Giardia lamblia.* – Science **243**: 75–77.

STEWART, K. D., MATTOX, K. R., 1975: Comparative cytology, evolution, and classification of the green algae with some considerations of the origin of the organisms with chlorophylls a and b. – Bot. Rev. **41**: 104–145.

STRIMMER, K., VON HAESELER, A., 1996: Quartet puzzling: a quartet maximum-likelihood method for reconstructing tree topologies. – Molec. Biol. Evol. **13**: 964–969.

SWOFFORD, D. L., 1993: PAUP: phylogenetic analysis using parsimony, v. 3.1.1. – Washington, D. C.: Smithsonian Institution.

– OLSEN, G. J., 1990: Phylogeny reconstruction. – In HILLIS, D. M., MORITZ, C., (Eds): Molecular systematics, pp. 411–501. – Sunderland: Sinauer.

– WADDELL, P. J., HILLIS, D. H., 1996: Phylogenetic inference. In HILLIS, D. M., MORITZ, C., MABLE, B. K., (Eds): Molecular systematics, pp. 407–425, 2nd edn. – Sunderland: Sinauer.

TAYLOR, F. J. R., 1976: Flagellate phylogeny: a study in conflicts. – J. Protozool. **23**: 28–40.

VAN DE PEER, Y., DE WACHTER, R., 1993: TREECON: a software package for the construction and drawing of evolutionary trees. – CABIOS **9**: 177–182.

– DE RIJK, P., DE WACHTER, R., 1996a: SSU rRNA database. Department of Biochemistry, University of Antwerpen http.//rrna.uia.ac.be/rrna/ssuform.html#Eukarya.

VAN DER AUWERA, G., DE WACHTER, R., 1996b: The evolution of stramenopiles and alveolates as derived by "Substitution Rate Calibration" of small ribosomal subunit RNA. – J. Molec. Evol. **42**: 201–210.

VAN DEN HOEK, C., JAHNS, H. M., MANN, D. G., 1993: Algen. 3rd edn. – New York: Thieme.

WAINRIGHT, P. O., HINKLE, G., SOGIN, M. L., STICKEL, S. K., 1993: Monophyletic origins of the *Metazoa*: an evolutionary link with the fungi. – Science **260**: 340–342.

WHATLEY, J. M., 1993: The endosymbiotic origin of chloroplasts. – Int. Rev. Cytol. **144**: 259–299.

WILKINSON, M., 1996: Majority-rule reduced consensus trees and their use in bootstrapping. – Molec. Biol. Evol. **13**: 437–444.

WOLF, P. G., SOLTIS, P. S., SOLTIS, D. E., 1994: Phylogenetic relationships of dennstaedtioid ferns: evidence from *rbc*L sequences. – Molec. Phyl. Evol. **3**: 383–392.

ZHARKIKH, A., LI, W.-H., 1995: Estimation of confidence in phylogeny: the complete and-partial bootstrap technique. – Molec. Phyl. Evol. **4**: 44–63.

Molecular systematics of oxygenic photosynthetic bacteria

SEÁN TURNER

Key words: *Cyanobacteria*, *Cyanophyta*, Blue-green algae, oxychlorobacteria, Prochlorophyta. – Plastid origins, chloroplast evolution, oxygenic photoautotrophy, ribosomal RNA, phylogenetics, molecular systematics, distance matrix, maximum likelihood, maximum parsimony.

Abstract: Molecular phylogenetic analyses of available small ribosomal subunit RNA sequences from cyanobacteria (Cyanophyta) and oxychlorobacteria (Prochlorophyta) were conducted to infer evolutionary relationships among oxygenic photoautotrophic bacteria and plastids. Matrices of pairwise evolutionary distances estimated under two models of molecular evolution provided a basis for phylogenetic analysis using two methods of tree inference. The sequences fell into ten discrete groups of taxa that were subsequently examined more thoroughly using maximum likelihood, maximum parsimony, and distance matrix methods. – Results confirmed those of previous studies, including the deep branching of *Pseudanabaena* strains and *Gloeobacter violaceus*, and the monophyletic grouping of heterocystous cyanobacteria. Several phenotypic characters traditionally used to classify cyanobacteria and oxychlorobacteria do not correlate with the phylogenetic relationships inferred from sequence analysis. In particular, the cyanobacterial genera *Synechococcus* and *Leptolyngbya* are clearly polyphyletic. Numerous other genera are paraphyletic. An unexpected result is an association of the pleurocapsalean cyanobacterium *Chroococcidiopsis thermalis* as a sister taxon to the heterocyst-forming nostocalean and stigonematalean cyanobacteria, distinctly apart from other pleurocapsalean cyanobacteria. – The plastids of photosynthetic eukaryotes constitute a monophyletic subtree within the cyanobacterial line of descent with no apparent relationship to any oxychlorobacteria which are themselves polyphyletic. No strong candidate for the sister taxon to plastids is indicated. Other molecular phylogenetic analyses are reviewed in the light of these results.

The earliest proposal that photosynthetic plastids may be derived from microorganisms of a bacterial nature was made more than a century ago (SCHIMPER 1883). The first explicit hypothesis that they are related to modern cyanobacteria (cyanophyta, "blue-green algae") is generally credited to MERESCHKOWSKY (1905, 1910). While proposals for an autogenous origin of such plastids continue to make an occasional appearance (JENSEN 1989, 1991), the weight of evidence is such that they are now generally accepted to be of xenogenous origin with a direct evolutionary link to modern cyanobacteria (GRAY & DOOLITTLE 1982; GRAY 1988, 1992, 1993; LOISEAUX-DE GOËR 1994). Consequently, a discussion of the origin of photosynthetic plastids and their derivatives must necessarily be based upon a discussion of evolutionary relationships among the cyanobacteria. Ideally, such relationships should be reflected in the systematics of these organisms, but for the cyanobacteria this poses particular difficulties.

The problematic nature of cyanobacterial systematics. The history of cyanobacterial systematics has been unusually tumultuous. The earliest taxonomic monographs describing cyanobacteria considered them to be a type of algal plant (for example, THURET 1875, BORNET & FLAHAULT 1886–1888, GOMONT 1892). The "modern" era of cyanobacterial systematics is usually acknowledged as beginning with the system proposed by GEITLER (1932). It has formed the basis of numerous revised systems proposed since then, including those of ELENKIN (1938, 1949), HOLLERBACH & al. (1953), DESIKACHARY (1959, 1973), FRITSCH (1959), STARMACH (1966), KONDRATEVA (1968), BOURELLY (1970), and GOLUBIC (1976). These systems share in common the view that the systematics of cyanobacteria should be based on traditional botanical criteria, a view sometimes referred to as the "Geitlerian" approach.

In the period from 1956–1981, an alternative system was developed by DROUET and DAILY that drastically reduced the number of genera and species of cyanobacteria (summarized in DROUET 1981). It was based on the hypothesis that the many morphological differences seen among field samples of cyanobacteria are ephemeral and that the numerous "species" of cyanobacteria are actually different "ecophenes" of a fairly small number of true taxa. However, it has since been shown that this system does not adequately reflect the true genetic diversity among cyanobacteria and it has consequently fallen into disuse.

A third approach, developed largely through the efforts of R. STANIER, R. RIPPKA, and their colleagues, is based on the recognition of the fact that the "blue-green algae" are unquestionably bacterial in nature and not simply a prokaryotic sister group to the other bacteria (STANIER & VAN NIEL 1962). On this basis, these investigators developed a provisional taxonomic system for cyanobacteria based on examination of strains in axenic culture using bacteriological rather than botanical criteria (HERDMANN & al. 1979a,b; RIPPKA & al. 1979; RIPPKA & COHEN-BAZIRE 1983; RIPPKA 1988; RIPPKA & HERDMAN 1992). In most cases, taxonomic assignments consisted of a genus name and a strain number, with some of the "genera" considered as provisional assemblages subject to future reassessment. Species names were assigned only for type strains. A proposal to formally place the systematics of cyanobacteria under the authority of the Bacteriological Code rather than the Botanical Code (STANIER & al. 1978) met with immediate and vigorous opposition from botanists and phycologists, particularly those specializing in "blue-green algae" (BOURRELLY 1979, GEITLER 1979, GOLUBIC 1979, LEWIN 1979, KONDRATYEVA [sic] 1982). Although the proposal was not adopted, the "Stanierian" system, with some modification, now forms the basis of cyanobacterial taxonomy as described in Bergey's Manual of Systematic Bacteriology, a recognized authority in bacterial systematics (BURGER-WIERSMA & MUR 1989; CASTENHOLZ 1989a,b,c; CASTENHOLZ & WATERBURY 1989; Lewin 1989; WATERBURY 1989; WATERBURY & RIPPKA 1989; WHITTON 1989).

In an effort to reconcile the differences between the botanical and bacteriological approaches, a compromise system was agreed to informally by a number of investigators in these disciplines (Friedmann & Borrowitzka 1982). More recently, K. Anagnostidis and J. Komárek have developed a formal system along the lines of this compromise. For historical reasons, its nomenclature is based mainly on botanical taxonomic criteria, but it also utilizes information obtained from bacteriological considerations using pure cultures where possible (Anagnostidis & Komárek 1985, 1988, 1990; Komárek & Anagnostidis 1986, 1989).

In this chapter, taxonomic nomenclature will follow mainly that of the Stanierian system but will also refer to the system of Anagnostidis and Komárek where appropriate. Readers seeking further information on the history of cyanobacterial systematics as well as discussion of species concepts for these organisms should consult the review of Castenholz (1992).

Molecular biology and cyanobacterial systematics. It is now recognized that modern approaches must be brought to bear on the problem of cyanobacterial systematics, including ultrastructural studies (Guglielmi & Cohen-Bazire 1984a, Jensen 1985, Hoffman 1988) and molecular biological methods (for example, Wilmotte & Golubic 1991). The two molecular methods most frequently used today in bacterial systematics are the determination of sequences of small ribosomal subunit ribonucleic acids (SSU rRNAs) and DNA-DNA hybridization. To a lesser extent, restriction fragment length polymorphism (RFLP) analysis of DNA has also been brought to bear on the problem.

The use of SSU rRNAs to determine the evolutionary relationships among bacteria was pioneered by Carl Woese and coworkers, initially through the comparison of ribonuclease T_1-generated oligonucleotide catalogs (Fox & al. 1977). The purpose was to establish a systematics of bacteria based on evolutionary relationships inferred by a direct comparison of homologous genes. The first such studies of cyanobacteria confirmed their bacterial nature as well as their link with photosynthetic plastids, although the exact nature of the relationship was, and continues to be, a matter of debate (Woese & al. 1975; Bonen & Doolittle 1975, 1976, 1978; Bonen & al. 1979; Seewaldt & Stackebrandt 1982; Stackebrandt & al. 1982; Van Valen 1982; Bremer & Bremer 1988).

Subsequently, the development of rapid nucleic acid sequencing techniques led to an explosive generation of data that has revolutionized our understanding of bacterial evolutionary relationships. Through analysis of SSU rRNA data, prokaryotic organisms are seen to fall into two distinct "Domains," the *Archaea* and the *Bacteria* (Woese & al. 1990). Within the latter group, the oxygenic phototrophs and derived organelles [cyanobacteria, oxychlorobacteria (prochlorophyta), and photosynthetic plastids] form a distinct, monophyletic assemblage, consistent with their unique phenotype of oxygenic photoautotrophy (Woese 1987). SSU rRNA sequence analysis has been used to explore further the evolutionary relationships within the cyanobacterial group (Tomioka & al. 1981; Giovannoni & al. 1988, 1990; Turner & al. 1989a; Douglas & Turner 1991; Weller & al. 1991, 1992; Nelissen & al. 1992, 1995a,b, 1996; Urbach & al. 1992; Wilmotte & al. 1992, 1993, 1994; Neilan & al. 1994, 1997; Ferris & al. 1996; Palinska & al. 1996).

Although SSU rRNA data are useful for determining differences among organisms at the genus level, they generally are not reliable for lower levels of taxonomic assignment. DNA-DNA hybridization studies have been used with success to determine evolutionary relationships among strains of cyanobacteria both between and within genera (KELLY & COWIE 1972, STAM & VENEMA 1977, STAM 1980, LACHANCE 1981, STULP & STAM 1984, WILMOTTE & STAM 1984, STAM & al. 1985). For even higher resolution, RFLP and allozyme analyses have been used to determine genetic differences among serotypes or strains of what might be single cyanobacterial species (DOUGLAS & CARR 1988, GOLDEN & al. 1989, WOOD & TOWNSEND 1990, ZIMMERMAN & BERGMAN 1990, ZIMMERMAN & ROSEN 1992). While SSU rRNA sequence analysis does not have the resolution of DNA-DNA hybridization or RFLP analysis, it does have the advantage of being a "single experiment" technique. That is, once an SSU rRNA sequence has been determined, it does not have to be redetermined to compare it with new sequences. The other methods require repeated experimentation with each new comparison. At present, then, the analysis of rRNA sequence data is the preferred method for investigating evolutionary relationships among the members of any particular bacterial group.

In this chapter, emphasis will be on a comprehensive examination of SSU rRNA data in order to explore relationships among the oxygenic photosynthetic bacteria and plastids and to bring up to date the conclusions that can be drawn from such analyses. Readers interested in a broader discussion that includes other methods of cyanobacterial chemotaxonomy should consult the review of WILMOTTE (1994). Where possible, cyanobacterial strains will be referred to by their corresponding designations from the Pasteur Culture Collection (PCC) or other publicly available culture collections. Sequence data obtained from strains coidentical with the publicly available strains will be so indicated in the text.

Procedures

Alignment. Ninety-four SSU rRNA sequences are included in this study and are derived from 78 strains of cyanobacteria, three oxychlorobacteria, eight clones of environmental DNA of putative cyanobacterial origin, three representative plastids, and two outgroup bacteria. Other partial or interrupted sequences that did not contain data distributed across the three major secondary structural domains of SSU rRNA were not included (GUTELL & al. 1985). Most of the sequence data are deposited in the GenBank/EMBL/DDBJ electronic data bases and/or the Ribosomal Data Base Project (MAIDAK & al. 1996). Others were provided by A. WILMOTTE, B. NEILAN, and D. DISTEL and J. WATERBURY. Several are unpublished data of the author.

The sequences were aligned on the basis of secondary structure analysis using the rRNA structure of *Synechococcus* PCC 6301 as template (GUTELL 1993). Outgroup taxa were the proteobacterium *Agrobacterium tumefaciens* and the Gram positive bacterium *Bacillus subtilus.* Ambiguous nucleotide identities were scored as "N." Hypervariable regions containing significant insertions/deletions were excluded from the position set used for the analysis. The resulting data set contained 1284 positions corresponding to nucleotides 29–68, 101–180, 220–451, 480–836, 850–1004, 1038–1129, 1144–1440, 1461–1491 of *Escherichia coli* SSU rRNA (BRÖSIUS & al. 1981). Positions for which any particular sequence lacked data within this set were treated as missing data for that sequence.

Complete tree analysis. Estimates of evolutionary distances between all pairs of sequences were determined under two models of evolution with the program DISTANCE of the PHYLIP package (version 3.5c) (FELSENSTEIN 1993). The first was the maximum likelihood model of FELSENSTEIN (1981) modified to incorportate a user-defined transition/ transversion ratio. This modified model has been included in the PHYLIP package since 1984 and is hereafter referred to as the F84 model. The second model was that of JIN & NEI (1990) in which variability of evolutionary rates among positions is assumed to follow a Gamma distribution with a user-defined value for the shape parameter "a." This model is hereafter referred to as the JN90 model.

The transition/transversion ratio for these models was determined in the following manner. A maximum parsimony analysis that excluded the two outgroup sequences was carried out with the program PAUP (version 3.1.1) (SWOFFORD 1993). Due to the large size of the data set, a heuristic search was limited to a single run with the input order of sequences "as is" and using the tree bisection-reconnection search procedure. This resulted in 108 equally most parsimonious trees. Of these, the "best among equals" was determined as those with the optimal parsimony likelihood index as defined by RODRIGO (1992). There were 16 equally optimal trees which were identical in topology except within a subtree of ten closely related strains of the genus *Microcystis.* Four of these trees were strictly bifurcating, and the average number of transitions and transversions for each of these was determined using the program MacClade (version 3.05) (MADDISON & MADDISON 1992). The respective values were the same for all four trees. As these values represent the minimum number of changes for the trees, the proportions of transitions and transversions were corrected to allow for additional, unseen changes (WAKELY 1996). The resulting transition/ transversion ratio, 1.8685, was used in all subsequent analyses.

The shape parameter "a" for distances estimated under the JN90 model was determined in two ways. In the first, the number of tree steps for each character (nucleotide position) was determined for each of the 16 optimal most parsimonious trees using the program PAUP. For each character, the values were found to be identical across all 16 trees. The mean and variance for the number of steps per character were computed and these values were then used in Equation (2) of TAMURA & NEI (1993) yielding a value of "a" of 0.366.

In the second method, the value was determined by numerical approximation using Equation (7) of KUMAR & RZHETSKY (1996). This equation incorporates the expected number of substitutions per site and the expected proportions of transitional and transversional differences. The latter were the corrected values as described above, and the former the mean number of changes per site as in the first method. The resulting value, 0.860, was sufficiently different from the first to suggest determining an intermediate value as well. Consequently, a third value was computed in the same manner as the second, but correcting the expected number of substitutions per site to take into account unseen changes. This third value was 0.581. The value of the corresponding "coefficient of variation" for each value of "a" was determined for estimation of the JN90 distances in the DISTANCE program of PHYLIP.

Each of the four matrices of pairwise distance estimates was used to infer phylogenetic trees in four ways with the PHYLIP package: the Neighbor-joining clustering algorithm (SAITOU & NEI 1987) of the NEIGHBOR program, and tree searches based on the least-squares optimality criterion of the FITCH program using three different weighting schemes. These included no weighting (CAVALLI-SFORZA & EDWARDS 1967), weighting by the inverse of the square of the observed distance estimate (FITCH & MARGOLIASH 1967), and an intermediate weighting using the inverse of the observed distance estimate. For trees inferred by the FITCH program, a single run was executed using sequence input order "as is" and without global rearrangement, as the large size of the data set precluded more exhaustive analyses.

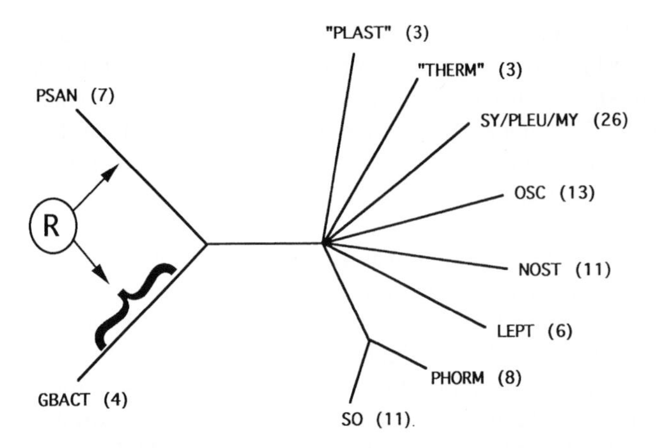

Fig. 1. Unrooted consensus cladogram of sixteen cyanobacterial trees inferred by distance matrix methods analyzing SSU rRNA sequence data. The two possible root (R) locations determined by outgroup taxa are indicated by arrows. The brace along the branch leading to the GBACT group indicates that the root location here falls within the group making it paraphyletic. Numbers in parentheses denote the number of taxa within each sequence group. Branch lengths are arbitrary. For further details, see text

Examination of the resulting 16 phylogenetic trees revealed a fairly consistent pattern of ten monophyletic groups as shown schematically in Fig. 1. For purposes of further discussion, the groups are given informal names that more or less reflect the organisms within them or some phenetic property thereof. Two of the groups, the *Gloeobacter* group (GBACT) and the *Pseudanabaena* group (PSAN), were consistently deepest in all the trees, but with the former often behaving as a paraphyletic rather than monophyletic group. The remaining eight groups showed no consistent intergroup branching order except for the *Synechococcus* (SO) and *Phormidium* (PHORM) groups which behaved as sister groups in all trees. This unresolved branching order is represented as a polytomy in Fig. 1. The remaining groups are the "plastid group" ("PLAST"), the "thermophilic unicellular group" ("THERM"), the *Synechocystis/Pleurocapsa/Microcystis* group (SY/PLEU/MY), the *Oscillatoria* (OSC) group, the *Nostoc* (NOST) group, and the *Leptolyngbya* (LEPT) group. Within each of the eight groups containing four or more sequences, there was no completely consistent pattern of internal branch order among the 16 trees.

For each set of trees inferred under the four models described above, the topologies of the four trees comprising the set were quantitatively compared by determining the log likelihood value for each topology using the program fastDNAml (OLSEN & al. 1994). This program and that of DNAML in PHYLIP incorporate the Kishino-Hasegawa test (KH test) to determine statistical significance of differences in log likelihood values among alternative trees (KISHINO & HASEGAWA 1989). Neighbor-joining trees were consistently better than trees inferred by the least-squares methods, but usually not significantly so at the 5% level of significance. The two instances in which there were significant differences involved the JN90 trees using the lower values of the "a" parameter (0.581 and 0.366) and the CAVALLI-SFORZA & EDWARDS weighting scheme. However, fastDNAml utilizes the F84 model of evolution and comparative evaluation of topologies inferred with the JN90 models with this method may be invalid. On the whole, there is no compelling reason to prefer any one of the 16 trees to the exclusion of all the others. Consequently, further phylogenetic analysis was limited to the seven individual subtrees containing six or more taxa (Fig. 1) in order to ascertain relationships within these smaller groups.

Subtree analysis. Each subtree containing six or more taxa was analyzed by maximum likelihood, distance, and parsimony methods. For each subtree, two cyanobacterial sequences from different subtrees were included as outgroup taxa.

Maximum likelihood analysis. Maximum likelihood trees were inferred using the fastDNAml program with ten different orders of addition of taxa. Local rearrangements during tree searches were allowed to cross one branch. Each distinct topology that resulted was further optimized with the global branch-swapping option. Reliability of the branching order in the final maximum likelihood tree was determined with the program PUZZLE (version 2.5) using the F84 model of evolution and approximate maximum likelihood values for all possible quartets of taxa (STRIMMER & VON HAESELER 1996). From these quartets, the program infers 1000 overall trees by random addition of taxa and reports the number of instances particular groupings of taxa occur in those trees. These "reliability values" are derived in a manner distinct from those obtained by bootstrap resampling but appear to be strongly correlated with bootstrap proportions (STRIMMER & VON HAESELER 1996). Additionally, a consensus tree from quartet puzzling analysis under the Hasegawa-Kishino-Yano model of evolution (HASEGAWA & al. 1985) using precise maximum likelihood values was derived using PUZZLE (version 2.3) and the CONSENSE program of PHYLIP.

Distance analysis. Subtrees were inferred by distance methods in the same manner as for the 16 larger trees discussed above, using the same values for the various parameters and with global branch-swapping during tree optimization, where applicable.

Parsimony analysis. Maximum parsimony trees were inferred using the branch-and-bound algorithm of PAUP (version 3.1.1) except for the SY/PLEU/MY group for which the number of taxa and inherent phylogenetic noise in the data precluded such an analysis. For that case, a heuristic search was done with 2000 random additions of taxa utilizing the tree bisection-reconnection search procedure. For the other subtrees, the branch-and-bound algorithm is guaranteed to find the most parsimonious tree(s).

For each subtree, all topologies inferred by these methods as well as the topologies for the pertinent subtree in each of the 16 larger trees were compared using the KH test in fastDNAml or DNAML (PHYLIP).

Molecular clock hypothesis. Each of the groups containing six or more taxa were examined to determine if the sequences comprising each group evolve with approximately clock-like behavior. This was done by imposing the assumption of a molecular clock during maximum likelihood tree inference using the program DNAMLK in the PHYLIP package. Analyses were carried out in the same manner as described for the F84 maximum likelihood trees except that outgroup taxa were not included. Analyses were then repeated in the absence of a clock assumption by "unrooting" these trees and determining the log likelihood values of the resulting topologies using the program DNAML. If an unrooted topology differed from that determined as best by a separate analysis of the same data with the DNAML or fastDNAml programs, it was verified as being not significantly worse by the KH test. The resulting log likelihood values of the rooted and corresponding unrooted trees were then compared by a likelihood ratio test (G test) (FELSENSTEIN 1993).

Results

Subtree phylogenetic analysis

***Gloeobacter* group.** This group, often the most deeply diverging in the trees discussed above, corresponds roughly to Branch B of WILMOTTE (1994). The group behaves paraphyletically in eight of the 16 trees. It consists in part of two rDNA clones derived from a thermal spring in Wyoming, USA, Octopus Spring type B and Octopus Spring type J, which behave as sister taxa in all the large trees. The group also includes a unicellular cyanobacterium recently isolated from Octopus Spring, *Synechococcus* sp. C9. The taxonomic assignment for strain C9 was provisional (WARD & al. 1994), and analysis of the molecular sequence data show it to be unrelated to the suggested type strain for the genus *Synechococcus* (see below). The organisms from which the Octopus Spring type B and Octopus Spring type J sequences were derived have yet to be identified, and at present there is no information on the ultrastructure or other features of strain C9. Its presence among the deepest branching cyanobacteria make it an obvious candidate for further studies.

The group also contains the type strain of the holotype species *Gloeobacter violaceus* PCC 7421, unique among cyanobacteria in having neither thylakoids nor phycobilisomes in the usual sense (RIPPKA & al. 1974, GUGLIELMI & al. 1981). Rather, the photosynthetic apparatus is located in the plasma membrane with phycobiliproteins arranged in rod-like arrays as an underlying cortical layer. This organism is also unique in lacking the lipid sulfoquinovosyl diacylglycerol that has been found in all other cyanobacteria and photosynthetic plastids examined (SELSTAM & CAMPBELL 1996). The lack of a highly developed photosystem coupled with its deep location in the cyanobacterial tree suggest that this organism may represent the most primitive cyanobacterium presently known. A compatibility analysis of the complete set of SSU rRNA sequences using the DNACOMP program of PHYLIP supports this conclusion, with *G. violaceus* being the deepest branching member of the tree so inferred (not shown). NELISSEN & al. (1995a) did an outgroup analysis in which a thousand iterations of a random selection of individual noncyanobacterial sequences were used to root a cyanobacterial tree and resulted in a consensus position of the root consistent with the results reported here. However, the multiple outgroup study did not include the sequences from strain C9 or the Octopus Spring rDNA clones, and included only a single strain of *Pseudanabaena*. A repetition of such a study using all the deeply branching cyanobacterial sequences is necessary to verify the lineage of *G. violaceus* as the deepest branch on the cyanobacterial tree, particularly since there is some evidence that the recently obtained sequences of thermophilic origin may prove to branch even deeper (FERRIS & al. 1996).

"Plastid group". This group, corresponding to a subgroup of Branch F of WILMOTTE (1994), is comprised of plastid sequences of which three representatives were used in the studies reported here. The chlorophyll *b*-containing "green" plastids were represented by the land plant *Marchantia polymorpha*, the phycobilin and/or chlorophyll *c*-containing "red" plastids by the "primitive" unicellular red alga *Galdieria sulphuraria* ALLEN 14-1-1, and the glaucocystophyte (glaucophyte) plastids by *Cyanophora paradoxa* UTEX 555. Analyses of SSU rRNA and other sequence data show the plastids as a monophyletic group within the cyanobacterial line of descent. However, the reliability of these results is a matter of continuing debate and further discussion of this group will be deferred until later in this chapter.

"Thermophilic unicellular group". This small group is made up of three sequences derived from thermophilic unicellular cyanobacteria. Two of these are closely related strains described as *Synechococcus lividus* (strains Y-7c-s and C1) (WARD & al. 1994, FERRIS & al. 1996). The third member has been identified as *Synechococcus elongatus*, but at the molecular level it is clearly distinct from the suggested type strain for this species, *S. elongatus* PCC 6301 (see below). In the large trees, this group sometimes behaves as a branch basal to the dichotomy leading to the *Synechococus* and *Phormidium* groups. In others, it branches almost as deeply as the *Gloeobacter* and *Pseudanabaena* groups. Because of this inconsistent placement, it is accorded separate group status here.

***Pseudanabaena* group.** This, the other of the two consistently deep-branching subtrees, corresponds closely to Branch C of Wilmotte (1994). It behaved as a monophyletic group in 12 of the 16 large trees discussed previously. The F84 maximum likelihood tree for this group is shown in Fig. 2/1. It was inferred using *Gloeobacter* PCC 7421 and *Synechococcus* C9 as outgroup taxa.

RIPPKA & HERDMAN (1992) divide the genus *Pseudanabaena* into five clusters of which four are represented among the seven sequences described here including the reference strains for cluster 2 (PCC 7367) and cluster 4 (PCC 7403). Clusters 1 and 3 are represented by strains PCC 6903 and PCC 7409, respectively. *Synechococcus* sp. B10, a thermophilic isolate from Octopus Spring that probably corresponds to rDNA clone Octopus Spring type P, is also in the group and its provisional taxonomic identification clearly needs revision. An organism described as *Limnothrix redekei* (strain MEFFERT 6705) completes the group. This species has been designated as the type species for the genus *Limnothrix* (MEFFERT 1988, ANAGNOSTIDIS & KOMÁREK 1988). Its presence among a cluster of *Pseudanabaena* strains suggests that the genus *Limnothrix* should be subsumed within the genus *Pseudanabaena* or that the individual *Pseudanabaena* clusters be accorded separate genus status.

Reliability values (RV) for the branches within the *Pseudanabaena* subtree are generally high (60%–100%) except for that connecting strains PCC 7367 and PCC 7403. Quartet puzzling analysis places PCC 7367 as the deepest-branching member of the group with strong support (RV>87%) and a tree in which such placement occurs has a log likelihood less than 0.02% lower than the maximum value. Maximum parsimony analysis found a single most parsimonious tree that differed in topology with the F84 maximum likelihood tree but was not significantly worse according the KH test. In all, the various methods of analysis resulted in nine distinct topologies for the *Pseudanabaena* subtree. The KH test showed none to be significantly worse than the F84 maximum likelihood tree.

***Synechocystis/Pleurocapsa/Microcystis* (SY/PLEU/MY) group.** This is the largest and most disparate of the subtrees, containing 26 strains of unicellular and filamentous cyanobacteria. In spite of this apparent diversity, they form a monophyletic group in all 16 of the large cyanobacterial trees. This group corresponds to Branch G of WILMOTTE (1994). Ten of the strains are closely related members of the genus *Microcystis* and form a monophyletic subgroup within this subtree. These will be treated separately below. The *Microcystis* subgroup was represented by the single strain PCC 7806 in the inference of the F84 maximum likelihood tree for the larger group. This tree was rooted using *Synechococcus* PCC 6301 and *Leptolyngbya* PCC 73110 as outgroup taxa and is displayed in Fig. 2/2.

In addition to *Microcystis,* unicellular members of this group include strains of *Synechococcus*, *Synechocystis*, *Gloeothece*, *Gloeocapsa*, and the marine, exosymbiotic oxychlorobacterium *Prochloron didemni.* The large assemblages of *Synechococcus* and *Synechocystis* strains are recognized as groups in need of further generic division rather than as genera sensu stricto (RIPPKA & HERDMAN 1992). These assemblages were originally differentiated on the basis of patterns of

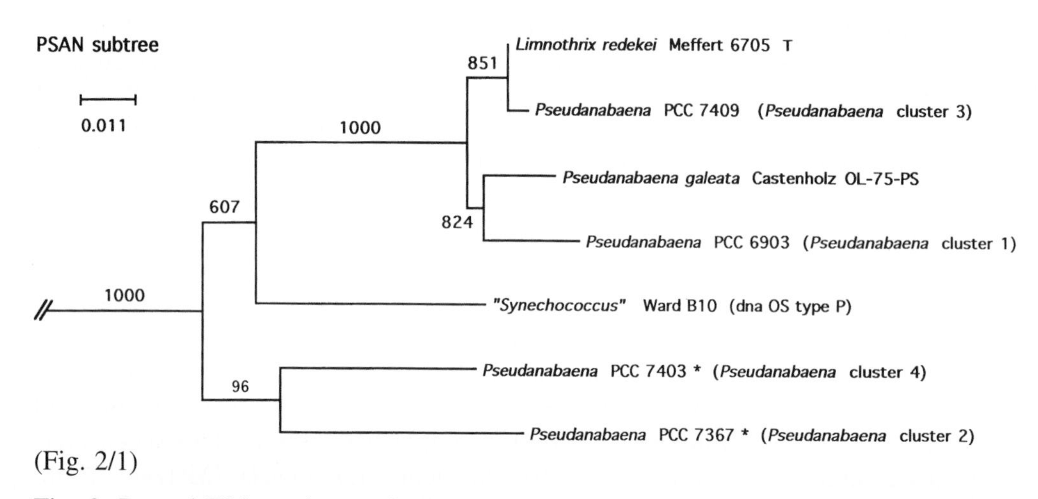

(Fig. 2/1)

Fig. 2. Rooted F84 maximum likelihood trees for individual cyanobacterial SSU rRNA sequence groups containing six or more taxa. Reference strains are denoted by an asterisk (*); (proposed) type strains are denoted by a capital T (T); (proposed) type species are denoted by a plus sign (+). Subgeneric assignments in the Stanierian taxonomic system sensu RIPPKA & HERDMAN (1992) or the 1989 edition of BERGEY'S Manual of Systematic Bacteriology are given in parentheses following the strain designations. Taxa liable to have been misidentified have their names in quotes. Alternative generic assignments for these are given in parentheses if available from the literature. Horizontal branch lengths are proportional to fixed point mutations per sequence position as indicated by scale bars. Note that scale bars are not uniform among trees. Vertical branches serve only to spread the trees for improved viewing. Numbers associated with internal branches indicate the number of times the subset of taxa defined by a particular branch were partitioned in 1000 quartet puzzled trees ("reliability value" × 10). Numbers corresponding to reliability values greater than 50% are shown in boldface. Approximate branch points of outgroup taxa are shown by a double slash (//). *Microcystis* PCC 7806 is boxed in the SY/PLEU/MY subtree to indicate that it represents the *Microcystis* subgroup. Abbreviations for strain designations are as follows: ACMM, Australian Collection of Marine Microorganisms (Australia); AWT, Australian Water Technologies (Australia); CCALA, Culture Collection of Autotrophic Organisms at Trebon (Czech Republic); CCAP, Culture Collection of Algae and Protozoa (UK); NIBB, National Institute of Basic Biology (Japan); NIES, National Institute for Environmental Studies (Japan); PCC, Pasteur Culture Collection (France); SAG, Sammlung von Algenkulturen (Germany); VRUC, 'Tor Vegata' University Collection of Rome (Italy). Strain designations not shown in this figure but given in the text are: ATCC, American Type Culture Collection (USA); CCMP, Provasoli-Gvillard National Center for Culture of Marine Phytoplankton (USA); NIVA, Norwegian Institute for Water Research (Norway); UTEX, University of Texas Culture Collection (USA). For further details, see text

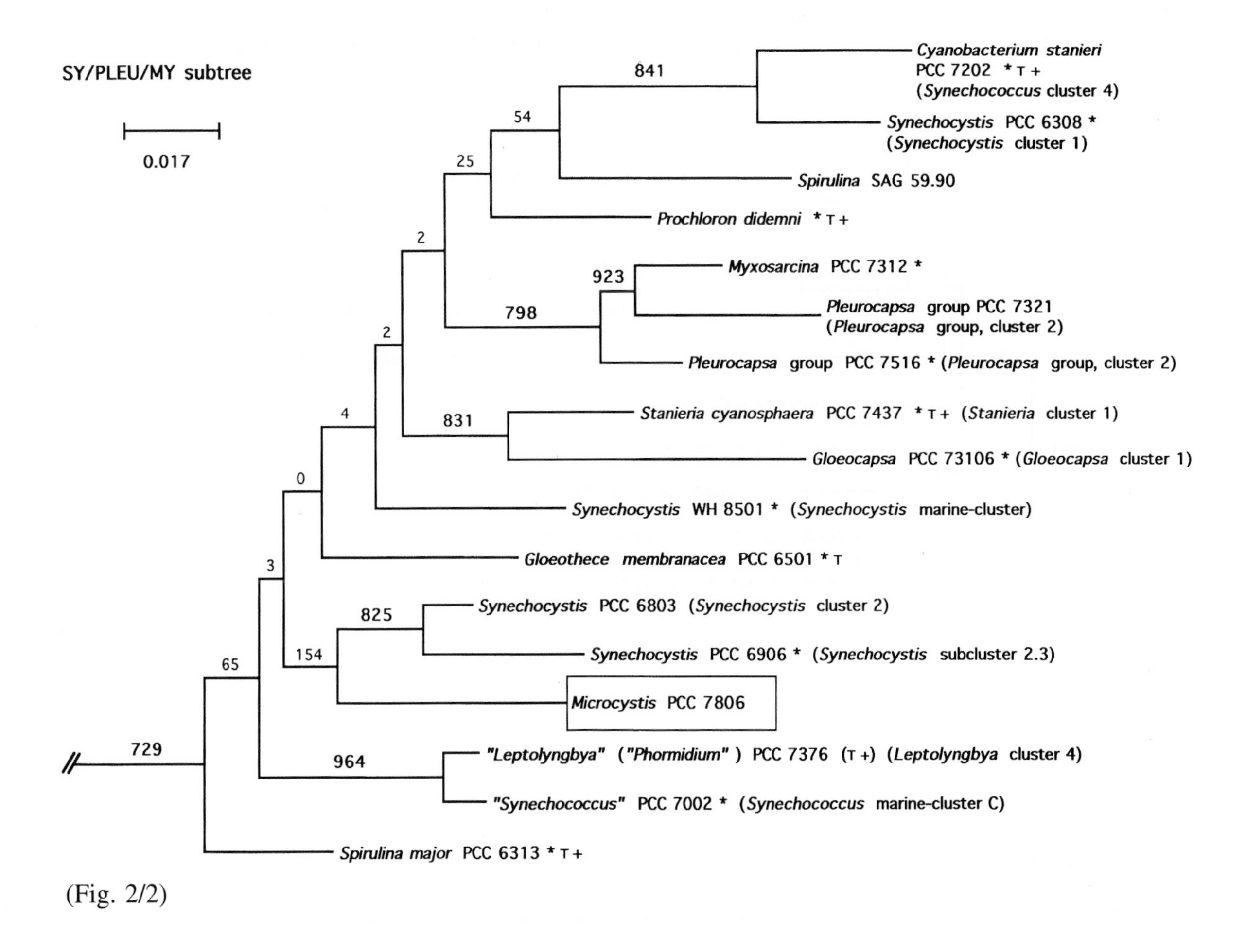
SY/PLEU/MY subtree
0.017
841
54
25
2
2
4
0
3
65
729
923
798
831
825
154
964
Cyanobacterium stanieri PCC 7202 * T + (Synechococcus cluster 4)
Synechocystis PCC 6308 * (Synechocystis cluster 1)
Spirulina SAG 59.90
Prochloron didemni * T +
Myxosarcina PCC 7312 *
Pleurocapsa group PCC 7321 (Pleurocapsa group, cluster 2)
Pleurocapsa group PCC 7516 * (Pleurocapsa group, cluster 2)
Stanieria cyanosphaera PCC 7437 * T + (Stanieria cluster 1)
Gloeocapsa PCC 73106 * (Gloeocapsa cluster 1)
Synechocystis WH 8501 * (Synechocystis marine-cluster)
Gloeothece membranacea PCC 6501 * T
Synechocystis PCC 6803 (Synechocystis cluster 2)
Synechocystis PCC 6906 * (Synechocystis subcluster 2.3)
Microcystis PCC 7806
"Leptolyngbya" ("Phormidium") PCC 7376 (T +) (Leptolyngbya cluster 4)
"Synechococcus" PCC 7002 * (Synechococcus marine-cluster C)
Spirulina major PCC 6313 * T +

(Fig. 2/2)

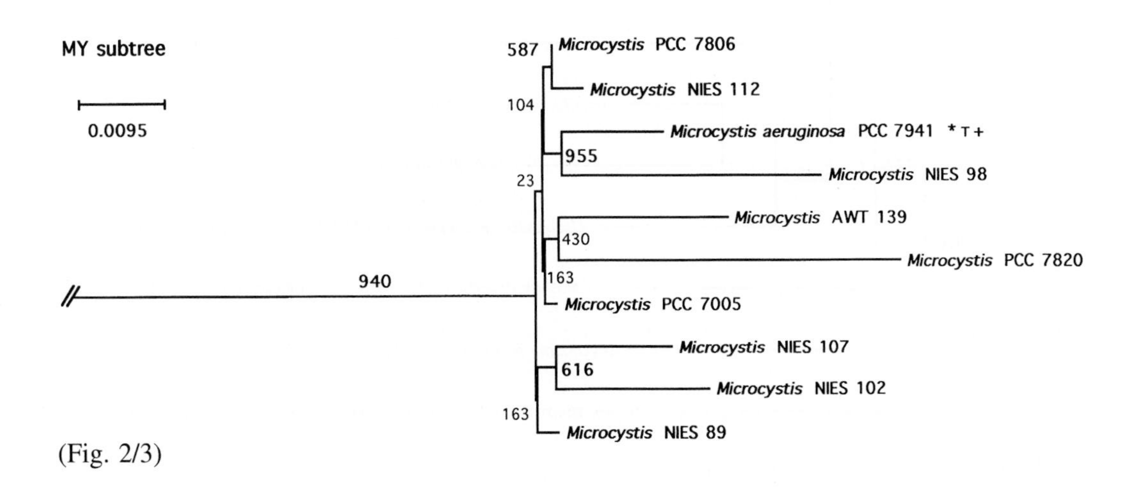
MY subtree
0.0095
940
587
104
23
955
430
163
616
163
Microcystis PCC 7806
Microcystis NIES 112
Microcystis aeruginosa PCC 7941 * T +
Microcystis NIES 98
Microcystis AWT 139
Microcystis PCC 7820
Microcystis PCC 7005
Microcystis NIES 107
Microcystis NIES 102
Microcystis NIES 89

(Fig. 2/3)

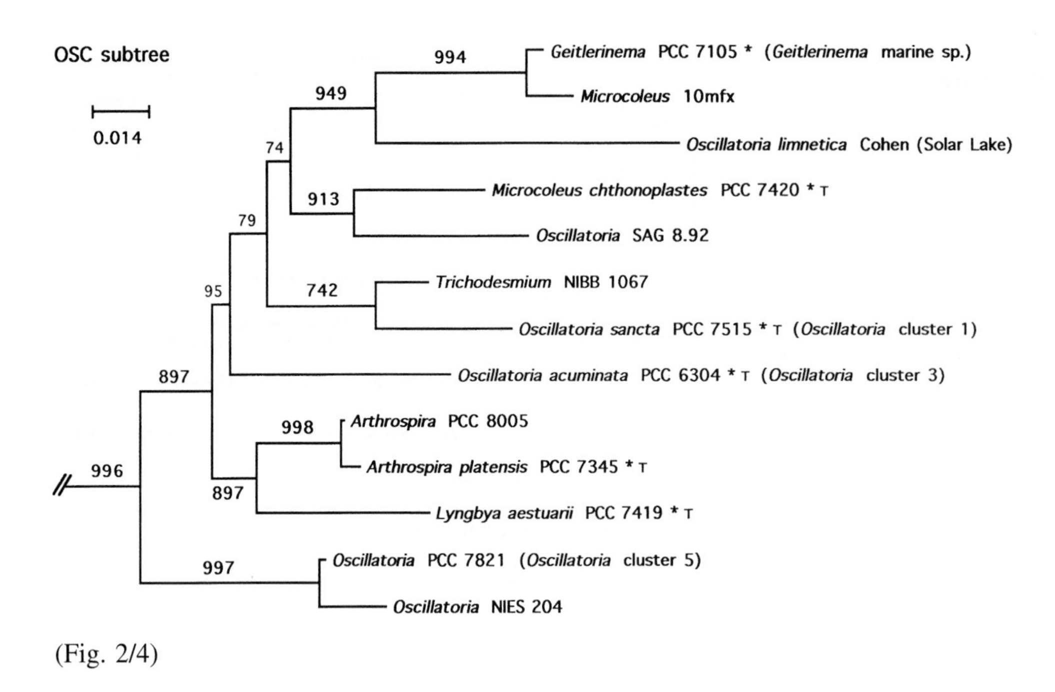
OSC subtree
0.014
994
949
74
913
79
95
742
897
998
996
897
997
Geitlerinema PCC 7105 * (Geitlerinema marine sp.)
Microcoleus 10mfx
Oscillatoria limnetica Cohen (Solar Lake)
Microcoleus chthonoplastes PCC 7420 * T
Oscillatoria SAG 8.92
Trichodesmium NIBB 1067
Oscillatoria sancta PCC 7515 * T (Oscillatoria cluster 1)
Oscillatoria acuminata PCC 6304 * T (Oscillatoria cluster 3)
Arthrospira PCC 8005
Arthrospira platensis PCC 7345 * T
Lyngbya aestuarii PCC 7419 * T
Oscillatoria PCC 7821 (Oscillatoria cluster 5)
Oscillatoria NIES 204

(Fig. 2/4)

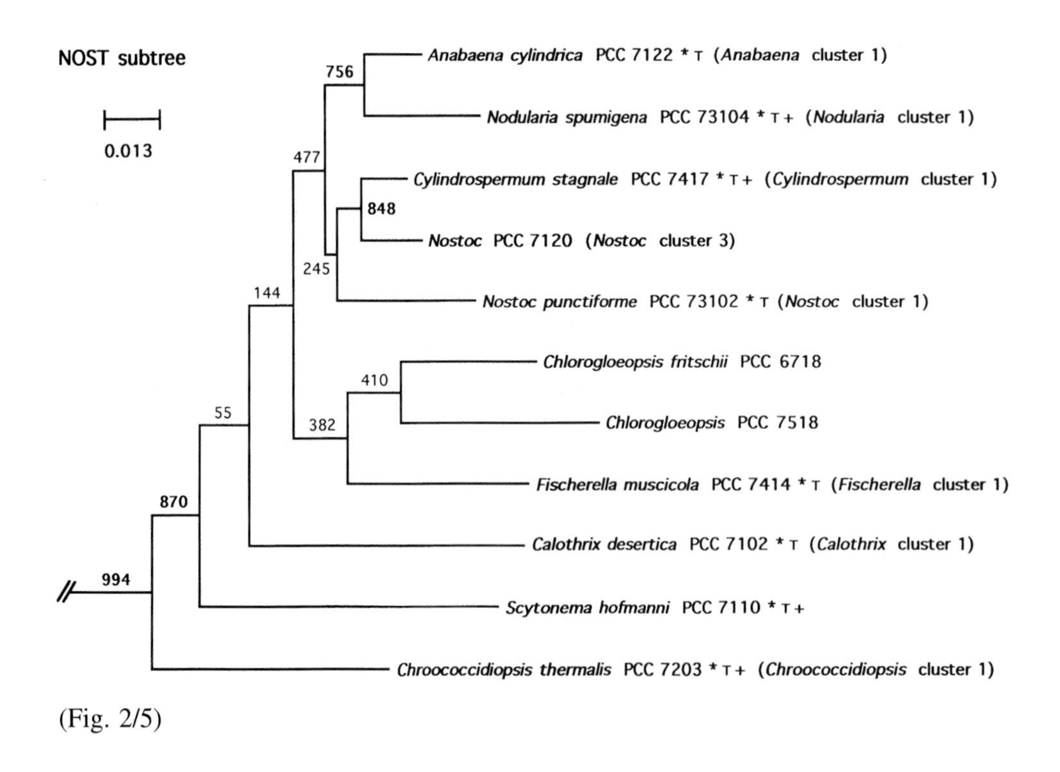
NOST subtree
0.013
756
477
848
245
144
410
55
382
870
994
Anabaena cylindrica PCC 7122 * T (Anabaena cluster 1)
Nodularia spumigena PCC 73104 * T + (Nodularia cluster 1)
Cylindrospermum stagnale PCC 7417 * T + (Cylindrospermum cluster 1)
Nostoc PCC 7120 (Nostoc cluster 3)
Nostoc punctiforme PCC 73102 * T (Nostoc cluster 1)
Chlorogloeopsis fritschii PCC 6718
Chlorogloeopsis PCC 7518
Fischerella muscicola PCC 7414 * T (Fischerella cluster 1)
Calothrix desertica PCC 7102 * T (Calothrix cluster 1)
Scytonema hofmanni PCC 7110 * T +
Chroococcidiopsis thermalis PCC 7203 * T + (Chroococcidiopsis cluster 1)

(Fig. 2/5)

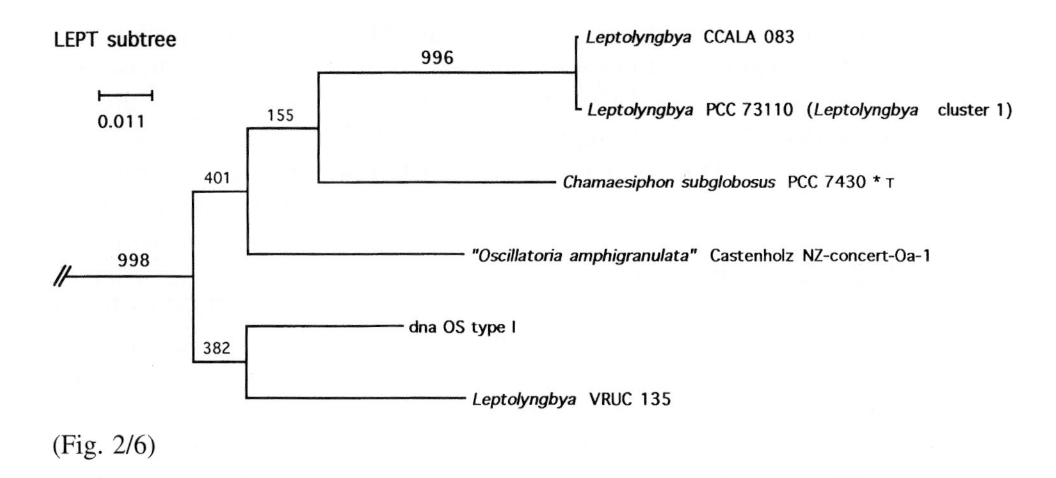

(Fig. 2/6)

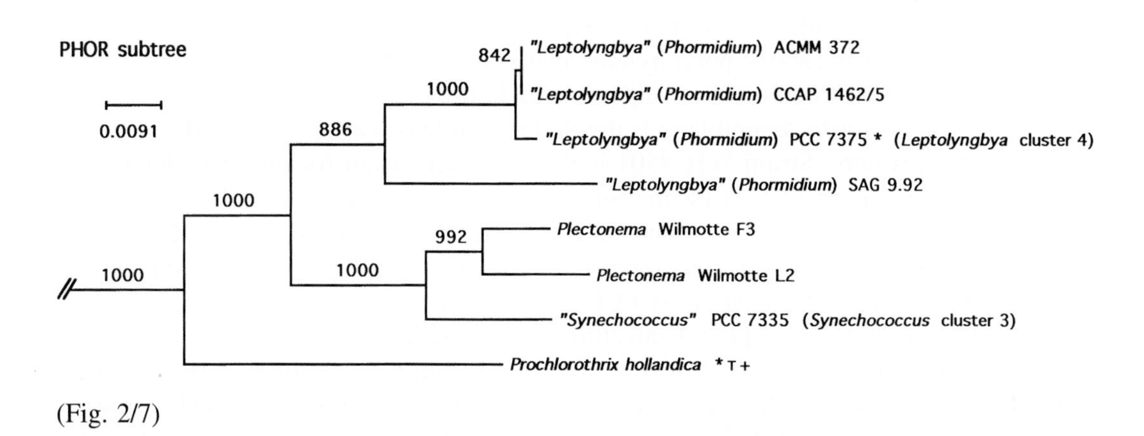

(Fig. 2/7)

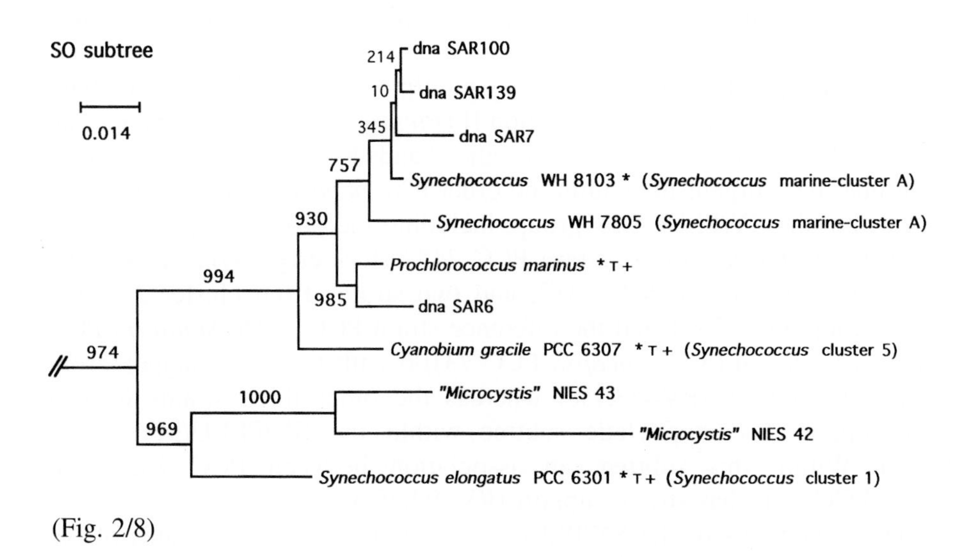

(Fig. 2/8)

cell division. Members of the *Synechococcus* assemblage undergo division in a single plane whereas members of the *Synechocystis* assemblage undergo division in two or three planes (RIPPKA & al.1979). The *Synechococcus* assemblage is divided into five clusters, two of which are represented here by PCC 7002 (cluster 3) and PCC 7202 (cluster 4). The former was designated the reference strain for *Synechococcus* marine-cluster C (WATERBURY & RIPPKA 1989) and the latter proposed as the type strain of the type species *Cyanobacterium stanieri* (RIPPKA & HERDMAN 1992). Strain PCC 7002 was originally designated as *Agmenellum quadruplicatum*, a taxonomic appellation that connotes multiplanar division. It was placed in the genus *Synechococcus* when the PCC culture did not exhibit this phenotype (RIPPKA & al. 1979). However, the suggested type species *Synechococcus elongatus* PCC 6301 branches distinctly apart from the *Synechococcus* strains that fall within the SY/PLEU/MY group (see below). The results presented in this study indicate that strains PCC 7002 and PCC 7202 appear to be more closely related to *Synechocystis* strains and may represent lines of descent that have lost the ability to divide in more than one plane. The *Synechococcus* assemblage as currently defined is clearly polyphyletic at the level of SSU rRNA with members scattered in at lest three distinct subtrees (see below).

The *Synechocystis* assemblage is divided into at least two clusters and has four representatives here. Strain WH 8501 is the reference strain for the *Synechocystis* marine-cluster, strain PCC 6308 the reference strain for cluster 1, and strain PCC 6906 the reference strain for subcluster 2.3 (WATERBURY & RIPPKA 1989, RIPPKA & HERDMAN 1992). Cluster 2 is also represented by strain PCC 6803. A phylogenetic association between *Synechocystis* PCC 6308 and *Cyanobacterium* PCC 7202 is moderately supported by quartet puzzling analysis (RV=84.1%).

Strains of the genus *Gloeocapsa* are divided into two clusters (RIPPKA & HERDMAN 1992). One of these, cluster 1, is represented in the SY/PLEU/MY group by the reference strain PCC 73106, and the genus *Gloeothece* by the proposed type strain of the neotype species *G. membranacea* PCC 6501. A discussion of *Prochloron* will be included in a section on the oxychlorobacteria elsewhere in this chapter.

Unicellular cyanobacteria that replicate by multiple rapid cell divisions without concomitant cell growth are placed in Section II (*Pleurocapsales*) of the Stanierian system (RIPPKA & HERDMAN 1992). On the other hand, KOMÁREK & ANAGNOSTIDIS (1986) subsume these organisms within the order *Chroococcales*. Here, four such strains fall within the SY/PLEU/MY group, including the proposed type strain of the type species *Stanieria cyanosphaera* PCC 7437 (formerly *Dermocarpa*), the reference strain *Myxosarcina* PCC 7312, and two strains from cluster 2 of the *Pleurocapsa* group, PCC 7321 and the reference strain PCC 7516. *Stanieria* PCC 7437 acts as sister taxon to *Gloeocapsa* PCC 73106 with moderate support from quartet puzzling analysis (RV=83.1%) whereas the other three strains form a moderately supported monophyletic subtree within the SY/PLEU/MY group (RV=79.8%). Within this subtree, an association between PCC 7321 and *Myxosarcina* PCC 7312 has strong support (RV=92.3%).

Filamentous members of the SY/PLEU/MY group include two strains of the genus *Spirulina* which has a very distinctive morphology. One is the proposed type strain of the type species *Spirulina major* PCC 6313 (RIPPKA & HERDMAN 1992).

The other is the marine epiphytic strain SAG 59.90 (sequence from strain WILMOTTE P7). The third filamentous member of the SY/PLEU/MY group is *Leptolyngbya* PCC 7376 which had been proposed as the type strain of the type species *Phormidium fragile* (GUGLIELMI & COHEN-BAZIRE 1984b) but is placed in cluster 4 of the genus *Leptolyngbya* by RIPPKA & HERDMAN (1992). However, it groups distinctly apart from the reference strain for this cluster, PCC 7375, as well as from other strains placed in different clusters of *Leptolyngbya*. On the other hand, strains formerly regarded as members of the genus *Phormidium* are closely related to PCC 7375 and not to PCC 7376 (see below). Moreover, PCC 7376 is strongly supported as being a relative of the unicellular strain *Synechococcus* PCC 7002 (RV= 96.4%). Its placement in the largely unicellular SY/PLEU/MY group may be an indication that it has evolved to a filamentous form from a unicellular ancestor.

Many of the internal branches of the SY/PEU/MY group have virtually no support at all (RV < 10%). This is due in large part to the disruption of an association of the two *Spirulina* strains that received moderate support from quartet puzzling analysis (RV= 72.2%). The *Spirulina* strains are placed far apart in the F84 maximum likelihood tree shown in Fig. 2 and as a consequence, the intermediate branches separating them are unsupported by the quartet puzzling analysis. Another moderately supported association not seen in the SY/PLEU/MY subtree in Fig. 2/2 is between *Gloeothece membrancacea* PCC 6501 and the *Microcystis* representative PCC 7806 (RV= 74.4%).

A heuristic maximum parsimony tree search resulted in two equally parsimonious trees one of which was identical in topology to the F84 maximum likelihood tree. Altogether, 21 alternate topologies were inferred with the various methods of analysis and only one of these was found to be significantly worse than the one shown in Fig. 2 according to the KH test. The relatively weak support for the monophyly of the group when analyzed apart from the other groups (RV= 72.9%) provides further evidence that the *Synechocystis/Pleurocapsa/Microcystis* group is the most unresolved of the subtrees examined in this study and requires much additional work.

Microcystis subgroup. Ten very closely related strains designated as *Microcystis* consistently formed a monophyletic subtree within the SY/PLEU/MY group even though no two SSU rRNA sequences were identical among these strains. The subgroup includes the proposed type strain of the type species *M. aeruginosa* PCC 7941 (RIPPKA & HERDMAN 1992). The F84 maximum likelihood tree is shown in Fig. 2/3 and was rooted using the sequences of *Pleurocapsa* PCC 7516 and *Synechococcus* PCC 6301 as outgroup taxa.

The *Microcystis* subtree displays a fanlike radiation but with some terminal branches notably longer than others, an indication that SSU rRNA genes have evolved at different rates among these strains. Quartet puzzling analysis reveals only one strongly supported phylogenetic relationship, that between the type strain PCC 7941 and NIES 98 (RV > 95%). Two other associations are weakly supported, one between strains NIES 107 and NIES 102 (RV= 61.6%), and the other between strains PCC 7806 and NIES 112 (RV= 58.7%). The lack of resolution among these sequences resulted in 22 topologies being produced by the various methods of analysis, not including maximum parsimony analysis. Of the 21 that differ from the maximum likelihood tree, only five were found to be significantly worse according to the KH test. In a branch-and-bound maximum parsimony tree search under the constraint that all trees be strictly bifurcating, a total of 208 equally most parsimonious trees were found, one of which was identical in topology to the F84 maximum likelihood tree, and none of the other 207 being significantly worse. One is unlikely to find a better example of the limitations of SSU rRNA data for discerning evolutionary relationships among such closely related taxa.

***Oscillatoria* group.** This group, composed of 13 filamentous strains, behaved as a monophyletic subtree in 12 of the 16 large cyanobacterial trees and corresponds closely to Branch F of WILMOTTE (1994). Its F84 maximum likelihood tree is shown in Fig. 2/4 and was inferred using *Synechococcus* PCC 6301 and *Leptolyngbya* PCC 73110 as outgroup taxa.

Six genera are represented in the *Oscillatoria* group. RIPPKA & HERDMAN (1992) divide the genus *Oscillatoria* into five clusters, three of which are represented here. Cluster 1 is represented by strain PCC 7515, the proposed type strain for *O. sancta*; cluster 3 by PCC 6304, the proposed type strain for *O. acuminata*; and cluster 5 by PCC 7821 (sequence from strain NIVA-CYA 18). The SSU rRNA sequence of this last strain is nearly identical with that of *Oscillatoria* NIES 204. Both strains have been described as *O. aghardii* for which ANAGNOSTIDIS & KOMÁREK (1988) propose the genus *Planktothrix*. The deep location of these two strains within the *Oscillatoria* subtree would be consistent with such an assignment.

Two closely related strains of *Arthrospira*, PCC 8005 and PCC 7345, the latter being the proposed type strain for *A. platensis*, are associated with PCC 7419, the proposed type strain for *Lyngbya aestuarii*. Two strains of *Microcoleus*, 10mfx and PCC 7420, do not appear as close relatives in this tree. The latter is the reference strain for *Microcoleus chthonoplastes* and shows a close relationship to the marine epiphyte *Oscillatoria* SAG 8.92. Strain 10mfx appears to be a close relative to *Geitlerinema* PCC 7105 which is the reference strain for marine species of *Geitlerinema*. This pair of sequences is itself closely associated with a strain of *Oscillatoria limnetica* that is capable of anaerobic anoyxgenic photosynthesis (COHEN & al. 1975). A proposal was made to reclassify this isolate of *O. limnetica* as *Phormidium hypolimneticum* (CAMPBELL 1985). However, its inclusion in the subtree that contains all other strains of *Oscillatoria* rather than in subtrees containing *Phormidium* strains (see below) argues for retention of the original taxonomic assignment.

The marine planktonic strain *Trichodesmium* NIBB 1067 acts as a sister taxon to *O. sancta* PCC 7515 in the subtree shown in Fig. 2/4. The situation is similar to that between *Limnothrix* and *Pseudanabaena* in that the result indicates a need to divide a large taxon, in this case *Oscillatoria*, into smaller units ("splitting") or to subsume one or more taxa into another ("lumping").

In most cases, the reliability values for the internal branches of the *Oscillatoria* subtree are quite high (RV $>$ 89%) although there are three branches notable for their lack of support (RV $<$ 10%). Maximum parsimony analysis resulted in a single most parsimonious tree identical in topology with the F84 maximum likelihood tree. In all, analyses gave rise to six distinct topologies of which only two were not significantly worse than the maximum likelihood/parsimony tree according to the KH test. This is another indication of the fairly robust topology for this group.

***Nostoc* group.** Filamentous strains capable of forming specialized cells for dinitrogen fixation (heterocysts or heterocytes) and as metabolically quiescent "resting" stages (akinetes) constitute the majority of this group, the F84 maximum likelihood subtree of which is shown in Fig. 2/5. The subtree contains eleven strains and was inferred using *Synechococcus* PCC 6301 and *Oscillatoria* PCC 7821 as outgroup taxa. It formed a monophyletic subtree in all 16 of the large trees, corresponding largely to Branch H of WILMOTTE (1994).

Among the heterocystous members of the *Nostoc* group are the type strains of the type species *Nodularia spumigena* PCC 73104, *Cylindrospermum stagnale* PCC 7417 (sequence from strain ATCC 29204), and *Scytonema hofmanni* PCC 7110. Also present are the type strains for the species *Anabaena cylindrica* PCC 7122, *Nostoc punctiforme* PCC 73102, *Calothrix desertica* PCC 7102, and *Fischerella muscicola* PCC 7414. RIPPKA & HERDMAN (1992) divide the genera *Anabaena*, *Cylindrospermum*, and *Nodularia* each into two clusters; the genus Nostoc into four; and the genus *Calothrix* into three. Cluster 1 of each of these genera is represented in the *Nostoc* subtree of Fig. 2/5. Cluster 3 of *Nostoc* is represented by strain PCC 7120. There has been some controversy as to whether this strain should be designated as *Anabaena* or *Nostoc* (RIPPKA 1988). Phylogenetic analysis of the SSU rRNA data indicates this strain as having a closer relationship with the proposed type species of the genus *Cylindrospermum* (RV = 84.8%) than with the proposed type strain of *Anabaena cylindrica*, and similarly, of *Anabaena* PCC 7122 with the proposed type species *Nodularia spumigena* PCC 73104 (RV = 75.6%). The proposed type strain *Nostoc punctiforme* PCC 73102 shows an affinity for the *Cylindorspermum*/PCC 7120 grouping but with no significant support (RV = 24.5%). However, in the quartet puzzling analysis an association including *Anabaena* PCC 7122, *Nodularia* PCC 7417, and strain PCC 7120 was found with even lower frequency (RV = 8.1%). While these results certainly do not settle the matter, they do favor at some level an assignment of strain PCC 7120 to the (paraphyletic) genus *Nostoc* rather than *Anabaena*.

Heterocystous, filamentous cyanobacteria that undergo cell division in more than one plane are placed in a separate section by RIPPKA & HERDMAN (1992) that would comprise in part the order *Stigonematales* sensu ANAGNOSTIDIS & KOMÁREK (1990). Three representatives are present in the *Nostoc* subtree of Fig. 2/5 including *Chlorogloeopsis fritschii* PCC 6718 and the proposed type strain *Fischerella muscicola* PCC 7414 which represents one of two clusters into which the genus *Fischerella* is divided by RIPPKA & HERDMAN (1992). Also present is *Chlorogloeopsis* PCC 7518, a strain that has lost its ability to form heterocysts as a result of prolonged maintenance in culture medium containing a source of fixed nitrogen. Its taxonomic characterization was made particularly difficult due to the resulting phenotype and was not resolved until recently by the use of SSU rRNA analysis (WILMOTTE & al. 1993). On the basis of molecular phylogenetic analysis, it would appear that the branching heterocystous cyanobacteria arose from within a larger assemblage made up of filamentous forms that divide in a single plane. The monophyly of the *Fischerella* and *Chlorogloeopsis* strains in Fig. 2/5 is not significantly supported (RV = 38.2%), and whether this phenotype arose more than once is unresolved by the present analysis. However, it is noteworthy that cyanobacteria in symbiotic association with the moss *Blasia pusilla* and assigned to the genus *Nostoc* demonstrate multiplanar cellular division that can be stably maintained when they are isolated and grown in culture (GORELOVA & al. 1996).

The remaining and deepest-branching member of the *Nostoc* group is of particular interest. It is the proposed type strain of the type species *Chroococcidiopsis thermalis* PCC 7203. *Chroococcidiopsis* is a genus that traditionally has been classified as a member of the order *Pleurocapsales* (Section II of RIPPKA & HERDMAN 1992). However, the SSU rRNA sequence data for other pleurocapsalean strains clearly place them in the SY/PLEU/MY group and separate from *Chroococcidiopsis* PCC 7203 (see above). There is no evidence to suggest that strain PCC 7203 was ever capable of differentiating heterocysts. However, this specialized cell type may have arisen after the divergence of *Chroococcidiopsis* and the progenitor(s) of the other members of the *Nostoc* group from a common ancestor. Intriguingly, under conditions of nitrogen starvation another *Chroococcidiopsis* strain is able to differentiate "survival" cells that are functionally comparable to akinetes (BILLI & GRILLI CAILOA 1996). Moreover, BRYANT (1982) demonstrated that strain PCC 7203 contains the phycobilin pigment phycoerythrocyanin which at that time was otherwise found only in heterocystous cyanobacteria. More recently, this pigment has also been found in a number of strains of the nonheterocystous, filamentous cyanobacterium *Microcoleus chthonoplastes* (PRUFERT-BEBOUT & GARCIA-PICHEL 1994). However, these isolates are likely to be distinct from the reference strain *Microcoleus chthonoplastes* PCC 7420 which lacks phycoerythrocyanin (RIPPKA & al. 1979) (see above). Molecular sequence data from the putative *Microcoleus* strains containing this pigment would be very helpful in determining its utility as a phylogenetic marker. The association of *Chroococcidiopsis* PCC 7203 with the heterocystous cyanobacteria and clearly apart from those strains near the root of the cyanobacterial tree contradicts the hypothesis suggested by others that this genus represents the most primitive cyanobacterial type (FRIEDMANN & al. 1994, GRILLI CAIOLA & al. 1996).

The topology of the *Nostoc* subtree shown in Fig. 2/5 is very poorly resolved with most branches having negligible support (RV < 50%). Aside from the associations discussed above involving *Anabaena* and *Nostoc* strains, the only other significantly supported branch places *Chroococcidiopsis* PCC 7203 as basal to the heterocystous cyanobacteria (RV = 87.0%). However, two groupings not present on the tree in Fig. 2/5 also received support in the quartet puzzling analysis. One is a pairing of *Calothrix* PCC 7102 with *Scytonema* PCC 7110 (RV = 64.1%) and the other a quartet made up of *Anabaena* PCC 7122, *Nodularia* PCC 73104, *Cylindrospermum* PCC 7417, and *Nostoc* PCC 7120 (RV = 52.5%). A branch-and-bound maximum parsimony tree search resulted in four equally most parsimonious trees, one of which was identical in topology to the F84 maximum likelihood tree.

In all, 17 topologies distinct from that with the maximum likelihood value were examined by the KH test and only six, including the three alternative parsimony trees, were not significantly worse. These results indicate that several subgroupings within the *Nostoc* group are unresolved, but the members of different subgroups cannot be freely exchanged without producing topologies that are considered statistically inferior.

***Leptolyngbya* group.** This small assemblage includes those sequences that make up Branch E of WILMOTTE (1994) as well as others. These include sequences from an environmental rDNA clone, a unicellular cyanobacterium, and four filamentous stains of cyanobacteria, three of which have been placed in the recently proposed genus *Leptolyngbya* (ANAGNOSTIDIS & KOMÁREK 1988). These sequences formed a monophyletic subtree in only three of the 16 large trees, specifically those inferred by the neighbor-joining algorithm using distances based on the JN90 model of evolution. Nonetheless, they are considered as a distinct group here on the basis of a corroborating result produced by a maximum likelihood analysis of a large subset of the complete cyanobacterial SSU rRNA sequence database (see below). The F84 maximum likelihood tree of the *Leptolyngbya* group is shown in Fig. 2/6. Outgroup taxa used in its inference were *Gloeobacter* PCC 7321 and *Pseudanabaena* PCC 7367.

The genus *Leptolyngbya* was created to accommodate a diverse set of cyanobacteria that had been placed in the provisional assemblage "LPP-group B" largely typified by growth as thin uniseriate trichomes enclosed within investing sheaths (RIPPKA & al. 1979, ANAGNOSTIDIS & KOMÁREK 1988). This taxonomic appellation has been accepted by RIPPKA & HERDMAN (1992) who divide the genus into four clusters, one of which, cluster 1, is represented by strain PCC 73110 in the tree shown in Fig. 2/6. This sequence is nearly identical with that of *Leptolyngbya* CCALA 083 (sequence from strain KOMÁREK 1964/112). The strain named *Oscillatoria amphigranulata* is likely to have been misidentified as it clearly groups separately from other strains of the genus *Oscillatoria* (see above). It shares in common with the environmental rDNA clone Octopus Spring type I a thermophilic phenotype, both being isolated from hot springs. *Leptolyngbya* VRUC 135 is likely to belong to a different cluster than PCC 73110 and CCALA 083. Additional sequence information from members of the other *Leptolyngbya* clusters is necessary to confirm this result.

The proposed type strain PCC 7430 of the unicellular species *Chamaesiphon subglobosus* also falls within this group. Members of this genus are unusual in that they replicate by budding (asymmetric cell division) while most other unicellular chroococcalean cyanobacteria produce daughter cells of equal size. Its presence is the cause for the low occurrence of monophyly of the *Leptolyngbya* group in the 16 large cyanobacterial trees. In 13 of these trees, PCC 7430 is placed as a basal member of the *Nostoc* group. However, an F84 maximum likelihood tree inference incorporating 76 cyanobacterial strains, the three plastid sequences, and the two outgroup sequences supports the placement of PCC 7430 within the *Leptolyngbya* subtree (not shown). Regardless of this result, there is no question that *Chamaesiphon* PCC 7430 is the most problematic strain in terms of determining its phylogenetic relationships with other cyanobacteria. Sequence data from other strains of *Chamaesiphon* are essential for resolution of this question.

Aside from the association between the *Leptolyngbya* strains PCC 73110 and CCALA 083, the support for the branches of the *Leptolyngbya* subtree in Fig. 2/6 is very poor (< 41%). However, a grouping of these two strains with *Oscillatoria amphigranulata* is weakly supported by quartet puzzling analysis (RV= 54.7%) and an F84 maximum likelihood tree with such a topology has log likelihood value that is less than 0.03% lower than that of the tree shown in Fig. 2.

A maximum parsimony analysis of the sequence data yielded a single most parsimonious tree identical in topology to that shown here. In all, four alternate topologies were examined and the KH test showed none to be significantly worse than the maximum likelihood tree.

***Phormidium* group.** This group contains sequences from six filamentous cyanobacterial strains, one unicellular strain, and the filamentous oxychlorobacterium *Prochlorothrix hollandica*. It is monophyletic in all 16 large cyanobacterial trees and corresponds in part to Branch D of Wilmotte (1994). Its F84 maximum likelihood tree is shown in Fig. 2/7. Strictly speaking, the name of the group might be considered a misnomer as it contains no strains presently recognized as being of the genus *Phormidium*. Four strains in the group, SAG 9.92 (sequence from strain Wilmotte D5), CCAP 1462/5, PCC 7375, and ACMM 372 (sequence from strain Wilmotte N182), were originally designated as *Phormidium* (Wilmotte 1991; 1994) but were subsequently reclassified as *Leptolyngbya* upon the acceptance of this genus within the Stanierian system (Nelissen & al. 1996). All are of marine origin and one of these, PCC 7375, is the reference strain for cluster 4 of *Leptolyngbya* (Rippka & Herdman 1992) but clearly branches apart from the other strains that comprise the *Leptolyngbya* group (see above). Hence, the name *Phormidium* is used here to distinguish between these two groups. Similar to the *Synechococcus* assemblage, analyses of SSU rRNA sequences of the *Leptolyngbya* strains scatter them among three separate groups. On this basis, it is suggested here that *Leptolyngbya* also be considered as an assemblage in need of further generic division rather than as a genus sensu stricto.

Two closely related marine strains, F3 and L2, are assigned to the genus *Plectonema* (Wilmotte 1991) and show a strong relationship to the unicellular marine strain *Synechococcus* PCC 7335 that is placed in cluster 3 of the *Synechococcus* group (Rippka & Herdman 1992). Similar to other unicellular strains designated as *Synechococcus*, its phylogenetic association separates it from the suggested type species *Synechococcus elongatus* PCC 6301 (see below), illustrating again the difficulty of inferring evolutionary relationships among morphologically simple cyanobacteria in the absence of molecular data. A discussion of *Prochlorothrix hollandica* will be included in a section on the oxychlorobacteria elsewhere in this chapter.

Quartet puzzling reliability values for all branches of the *Phormidium* subtree are high (84–100%) and the topology shown in Fig. 2/7 occurred in the large majority of the various analyses including the single maximum parsimony tree. In all, only three alternate topologies were produced, none of which were significantly worse than the one shown here according to the KH test.

***Synechococcus* group.** Eleven sequences make up this group, four of which are derived from marine environmental rDNA clones and the remaining seven being from unicellular strains, including the marine planktonic oxychlorobacterium *Prochlorococcus marinus*. A number of these sequences make up Branch A and part of Branch D of Willmotte (1994). The *Synechococcus* group behaved as a monophyletic subtree in all 16 of the large cyanobacterial trees, and its F84 maximum likelihood tree is shown in Fig. 2/8. It was rooted using *Leptolyngbya* PCC 73110 and *Plectonema* F3 as outgroup taxa.

The rDNA clones are from studies of picoplankton in the Sargasso Sea and as such are almost certainly derived from unicellular organisms (Giovannoni & al. 1990, Britschgi & Giovannoni 1991). Three of these, SAR7, SAR100, and

SAR139, fall within a subgroup that includes the marine unicellular cyanobacteria *Synechococcus* WH 7805 and *Synechococcus* WH 8103. The latter has been suggested as the reference strain for marine-cluster A of the *Synechococcus* assemblage (WATERBURY & RIPPKA 1989). The fourth rDNA clone, SAR6, behaves as a close relative of the oxychlorobacterium *Prochlorococcus marinus* (RV= 98.5%) and may well be derived from a similar organism. These seven strains are associated with the proposed type strain of the type species *Cyanobium gracile* PCC 6307 (RV= 93.0%) which is placed in cluster 5 of the *Synechococcus* group (RIPPKA & HERDMAN 1992).

Synechococcus cluster 1 is represented by the proposed type strain of the type species *Synechococcus elongatus* PCC 6301 which is associated in the tree with two strains described as *Microcystis*, NIES 42 and NIES 43. However, the position of these two strains within the large cyanobacterial trees is clearly different from other strains of *Microcystis* including the proposed type species *M. aeruginosa* PCC 7941 (see above). Hence, they are likely to have been misidentified.

Quartet puzzling analysis showed most of the groupings within the *Synechococcus* subtree to be robust (RV= 75.7–100%). Only the branches involving *Synechococcus* WH 8103 and three of the environmental rDNA clones were without support (RV < 35%). Different arrangements of these taxa were supported by quartet puzzling analysis and not present in the tree shown in Fig. 2/8. These are a pairing of rDNA clone SAR139 with *Synechococcus* WH 8103 (RV = 58.6%) and an association involving these two sequences and rDNA clone SAR100 (RV = 61.9%). A tree including these topological features and otherwise unchanged from that shown in Fig. 2/8 had a log likelihood value that was less than 0.006% lower than the maximum likelihood value.

A maximum parsimony analysis of these sequence data was not possible with the same outgroup taxa because under such circumstances the ingroup behaved paraphyletically. The relatively long branch lengths of the two putative *Microcystis* strains caused *Plectonema* F3 to be drawn into the *Synechococcus* group. Substitution of strain F3 by *Pseudanabaena* PCC 7367 alleviated this problem and resulted in no change of ingroup topology under F84 maximum likelihood tree inference. Three equally most parsimonious trees were inferred by maximum parsimony in this way, one of which was identical in topology to the maximum likelihood tree and the other two of which wre not significantly worse according to the KH test. The anomalous behavior of *Plectonema* F3 did not occur with the other methods of analysis except in a single instance, showing that the choice of outgroup can have a disproportionately adverse impact on phylogenetic analyses when parsimony methods are employed (BAUM & ESTABROOK 1996). This serves as a cautionary example against the indiscriminate use of such methods with some of these data. In all, 18 alternate topologies were examined and only nine were found not to be significantly worse than that of the maximum likelihood subtree.

Molecular clock hypothesis

Examination of the molecular clock hypothesis for the individual groups led to mixed results. This hypothesis was not rejected at the 5% significance level for the *Pseudanabaena*, *Leptolyngbya*, and *Phormidium* groups ($p > 0.05$). It is noteworthy that in the case of the *Leptolyngbya* group, the tree inferred under the assumption of a molecular clock had a rooted topology that differed from that of the subtree rooted by the inclusion of outgroup taxa, but only by less than 0.03% in log likelihood value. Similar to maximum parsimony analysis of the *Synechococcus* group, this demonstrates that in some instances the choice of outgroup taxa may influence the topology of the ingroup, although much less dramatically in this case.

The molecular clock hypothesis was rejected for the *Nostoc* group ($p < 0.01$), as well as the *Synechococcus*, the *Oscillatoria*, the *Synechocystis/ Pleurocapsa/ Microcystis* group, and the *Microcystis* subgroup ($p \ll 0.001$). In the case of the *Synechococcus* group, this result may explain in part the unusual root location in the cyanobacterial tree inferred by WILMOTTE (1994). In that instance, a form of midpoint rooting was used, which implicitly assumes clock-like behavior, resulting in the SAR rDNA and *Prochlorococcus* sequences being partitioned from the other *Synechococcus/Phormidium* supergroup sequences and placed as the deepest branch on the cyanobacterial tree with some statistical support (bootstrap resampling proportion > 50%). The assumption of a molecular clock implicit in midpoint rooting is liable to be invalid for these data. However, it must be emphasized that in the study reported here, the molecular clock hypothesis was examined only within individual groups and not among them. It remains to be verified that groups lacking internal clock-like behavior will also fail to show such behavior when compared to each other.

Evolutionary relationships among cyanobacterial groups

Among the sixteen large cyanobacterial trees inferred by distance methods, the topologies are inconsistent in terms of the relative branching orders among the ten principal sequence groups. The result of an F84 maximum likelihood analysis on a subset of the complete sequence data set is shown schematically in Fig. 3 (see also above). The tree is highly asymmetric, with only *Synechococcus* and *Phormidium* subtrees, and the *Oscillatoria* and *Synechocystis/Pleurocapsa/Microcystis* subtrees behaving as sister groups. This tree also indicates the *Nostoc* and *Leptolyngbya* subtrees as being nearest-neighbors among the major groups of cyanobacteria. This result may help explain the tendency of *Chamaesiphon* PCC 7204 to switch its association between these two groups in the 16 large trees inferred by distance methods. As in some of the distance trees, the thermophilic unicellular group branches quite deeply. If this accurately reflects the phylogenetic position of these organisms, they may be considered as part of the paraphyletic grade made up of the members of the *Gloeobacter* group which also contains sequences of thermophilic origin.

Maximum likelihood methods have been shown to be particularly robust to violations of the assumptions underlying various models of molecular evolution used in phylogenetic tree inference (HILLIS & al. 1994, KUHNER & FELSENSTEIN 1994, YANG 1994). As such, the tree shown in Fig. 3 may be considered reasonably reliable in terms of the branching order among the cyanobacterial groups. However, it is the result of a preliminary analysis and should by no means be considered definitive. Much additional work remains to be done before the

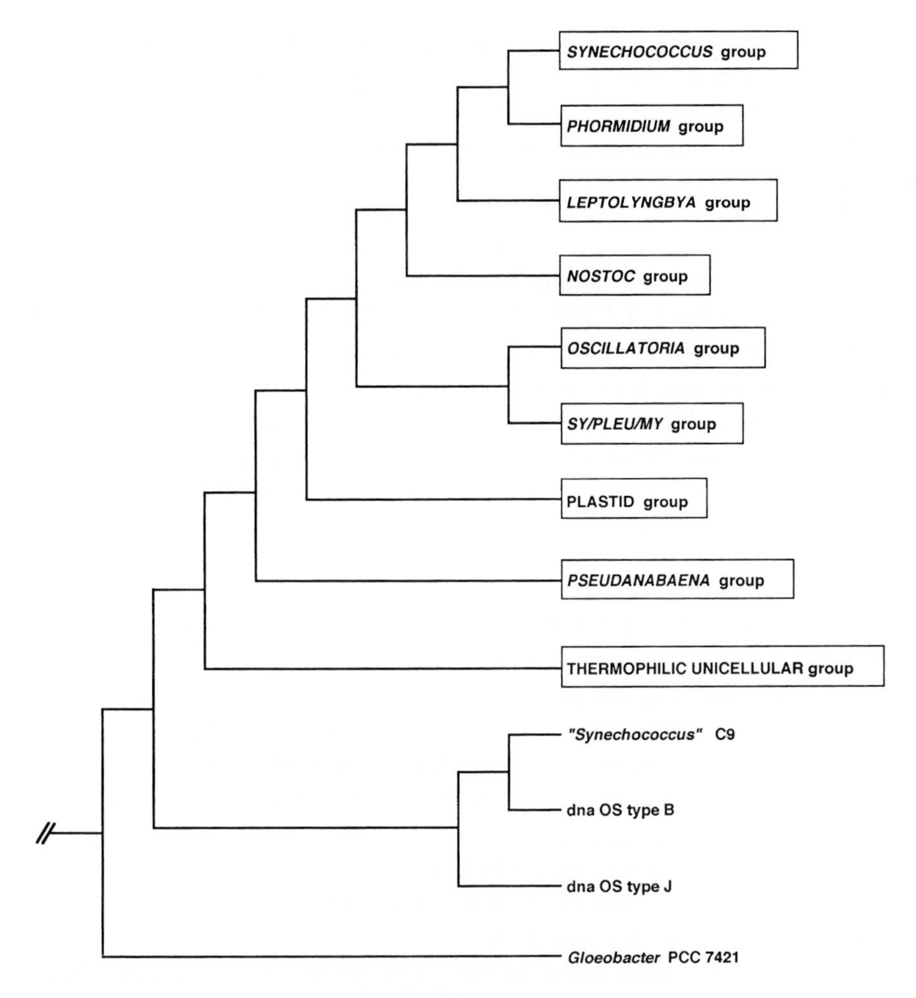

Fig. 3. Rooted cladogram depicting relationships among the ten cyanobacterial and plastid SSU rRNA sequence groups inferred by F84 maximum likelihood analysis. The tree on which the cladogram is based was inferred with the program fastDNAml using a single order of taxon input "as is." Local rearrangements during tree building were allowed to cross one branch and global rearrangement followed addition of the final taxon. Members of the *Gloeobacter* group are shown individually as location of the root made the group paraphyletic. For further details, see text and the legends to Figs. 1 and 2

relationships among the cyanobacterial SSU rRNA sequence groups are clearly defined.

Discussion

Utility of molecular sequence data analysis

The analyses of cyanobacterial SSU rRNA sequence data reported here and elsewhere reveal that a number of diacritical features traditionally used in the taxonomy and systematics of these organisms do not accurately reflect the evolutionary relationships among many strains. In particular, whether a strain is unicellular or filamentous is often poorly correlated with its phylogenetic associations. This is not really surprising as there are a number of reports in which unicellular strains of cyanobacteria are easily mutated into filamentous phenotypes (INGRAM & BLACKWELL 1974, DOLGANOV & GROSSMAN 1993), and cell size can be changed readily as well (INGRAM & al. 1975). From results presented here and elsewhere, the number of planes of division among unicellular and heterocystous strains would also appear to be a feature of some plasticity.

The clonal nature of bacterial reproduction can lead to the phenomenon of "periodic selection" in which a single mutation conferring a selective advantage in a particular environment can lead to the complete replacement of the original phenotype by the mutant. An example is the inadvertant selection for mutants of nostocalean or stigonematalean cyanobacteria no longer able to produce heterocysts due to prolonged serial culture in media containing a source of fixed nitrogen (see above). A ubiquitous "house-keeping" gene such as SSU rRNA accumulates fixed mutations in a neutral or nearly neutral manner and is less likely to be influenced by environmental factors that can give rise to phenotypic differences among strains (WOESE 1987; COHAN 1994, 1996). Consequently, it can serve as a stable measure of genetic relatedness that can help resolve uncertainties raised by other, conflicting data. On the other hand, it often cannot be relied upon to distinguish among populations that are clearly different on the basis of one or more ecological criteria. At present, the depth to which SSU rRNA sequence analysis can reliably answer questions of bacterial evolutionary systematics is unclear (FOX & al. 1992, STACKEBRANDT & GOEBEL 1994). In the case of cyanobacteria, a conservative approach is presently best and SSU rRNA data should not be relied upon for purposes of identification below the level of genus, such as for the *Microcystis* subgroup (Fig. 2/2 and 2/3).

The lack of resolution within many of the phylogenetic trees presented here may seem disappointing to some, but it should be viewed as a reflection of the probable evolutionary history of these organisms. The various groups of cyanobacteria appear to be the result of one or more explosive episodes of evolutionary radiation resulting in an array of morphotypes, the breadth of which is not generally seen among other bacterial groups but which is not strongly reflected at the level of SSU rRNA gene sequences (Fig. 1).

Among bacteria, phylogenetic analyses of the large ribosomal subunit RNA (LSU rRNA) and translational elongation factor Tu (EF-Tu) support the results obtained with SSU rRNA data (LUDWIG & al. 1993, LUDWIG & SCHLEIFER 1994). This might reflect concerted evolution among the three gene products since they must interact either directly or indirectly in the ribosomal macromolecular complex during protein synthesis. Interaction in large macromolecular assemblies would be expected to select against fixation of laterally transferred genes from other organisms since foreign gene products would be unlikely to function as well as native ones under such circumstances. For LSU rRNA, its greater number of sequence positions and greater rate of change relative to SSU rRNA would make it a useful adjunct in phylogenetic analysis of cyanobacteria, Moreover, constraints imposed by the genetic code often make protein gene sequences such as EF-Tu less susceptible than rRNAs to artifacts of base composition bias or tachytely provided that they are analyzed at the amino acid level. However, they are more liable to artifacts of paralogy due to lateral transfer or gene conversion if they function as "stand-alone" enzymes or otherwise have little interaction with at least several other macromolecules. If the components of a macromolecular complex are encoded on a single operon, the possibility of lateral transfer and gene replacement for the full complement of gene products increases significantly. For these reasons, an approach utilizing multiple gene sequences is necessary to ensure that the phylogeny of the genes truly mirrors that of the organisms of which they are a part, and separate suites of gene sequences that encode interactive products are generally to be preferred over those encoding noninteractive ones.

Oxychlorobacteria (Prochlorohyta) and green plastid phylogeny

Cyanobacteria are phenetically defined as oxygenic photosynthetic prokaryotes containing the photosynthetic pigments chlorophyll *a*, phycocyanin, and allophycocyanin. Many strains also contain the photopigment phycoerythrin. The photosynthetic plastids of glaucocystophytes, rhodophytes, and cryptophytes also contain this primary complement of pigments and an evolutionary relationship between cyanobacteria and these plastid types formed a practical working hypothesis in the context of the serial endosymbiosis theory, although cryptophytes also contain the pigment chlorophyll *c*. The plastids of most green algae and land plants, on the other hand, contain the photopigments chlorophylls *a* and *b*. The different pigment compositions of various plastid types led some investigators to propose multiple independent origins of plastids involving endosymbioses between different photosynthetic organisms containing the appropriate suite of pigments and their respective host organisms (Sagan 1967, Raven 1970). The subsequent discovery of oxygenic photosynthetic prokaryotes containing chlorophylls *a* and *b* understandably led to great excitement as the "missing link" in the history of green plastids apparently had been found. At present, three distinct oxychlorobacteria have been characterized and there are reports of the existence of others (Matthijs & al. 1994).

The first to be discovered, *Prochloron didemni*, is a unicellular exosymbiont of didemnid ascidians in tropical and subtropical marine waters. To date, it has proved intractable to cultivation under laboratory conditions, a situation that greatly limited investigations on this organism. The second, *Prochlorothrix hollandica*, is a filamentous, planktonic bacterium found in shallow lakes in the Netherlands. It is readily cultured in defined mineral media thus providing a steady supply of research material. Not surprisingly, it is the most well characterized of the oxychlorobacteria. The third, *Prochlorococcus marinus*, is an abundant unicellular picoplankter in the euphotic zone of temperate and tropical marine waters. It is phenotypically distinct from the other two oxychlorobacteria in having divinyl forms of chlorophylls *a* and *b*. It too can be cultivated in appropriate media but as yet there have been no reports of axenic cultures having been established. In addition to chlorophylls *a* and *b*, both *Prochloron* and *Prochlorococcus* contain the chlorophyll *c*-like pigment Mg 3, 8-divinyl-pheoporphyrin a_5, a derivative of which might also be present in *P. hollandica* (Goericke & Repeta 1992, Partensky & al. 1993, Larkum & al. 1994).

The history, biochemistry, ultrastructure, and physiology of oxychlorobacteria have been well reviewed (BULLERJAHN & POST 1993, MATTHIJS & al. 1994, POST & BULLERJAHN 1994). Consequently, discussion here will be limited to the analysis of molecular sequence data in regard to possible evolutionary relationships with green plastids with an emphasis on the more controversial results.

SSU rRNA. As reported here and elsewhere, SSU rRNA sequence analysis indicates the lack of a close evolutionary relationship both between the different oxychlorobacteria and green plastids and among the oxychlorobacteria themselves (TURNER & al. 1989, URBACH & al. 1992). These results have been called into question based on the hypothesis that there has been such a loss of phylogenetic signal over time in the corresponding genes that inferring presumably true evolutionary relationships between green plastids and one or more oxychlorobacteria may not be possible with present techniques. It was suggested that plastids in inferred phylogenetic trees artifactually group together and apart from oxychlorobacteria, *P. hollandica* in particular, as a result of the plastid sequences having converged to similar guanine + cytosine content (LOCKHART & al. 1993). This criticism merits close inspection as it bears on all phylogenetic analyses involving rRNA data.

In the study of LOCKHART & al. (1993), rRNA sequences were treated as binary data in a series of quartet parsimony analyses with a correction for apparent signal arising from random noise. The authors purport to show that there is an increasing loss in phylogenetic signal with increasing evolutionary distance among plastids. What the authors actually demonstrate is that there is a lack of resolution within the "red" plastid line of descent made up of plastids from rhodophytes, cryptophytes, and chromophytes sensu lato. This phenomenon previously had been observed in a bootstrap analysis of a nearly identical data set (DOUGLAS & TURNER 1991). Thus, for a quartet that includes one "green" plastid and three "red" plastids none of the three possible unrooted trees is likely to be favored. The authors fail to justify how a lack of resolution within a subset of taxa can be extended to question results of the analysis involving other taxa. Regardless of this, their analysis confirmed earlier results based on SSU rRNA analysis in that *P. hollandica* grouped with a cyanobacterium and apart from a green plastid with the same degree of support as a green algal plastid grouped with a land plant plastid to the exclusion of a cyanobacterium. The authors suggested this to be artifactual because another unexpected result also occurred, specifically as association between the phycobilin-containing plastid of the red alga *Cyanidium caldarium* (*Galdieria sulfuraria*) with that of the chromophyte *Olisthodiscus luteus* (*Heterosigma akashiwo*, *H. carterae*), which contains chlorophyll *c*, rather than with the phycobilin-containing cyanobacterium *Anacystis nidulans* (*Synechococcus* sp.). But the plastids of *C. caldarium* and *O. luteus* both contain *rbcL* genes that are clearly more closely related to those of chemoautotrophic proteobacteria than to cyanobacteria or green plastids, a situation likely due to a common ancestral event involving a lateral transfer of genetic material (DELWICHE & PALMER 1996). On this basis, these taxa would be expected to share a common evolutionary history that would group them together in phylogenetic trees inferred from SSU rRNA and other molecular data (MORDEN & al. 1992). Furthermore, the argument that true historical patterns between cyanobacteria and *P. hollandica* SSU rRNAs are likely to be lost due to the greater age of divergence of these organisms relative to those of green plastids implicitly assumes universal clock-like behavior in the evolution of SSU rRNA genes, a view to be contrasted with that of probable evolutionary hypobradytely among cyanobacteria relative to other organisms such as plastid-bearing eukaryotes (SCHOPF 1994). In short, these investigators define the "correct" phylogeny as that grouping taxa with similar light-harvesting pigments and when their results contradict this model they are summarily dismissed. Moreover, the extrapolation of results for four-taxon trees to comment on results derived from multitaxon studies is of questionable validity since the case of four taxa is guaranteed to have the greatest amount of noise (highest probability that by chance alone two taxa will share a character state in common to the exclusion of the other taxa) and the least amount of signal (smallest number of "phylogenetically informative" positions). This is true whether nucleic acid sequence data are scored as binary or quaternary characters.

psbA. This gene codes for the D1 protein that is an integral part of the reaction center of photosystem II. It is located in the plastid genome of all algae and land plants so far examined. The immature D1 protein of *P. hollandica* shares with those of green algae and land plants a gap (indel) of seven amino acids near the carboxy terminus relative to rhodophyte, glaucocystophyte, and chromophyte plastids as well as cyanobacteria and the oxychlorobacteria *Prochloron didmeni* and *Prochlorococcus marinus* (MORDEN & GOLDEN 1989a, LOCKHART & al. 1993, HESS & al. 1995). However, phylogenetic analyses using complete sequences do not support an association of *P. hollandica* with green plastids (MORDEN & GOLDEN 1989a,b; KISHINO & al. 1990; HESS & al. 1995). Nonetheless, some investigators have expressed the view that the probability of such an indel event of the same size and in the same location is so unlikely to have occurred in two independent instances that it constitutes strong evidence for an evolutionary relationship between *P. hollandica* and green plastids (LLOYD & CALDER 1991, LOCKHART & al. 1993). A priori, one might be inclined to agree, but when viewed in the light of all available evidence, this argument is far less compelling. A number of features related to this indel make its use as a phylogenetic marker problematic: (1) the indel occurs in a hypervariable region of the protein that is post-translationally cleaved (MARDER & al. 1984, TAKAHASHI & al. 1990). This cleavage is necessary for proper formation of the oxygen-evolving complex (DINER & al. 1988). Within the cleaved oligopeptide, additional single amino acid deletions are found for the unicellular green algae *Chlorella ellipsoidea* and *Chlamydomonas reinhardtii* as well as in the oxychlorobacterium *Prochlorococcus marinus* (ERICKSON & al. 1984, YAMADA 1991, HESS & al. 1995). The *psbA* gene in the plastid of *Euglena gracilis* does not even encode that part of the D1 protein that is otherwise cleaved in other organisms (KARABIN & al. 1984). Collectively, these facts indicate strongly that this segment of the *psbA* gene is hypermutable. (2) In some cyanobacterial *psbA* genes, the corresponding indel region is flanked by short direct repeats (MORDEN & GOLDEN 1991). It has been empirically established that the frequency of DNA excision events is inversely proportional to the distance between direct repeats (CHÉDIN & al. 1994). Consequently, an event involving a stretch of DNA flanked by direct repeats and as short as that corresponding to the seven amino acid indel would have a relatively high probability of occurring more than once, particularly in such a hypermutable region as this. (3) Recently, the *psbA* gene sequence of the cyanobacterium *Cyanothece* ATCC 51142 has been determined (Accession U39610). In this strain there appears to be an indel of four amino acids in the same location as the seven amino acid indel of *P. hollandica* and green plastids. However, in this case alignment at both the amino acid and nucleotide levels argues for two separate deletion events involving noncontiguous dipeptides (not shown). Regardless of the number of DNA excision events that may have occurred in this organism, this constitutes further evidence that indel mutations may have occurred on several independent occasions at this location of the *psbA* gene. The conclusion drawn here is that indel analysis within a post-translationally cleaved hypermutable oligopeptide has little to recommend it as a reliable indicator of evolutionary relationships.

Light-harvesting complexes (LHCs). The principle assumption underlying the presumptive evolutionary relationship between oxychlorobacteria and chlorophyll *b*- containing plastids is that similar light-harvesting systems are unlikely to have evolved independently two or more times (see GREEN & DURNFORD (1996) for a comprehensive review of chlorophyll-binding proteins). Comparison of the sequences encoding the LHC proteins for both green plastids and oxychlorobacteria has raised strong doubts as to the validity of this model. Examination of the genes encoding photosystem II LHC proteins of oxychlorobacteria indicates a greater similarity with the *isiA* gene of cyanobacteria than with the LHC-protein genes of green plastids (VAN DER STAAY & al. 1995, LA ROCHE & al. 1995). Expression of the *isiA* gene occurs in cyanobacteria subjected to iron depletion, producing a chlorophyll-binding protein, CP43′, that is related to the constitutively expressed CP43 protein (encoded by *psb*C) and CP47 protein (encoded by *psbB*) that comprise part of the photosystem II core complex. Subsequent phylogenetic analysis of the sequence data confirms a close evolutionary link between the oxychlorobacterial chlorophyll-binding protein genes and the *isiA* genes of cyanobacteria (LA ROCHE & al. 1996). The eukaryotic light-harvesting complex chlorophyll *a/b*-binding proteins, on the other hand, are similar to the early light induced proteins (ELIPs) of plants and the high light inducible protein (HLIP, encoded by *hliA*) of cyanobacteria, both of which appear to be involved in the acclimation of the photosynthetic apparatus to conditions of high light intensity (MIROSHNICHENKO DOLGANOV & al. 1995). Thus, the light-harvesting antenna proteins of oxychlorobacteria are of a separate evolutionary origin than those of the plastids of chlorophytes and land plants.

The phenotypic similarity in photosynthetic pigments may also be misleading as the assumption here is that chlorophyll *b* in oxychlorobacteria and green plastids is the product of a common synapomorphic metabolic pathway (LARKUM 1991). However, there is precedent for production of identical chlorophyll isomers by closely related but different organisms utilizing different biosynthetic pathways. In *Rhodobacter sphaeroides*, formation of ring E in bacteriochlorophyll *a* involves a hydratase mechanism while in *Roseobacter denitrificans* the production of an identical structure proceeds via a monoxygenase reaction (PORRA & al. 1996). It is possible that chlorophyll *b* synthesis in oxychlorobacteria and "green" photosynthetic eukaryotes may also involve independently derived pathways.

The vast majority of molecular sequence analyses favors the hypothesis that oxychlorobacteria and green plastids are not close relatives (for further review, see the chapter by DELWICHE & PALMER 1997 in this volume). However, this obviously does not preclude the future discovery of one or more oxychlorobacteria that may prove to be the sister taxon or taxa to green plastids. The recent discovery of *Acaryochloris marina*, a unicellular oxygenic photosynthetic bacterium whose primary photosynthetic pigment is chlorophyll *d*, is an indication of how little is actually known of present day oxygenic photosynthetic life (MIYASHITA & al. 1996a,b). As of this writing, no molecular sequence data are available for *A. marina* so no comment can be made concerning its phylogenetic position. However, it does contain a water soluble phycobilin-like pigment (MIYASHITA & al. 1996b) which strongly suggests that it falls within the cyanobacterial line of descent although its thylakoids are apressed and lack obvious phycobilisomes (MIYASHITA & al. 1996a). Remarkably, it also is found as an exosymbiont of clonal ascidians yet is phenotypically distinct from *Prochloron didemni*. This raises the possibility that *A. marina* may be a close relative of the ascidian-associated cyanobacterium *Synechocystis trididemni* which contains chlorophyll *a* and phycobilin pigments but has few or no obvious phycobilisomes when examined by electron microscopy (COX & al. 1985), and which has itself been suggested to be a close relative of *P. didemni* (COX 1986, COX & DIBBAYAWAN 1987). Unfortunately, there are no gene or protein sequence data for *S. tridemni* either, a situation that makes such speculations all the more tentative.

The discovery of *A. marina* as well as the demonstration of functional phycobiliprotein genes in the oxychlorobacterium *Prochlorococcus marinus* CCMP 1375 (HESS & al. 1996) serves to blur further the phylogenetically artificial distinction between cyanobacteria and oxychlorobacteria. Among extant photosynthetic bacteria, *P. marinus* comes closest in terms of pigment composition to a proposed ancestral organism that contained chlorophylls *a* and *b* as well as phycobilins (BRYANT 1992). If such a precursor organism did in fact exist and if the results of molecular phylogenetic analyses are assumed to be even partly correct, such as the consistently deep placement of the cyanobacterium *Gloeobacter violaceus* PCC 7421, then the numerous differences in photosynthetic pigment composition among cyanobacteria, oxychlorobacteria, and plastids may represent multiple independent losses of these pigments in addition to, or in place of, multiple independent gains.

Plastid monophyly: "ancestry" versus "origin"

Phylogenetic trees inferred from molecular sequence analyses generally support the monophyly of plastids within the cyanobacterial line of descent (reviewed by DELWICHE & PALMER 1997 in this volume). Many investigators view this as evidence for a single primary origin of photosynthetic plastids wherein a eukaryote endocytosed an oxygenic photosynthetic bacterium that consequently became the progenitor of all modern plastids. This initial event is posited to have occurred in the interval represented in phylogenetic trees by the branch immediately leading to a single node from which paths to the terminal plastid nodes can be traced. Descent from this single founder plastid is thought to have occurred both by direct descent from the newly photosynthetic eukaryotic host and indirectly by means of one or more secondary endosymbioses, each involving endocytosis of a descendent plastid-bearing eukaryote by an aplastidic eukaryote.

While plastid monophyly in inferred trees is a necessary condition for the single primary origin hypothesis to be true, it is not sufficient to establish it unequivocally even if all independently derived trees are consistent and there is no uncertainty as to the validity of the data or analytical methods. This is due to the fact that the root node of a plastid subtree may represent either a plastid, thus supporting the hypothesis as above, or it may represent an oxygenic photosynthetic bacterium, two or more descendants of which subsequently established closely related but separate lines of plastid descent via independent primary endosymbioses. That is, primary endosymbiotic events may have occurred not in the interval leading to the root node of the plastid subtree, but rather in subsequent intervals represented by branches between the root node and the terminal plastid nodes. Barring extinction of all free-living relatives of these putative plastid founders, the latter interpretation predicts the existence of modern oxyphotobacteria of which inclusion in phylogenetic analyses would sunder the monophyly of plastids seen in current trees. As yet, no such organisms are known, but in regard to molecular sequence data the available sample size of oxyphotobacteria is minuscule. Currently, then, the more parsimonious interpretation of most available results from molecular phylogenetic analysis favors the hypothesis of a single primary origin of photosynthetic plastids. However, it is suggested here that to avoid semantic confusion, investigators take care to distinguish between a monophyletic origin of plastids, referring to a single primary endosymbiosis as discussed above, and a monophyletic ancestry of plastids, referring to the fact that plastids, and only plastids, comprise the terminal nodes of a monophyletic clade within the cyanobacterial line of descent irrespective of the number of primary endosymbioses that may have occurred.

A second necessary condition for the single primary origin hypothesis to be true is that phylogenetic trees of photosynthetic eukaryotes whose plastids are thought to be directly descended from the original primary endosymbiosis, as opposed to secondary endosymbioses, should be congruent with trees inferred from plastid sequence data for those same taxa. At present, this point is unresolved as is discussed in more detail in the following chapter by DELWICHE & PALMER (1997) and to which the reader is directed for a comprehensive review of this most interesting subject of plastid phylogeny.

I am grateful to A. WILMOTTE, B. NIELAN, and D. DISTEL and J. WATERBURY for providing sequence data prior to publication; to R. G. HILLER and B. R. GREEN for bringing several references to my attention; and to B. R. GREEN and B. NIELAN for providing manuscripts prior to publication. Thanks are also due J.D. PALMER for helpful discussions.

References

ANAGNOSTIDIS, K., KOMÁREK, J., 1985: Modern approach to the classification system of cyanophytes 1 – Introduction. – Arch. Hydrobiol. [Suppl.] **71**, Algol. Stud. **38/39**: 291–302.

– – 1988: Modern approach to the classification system of cyanophytes 3-*Oscillatoriales*. – Arch. Hydrobiol. [Suppl.] **80**, Algol. Stud. **50/53**: 327–472.

– – 1990: Modern approach to the classification system of cyanophytes 5-*Stigonematales*. – Arch. Hydrobiol. [Suppl.] **86**, Algol. Stud. **59**: 1–73.

BAUM, B. R., ESTABROOK, G. F., 1996: Impact of outgroup inclusion on estimates by parsimony of undirected branching of ingroup phylogenetic lines. – Taxon **45**: 243–257.

BILLI, D., GRILLI CAIOLA, M., 1996: Effects of nitrogen limitation and starvation on *Chroococcidiopsis* sp. (*Chroococcales*). – New Phytol. **133**: 563–571.

BONEN, L., DOOLITTLE, W. F., 1975: On the prokaryotic nature of red algal chloroplasts. – Proc. Natl. Acad. Sci. USA **72**: 2310–2314.

– – 1976: Partial sequences of 16S rRNA and the phylogeny of blue-green algae and chloroplasts. – Nature **261**: 669–673.

– – 1978: Ribosomal RNA homologies and the evolution of the filamentous blue-green bacteria. – J. Molec. Evol. **10**: 283–291.

– – FOX, G. E., 1979: Cyanobacterial evolution: results of 16S ribosomal ribonucleic acid sequence analyses. – Canad. J. Biochem. **57**: 879–888.

BORNET, E., FLAHAULT, C., 1886–1888. Revision des Nostocacees heterocystees. – Ann. Sci. Nat., Bot. **7**, Ser. **3**: 323–381; **4**: 343–373 **5**: 52–129; **7**: 178–262.

BOURELLY, P., 1970: Les algues d'eau douce. Initiation a la systematique. 3. Les algues bleues et rouges. – Paris: Boubee.

– 1979: Les cyanophycees, algues ou bacteries? – Rev. Algol., N.S., **14**: 5–9.

BREMER, B., BREMER, K., 1988: Cladistic analysis of blue-green procaryote interrelationships and chloroplast origin based on 16s rRNA oligonucleotide catalogues. – J. Evol. Biol. **2**: 13–30.

BRITSCHGI, T. B., GIOVANNONI, S. J., 1991: Phylogenetic analysis of a natural marine bacterioplankton population by rRNA cloning and sequencing. – Appl. Environ. Microbiol. **57**: 1707–1713.

BRÖSIUS, J., DULL, T. J., SLEETER, D. D., NOLLER, H. F., 1981: Gene organization and primary structure of a ribosomal RNA operon from *Escherichia coli*. – J. Molec. Biol. **148**: 107–127.

BRYANT, D. A., 1982: Phycoerythrocyanin and phycoerythrin: properties and occurrence in cyanobacteria. – J. Gen. Microbiol. **128**: 835–844.

– 1992: Puzzles of chloroplast ancestry. – Curr. Biol. **2**: 240–242.

BULLERJAHN, G. S., POST, A. F., 1993: The prochlorophytes: are they more than just chlorophyll *a/b*-containing cyanobacteria?. – Crit. Rev. Microbiol. **19**: 43–59.

BURGER-WIERSMA, T., MUR, L. R., 1989: Genus "*Prochlorothrix*" BURGER-WIERSMA, VEENHUIS, KORTHALS, VAN DE WIEL and MUR 1986. – In STALEY, J. T., BRYANT, M. P., PFENNIG, N., HOLT, J. G., (Eds): BERGEY'S manual of systematic bacteriology. **3**, pp. 1805–1806. – Baltimore: Williams & Wilkins.

CAMPBELL, S. E., 1985: "*Oscillatoria limnetica*" from Solar Lake, Sinai, is a *Phormidium* (*Cyanophyta* or *Cyanobacteria*). – Arch. Hydrobiol. [Suppl.] **71**, Algol. Stud. **38/39**: 175–190.

CASTENHOLZ, R. W., 1989a: Subsection III, order *Oscillatoriales*. – In STALEY, J.T., BRYANT, M. P., PFENNIG, N., HOLT, J. G., (Eds): BERGEY'S manual of systematic bacteriology. **3**, pp. 1771–1780. – Baltimore: Williams & Wilkins.

– 1989b: Subsection IV, order *Nostocales*. – In STALEY, J. T., BRYANT, M. P., PFENNIG, N., HOLT, J. G., (Eds): BERGEY'S manual of systematic bacteriology. **3**, pp. 1780–1793. – Baltimore: Williams & Wilkins.

– 1989c: Subsection V, order *Stigonematales*. – In STALEY, J. T., BRYANT, M.P., PFENNIG, N.,

Holt, J. G., (Eds): Bergey's manual of systematic bacteriology. **3**, pp. 1794–1799. – Baltimore: Williams & Wilkins.
– 1992: Species usage, concept, and evolution in the cyanobacteria (blue-green algae). – J. Phycol. **28**: 737–745.
– Waterbury, J. B., 1989: Oxygenic photosynthetic bacteria (sect. 19), group I. *Cyanobacteria.* – In Staley, J. T., Bryant, M. P., Pfennig, N., Holt, J. G., (Eds): Bergey's manual of systematic bacteriology. **3**, pp. 1710–1728. – Baltimore: Williams & Wilkins.
Cavalli-Sforza, L. L., Edwards, A. W. F., 1967: Phylogenetic analysis: models and estimation procedures. – Evolution **32**: 550–570.
Chédin, F., Dervyn, E., Dervyn, R., Ehrlich, S. D., Noirot, P., 1994: Frequency of deletion formation decreases exponentially with distance between short direct repeats. – Molec. Microbiol. **12**: 561–569.
Cohan, F. M., 1994: Genetic exchange and evolutionary divergence in prokaryotes. – Trends Ecol. Evol. **9**: 175–180.
– 1996: The role of genetic exchange in bacterial evolution. – A.S.M. News **62**: 631–636.
Cohen, Y., Padan, E., Shilo, M., 1975: Facultative anoxygenic photosynthesis in the cyanobacterium *Oscillatoria limnetica.* – J. Bacteriol. **123**: 855–861.
Cox, G., 1986: Comparison of *Prochloron* from different hosts. I. Structural and ultrastructural characteristics. – New Phytol. **104**: 429–445.
– Dibbayawan, T., 1987: A chloroplast-like DNA arrangement in *Synechocystis trididemni* (*Chrooccales, Cyanophyta*). – Phycologia **26**: 148–151.
– Hiller, R.G., Larkum, A. W. D., 1985: An unusual cyanophyte, containing phycourobilin and symbiotic with ascidians and sponges. – Mar. Biol. **89**: 149–163.
Delwiche, C. F., Palmer, J. D., 1996: Rampant horizontal transfer and duplication of rubisco genes in eubacteria and plastids. – Molec. Biol. Evol. **13**: 873–882.
– – 1997: The origin of plastids and their spread via secondary symbiosis. – Pl. Syst. Evol., [Suppl.] **11**: 53–86.
Desikachary, T. V., 1959: *Cyanophyta.* – New Delhi: Indian Council of Agricultural Research.
– 1973: Status of classical taxonomy. – In Carr, N. G., Whitton, B. A., (Eds): The biology of blue-green algae, pp. 473–486. – Oxford: Blackwell.
Diner, B. A., Ries, D. F., Cohen, B. N., Metz, J. G., 1988: COOH-terminal processing of polypeptide D1 of the photosystem II reaction center of *Scenedesmus obliquus* is necessary for the assembly of the oxygen-evolving complex. – J. Biol. Chem. **263**: 8972–8980.
Dolganov, N., Grossman, A. R., 1993: Insertional inactivation of genes to isolate mutants of *Synechococcus* sp. strain PCC 7942: isolation of filamentous strains. – J. Bacteriol. **175**: 7644–7651.
Douglas, S. E., Carr, N., 1988: Examination of genetic relatedness of marine *Synechococcus* spp. by using restriction fragment length polymorphisms. – Appl. Environm. Microbiol. **54**: 3071–3078.
– Turner, S., 1991: Molecular evidence for the origin of plastids from a cyanobacterium-like ancestor. – J. Molec. Evol. **33**: 267–273.
Drouet, F., 1981: Revision of the *Stigonemataceae* with a summary of the classification of the blue-green algae. – Beih. Nova Hedwigia, **66**. – Vaduz: Cramer.
Elenkin, A. A., 1938 & 1949: Sinezelenye vodorosli SSSR (Monographia algarum cyanophycearum aquidulcium et terrestrium in finibus URSS inventarum). Systematic part, **1 & 2.** – Moscow, Leningrad: Akad. Nauk. SSSR.
Erikson, J. M., Rahire, M., Rochaix, J.-D., 1984: *Chlamydomonas reihnardii* [sic] gene for

the 32 000 mol. wt. protein of photosystem II contains four large introns and is located entirely within the chloroplast inverted repeat. – E.M.B.O. J. **3**: 2753–2762.

FELSENSTEIN, J., 1981: Maximum likelihood and minimum-steps methods for estimating evolutionary trees from data on discrete characters. – Syst. Zool. **22**: 240–249.

– 1993: PHYLIP (Phylogeny Inference Package) version 3.5c.– Distributed by the author. – Seattle: Department of Genetics, University of Washington.

FERRIS, M. J., RUFF-ROBERTS, A. L., KOPCZYNSKI, E. D., BATESON, M. M., WARD, D. M., 1996: Enrichment culture and microscopy conceal diverse thermophilic *Synechococcus* populations in a single hot spring microbial mat habitat. – Appl. Environm. Microbiol. **62**: 1045–1050.

FITCH, W. M., MARGOLIASH, E., 1967: Construction of phylogenetic trees. – Science **155**: 279–284.

FOX, G. E., PECHMAN, D. R., WOESE, C. R., 1977: Comparative cataloging of the 16S ribosomal ribonucleic acid: molecular approach to procaryotic systematics. – Int. J. Syst. Bacteriol. **27**: 44–57.

– WISOTZKEY, J. D., JURTSHUK, P., 1992: How close is close: 16S rRNA sequence identity may not be sufficient to guarantee species identity. – Int. J. Syst. Bacteriol. **42**: 166–170.

FRIEDMANN, E. I., BOROWITZKA, L. J., 1982: The symposium on taxonomic concepts in blue-green algae: towards a compromise with the bacteriological code? –Taxon **3**: 673–683.

– OCAMPO-FRIEDMANN, R., HUA, M., 1994: *Chroococcidiopsis*, the most primitive living cyanobacterium? – Origins Life Evol. Biosphere **24**: 269–270.

FRITSCH, F. E., 1959: The structure and reproduction of the algae. **1**. – Cambridge: Cambridge University Press.

GEITLER, L., 1932: *Cyanophyceae*. – RABENHORST'S Kryptogamenflora von Deutschland, Österreich und der Schweiz, **14**. – Leipzig: Akademische Verlagsgesellschaft (Repri. 1971; New York: Johnson.)

– 1979: Einige kritische Bemerkungen zu neuen zusammenfassenden Darstellungen der Morphologie und Systematik der Cyanophyceen. – Pl. Syst. Evol. **132**: 153–160.

GIOVANNONI, S. J., TURNER, S., OLSEN, G. J., BARNS, S., LANE, D. J., PACE, N. R., 1988: Evolutionary relationships among cyanobacteria and green chloroplasts. – J. Bacteriol. **170**: 3584–3592.

– BRITSCHGI, T. B., MOYER, C. L., FIELD, K. G., 1990: Genetic diversity in Sargasso Sea bacterioplankton. – Nature **345**: 60–63.

GOERICKE, R., REPETA, D. J., 1992: The pigments of *Prochlorococcus marinus*: the presence of divinyl chlorophyll *a* and *b* in a marine procaryote. – Limnol. & Oceanogr. **37**: 425–433.

GOLDEN, S. S., NALTY, M. S., CHO, D.-S. C., 1989: Genetic relationship of two highly studied *Synechococcus* strains designated *Anacystis nidulans*. – J. Bacteriol. **171**: 24–29.

GOLUBIC, S., 1976: Taxonomy of extant stromatolite building cyanophytes. – In WALTER, M. R., (Ed.): Stromatolites. – New York: Elsevier Developments in Sedimentology **20**: 127–140.

– 1979: Cyanobacteria (blue-green algae) under the bacteriological code? An ecological objection. – Taxon **28**: 387–389.

GOMONT, M., 1892: Monographie des Oscillariees. – Ann. Sci. Nat., Bot. **7**, Ser. **15**: 263–368, **16**: 91–264. (Reprinted 1962; Weinheim: Cramer.)

GORELOVA, O. A., BAULINA, O. I., SHCHELMANOVA, A. G., KORZHENEVSKAYA, T. G., GUSEV, M. V., 1996: Heteromorphism of the cyanobacterium *Nostoc* sp., a microsymbiont of the *Blasia pusilla* moss.–Microbiology **65**: 791–726 (English translation of Mikrobiologiya **65**: 824–832).

GRAY, M. W., 1988: Organelle origins and ribosomal RNA. – Biochem. Cell Biol. **66**: 325–348.
– 1992: The endosymbiont hypothesis revisited. – Int. Rev. Cytol. **141**: 233–357.
– 1993: Origin and evolution of organelle genomes. – Curr. Opin. Genet. Developm. **3**: 884–890.
– DOOLITTLE, W. F., 1982: Has the endosymbiont hypothesis been proven? – Microbiol. Rev. **46**: 1–42.
GREEN, B. R., DURNFORD, D. G., 1996: The chlorophyll-carotenoid proteins of oxygenic photosynthesis. – Annual Rev. Pl. Physiol. Pl. Molec. Biol. **47**: 685–714.
GRILLI CAIOLA, M., CANNI, A., OCAMPO-FRIEDMANN, R., 1996: Iron superoxide dismutase (Fe-SOD) localization in *Chroococcidiopsis* sp. (*Chroococcales, Cyanobacteria*). – Phycologia **35**: 90–94.
GUGLIELMI, G., COHEN-BAZIRE, G., 1984a: Étude taxonomique d'un genre de cyanobactérie oscillatoriacée: le genre *Pseudanabaena* LAUTERBORN. I. Étude ultrastructurale. – Protistologica **20**: 377–391.
– – 1984b: Étude taxonomique d'un genre de cyanobactérie oscillatoriacée: le genre *Pseudanabaena* LAUTERBORN. II. Analyse de la composition moléculaire et de al structure des phycobilisomes. – Protistologica **20**: 393–413.
– – BRYANT, D. A., 1981: The structure of *Gloeobacter violaceus* and its phycobilisomes. – Arch. Microbiol. **129**: 181–189.
GUTELL, R. R., 1993: Collection of small subunit (16S- and 16S-like) ribosomal RNA structures. – Nucl. Acids Res. **21**: 3051–3054.
– WEISER, B., WOESE, C. R., NOLLER, H. F., 1985: Comparative anatomy of 16S-like ribosomal RNA. – Prog. Nucl. Acid Res. Molec. Biol. **32**: 155–216.
HASEGAWA, M., KISHINO, H., YANO, T., 1985: Dating of the human-ape splitting by a molecular clock of mitochondrial DNA. – J. Molec. Evol. **22**: 132–147.
HESS, W. R., WEIHE, A., LOISEAUX-DE GOËR, S., PARTENSKY, F., VAULOT, D., 1995: Characterization of the single *psbA* gene of *Prochlorococcus marinus* CCMP 1375 (*Prochlorophyta*). – Pl. Molec. Biol. **27**: 1189–1196.
– PARTENSKY, F., VAN DER STAAY, G.W. M., GARCIA-FERNANDEZ, J. M., BÖRNER, T., VAULOT, D., 1996: Coexistence of phycoerythrin and a chlorophyll *a*/*b* antenna in a marine prokaryote. – Proc. Natl. Acad. Sci. USA **93**: 11126–11130.
HERDMAN, M., JANVIER, M., WATERBURY, J. B., RIPPKA, R., STANIER, R. Y., MANDEL, M., 1979a: Deoxyribonucleic acid base composition of cyanobacteria. – J. Gen. Microbiol. **111**: 63–71.
– – RIPPKA, R., STANIER, R. Y., 1979b: Genome size of cyanobacteria. – J. Gen. Microbiol. **111**: 73–85.
HILLIS, D. M., HUELSENBECK, J. P., SWOFFORD, D. L., 1994: Hobglobin of phylogenetics? – Nature **369**: 363–364.
HOFFMANN, L., 1988: Criteria for the classification of blue-green algae (cyanobacteria) at the genus and at the species level. – Arch. Hydrobiol. [Suppl.] **80**, Algol. Stud. **50–53**: 131–139.
HOLLERBACH, M. M., KOSINSKAJA, E. K., POLJANSKIJ, V. I., 1953: Sinezelenye vodorosli. [Blue–green algae.]. – In: Opred. presnov. vodoros. SSSR, **2.** – Moskva: Sov. Nauka.
INGRAM, L. O., BLACKWELL, M. M., 1974: Macromolecular requirements for the physical initiaion of cell division in a filamentous mutant of *Agmenellum quadruplicatum*. – Biochim. Biophys. Acta **335**: 295–302.
– OLSON, G. J., BLACKWELL, M. M., 1975: Isolation of a small-cell mutant in the blue-green bacterium *Agmenellum quadruplicatum*. – J. Bacteriol. **123**: 743–746.

JENSEN, T. E., 1985: Cell inclusions in the *Cyanobacteria.* – Arch. Hydrobiol. [Suppl.] **71**, Algol. Stud. **38/39**: 33–73.

– 1989: Thylakoids in aged cyanobacterial cells suggest origin of eukaryotic nuclear membranes. – Cytobios **60**: 47–61.

– 1991: Autogenous bacterial origin of the eukaryotic cell. – Endocytobiosis Cell Res. **8**: 1–16.

JIN, L., NEI, M., 1990: Limitations of the evolutionary parsimony method of phylogenetic analysis. – Molec. Biol. Evol. **7**: 82–102.

KARABIN, G. D., FARLEY, M., HALLICK, R. B., 1984: Chloroplast gene for M_r 32,000 polypeptide of photosystem II in *Euglena gracilis* is interrupted by four introns with conserved boundary sequences. – Nucl. Acids Res. **12**: 5801–5812.

KELLY, M. L., COWIE, D. B., 1972: DNA-DNA hybridization studies of blue-green algae. – Carnegie Inst. Wash. Yearbook **71**: 276–281.

KISHINO, H., HASEGAWA, M., 1989: Evaluation of the maximum likelihood estimate of the evolutionary tree topologies from DNA sequence data, and the branching order in *Hominoidea.* – J. Molec. Evol. **29**: 170–179.

– MIYATA, T., HASEGAWA, M., 1990: Maximum likelihood inference of protein phylogeny and the origin of chloroplasts. – J. Molec. Evol. **31**: 151–160.

KOMÁREK, J., ANAGNOSTIDIS, K., 1986: Modern approach to the classification system of cyanophytes 2-*Chroococcales.* – Arch. Hydrobiol. [Suppl.] **73**, Algol. Stud. **43**: 157–226.

– – 1989: Modern approach to the classification system of cyanophytes 4-Nostocales. – Arch. Hydrobiol. [Suppl.] **82**, Algol. Stud. **56**: 247–345.

KONDRATEVA, N.V., 1968: *Cyanophyta.* – In Vizn. Prisnov. Vodorost. Ukr. RSR 1.2. – Kiev: Vid. "Naukova dumka".

KONDRATYEVA [sic], N. V., 1982: On difference of opinions of phycologists and bacteriologists concerning the nomenclature of cyanophyta. – Arch. Protistenk. **126**: 247–259.

KUHNER, M. K., FELSENSTEIN, J., 1994: A simulation comparison of phylogeny algorithms under equal and unequal evolutionary rates. – Molec. Biol. Evol. **11**: 459–468.

KUMAR, S., RZHETSKY, A., 1996: Evolutionary relationships of eukaryotic kingdoms. – J. Molec. Evol. **42**: 183–193.

LACHANCE, M.-A., 1981: Genetic relatedness of heterocystous cyanobacteria by deoxyribonucleic acid-deoxyribonucleic acid reassociation. – Int. J. Syst. Bacteriol. **31**: 139–147.

LARKUM, A. W. D., 1991: The evolution of chlorophylls. – In SCHEER, H., (Ed.): Chlorophylls, pp. 367–383. – Boca Raton: CRC Press.

– SCARAMUZZI, C., COX, G. C., HILLER, R. G., TURNER, A. G., 1994: Light-harvesting chlorophyll *c*-like pigment in *Prochloron.* – Proc. Natl. Acad. Sci. USA **91**: 679–683.

LA ROCHE, J., PARTENSKY, F., FALKOWSKI, P., 1995: The major light-harvesting antenna of *Prochlorococcus marinus* is similar to CP43′, a chl binding protein induced by iron limitation in cyanobacteria. – In MATHIS, P., (Ed.): Photosynthesis: from light to biosphere, 1, pp. 171–174. – Dordrecht: Kluwer.

– VAN DER STAAY, G. W. M., PARTENSKY, F., DUCRET, A., AEBERSOLD, R., LI, R., GOLDEN, S. S., HILLER, R. G., WRENCH, P. M., LARKUM, A. W. D., GREEN, B. R., 1996: Independent evolutionary of the prochlorophyte and green plant chlorophyll *a/b* light-harvesting proteins. – Proc. Natl. Acad. Sci. USA **93**: 15244–15248.

LEWIN, R. A., 1979: Formal taxonomic treatment of cyanophytes. – Int. J. Syst. Bacteriol. **29**: 411–412.

– 1989: Group II. Order *Prochlorales* LEWIN 1977. – In STALEY, J. T., BRYANT, M. P., PFENNIG, N., HOLT, J. G., (Eds): BERGEY's manual of systematic bacteriology. **3**, pp. 1805–1806. – Baltimore: Williams & Wilkins.

LLOYD, D. G., CALDER, V. L., 1991: Multi-residue gaps, a class of molecular characters with exceptional reliability for phylogenetic analyses. – J. Evol. Biol. **4**: 9–21.

LOCKHART, P. J., PENNY, D., HENDY, M. D., LARKUM, A. D. W., 1993: Is *Prochlorothrix hollandica* the best choice as a prokaryotic model for higher plant chl *a/b* photosynthesis? – Photosyn. Res. **37**: 61–68.

LOISEAUX-DE GOËR, S., 1994: Plastid lineages. – Progr. Phycol. Res. **10**: 137–177.

LUDWIG, W., SCHLEIFER, K. H., 1994: Bacterial phylogeny based on 16S and 23S rRNA sequence analysis. – F.E.M.S. Microbiol. Rev. **15**: 155–173.

– NEUMAIER, J., KLUGBAUER, N., BROCKMANN, E., ROLLER, C., JILG, S., REETZ, K., SCHACHTNER, I., LUDVIGSEN, A., BACHLEITNER, M., FISCHER, U., SCHLEIFER, K. H., 1993: Phylogenetic relationships of *Bacteria* based on comparative sequence analysis of elongation facter Tu and ATP-synthase β-subunit genes. – Antonie van Leeuwenhoek **64**: 285–305.

MADDISON, W. P., MADDISON, D. R., 1992: MacClade: Analysis of phylogeny and character evolution. Version 3.0. – Sunderland, MA: Sinauer.

MAIDAK, B. L., OLSEN, G. J., LARSEN, N., OVERBEEK, R., MCCAUGHEY, M. J., WOESE, C. R., 1996: The ribosomal data base project (RDP). – Nucl. Acids Res. **24**: 82–85.

MARDER, J. B., GOLOUBINOFF, P., EDELMAN, M., 1984: Molecular architecture of the rapidly metabolized 32-kilodalton protein of photosystem II. – J. Biol. Chem. **259**: 3900–3908.

MATTHIJS, H. C. P., VAN DER STAAY, G. W. M., MUR, L. R., 1994: Prochlorophytes: the 'other' cyanobacteria?. – In BRYANT, D. A., (Ed.): The molecular biology of cyanobacteria, pp. 49–64. – The Netherlands: Kluwer.

MEFFERT, M.-E., 1988: *Limnothrix* MEFFERT nov. gen. The unsheathed planktic cyanophycean filaments with polar and central gas vacuoles. – Arch. Hydrobiol. [Suppl.] **80**, Algol. Stud. **50–53**: 269–276.

MERESCHKOWSKY, C., 1905: Über die Natur und den Ursprung der Chromatophoren im Pflanzenreiche. – Biol. Centralbl. **25**: 593–604.

– 1910: Theorie der zwei Plasmaarten als Grundlage der Symbiogenesis, einer neuen Lehre von der Entstehung der Organismen. – Biol. Centralbl. **30**: 278–303, 321–347, 353–367.

MIYASHITA, H., IKEMOTO, H., KURANO, N., ADACHI, K., CHIHARA, M., MIYACHI, S., 1996a: Chlorophyll *d* as a major pigment. – Nature **383**: 402.

– KURANO, N., MIYACHI, S., 1996b: A new oxygenic photosynthetic prokaryote containing chlorophyll *d* as a major pigment. – Pl. Cell Physiol. **37** [Suppl.]: s48.

MIROSHNICHENKO DOLGANOV, N. A., BHAYA, D., GROSSMAN, A. R., 1995: Cyanobacterial protein with similarity to the chlorophyll *a/b* binding proteins of higher plants: evolution and regulation. – Proc. Natl. Acad. Sci. USA **92**: 636–640.

MORDEN, C. W., GOLDEN, S. S., 1989a: *psbA* genes indicate common ancestry of prochlorophytes and chloroplasts. – Nature **337**: 382–385.

– – 1989b: Corrigendum: *psbA* genes indicate common ancestry of prochlorophytes and chloroplasts. – Nature **339**: 400.

– – 1991: Sequence analysis and phylogenetic reconstruction of the genes encoding the large and small subunits of ribulose-1, 5-bisphosphate carboxylase/oxygenase from the chlorophyll *b*-containing prokaryote *Prochlorothrix hollandica*. – J. Molec. Evol. **32**: 379–395.

– DELWICHE, C. F., KUHSEL, M., PALMER, J. D., 1992: Gene phylogenies and the endosymbiotic origin of plastids. – BioSystems **28**: 75–90.

NEILAN, B. A., COX, P. T., HAWKINS, P. R., GOODMAN, A. E., 1994: 16S ribosomal RNA gene sequence and phylogeny of toxic *Microcystis* sp. (cyanobacteria). – DNA Sequence **4**: 333–337.

– JACOBS, D., DEL DOT, T., BLACKALL, L. L., HAWKINS, P. R., COX, P. T., GOODMAN, A. E., 1997: rRNA sequences and evolutionary relationships among toxic and nontoxic cyanobacteria of the genus *Microcystis*. Int. J. Syst. Bacteriol. **47**: 693–697.

NELISSEN, B., WILMOTTE, A., DE BAERE, R., HAES, F., VAN DE PEER, Y., NEEFS, J.-M., DE WACHTER, R., 1992: Phylogenetic study of cyanobacteria on the basis of 16S ribosomal RNA sequences. – Belg. J. Bot. **125**: 210–213.

– – NEEFS, J. -M., DE WACHTER, R., 1995a: Phylogenetic relationships among filamentous helical cyanobacteria investigated on the basis of 16S ribosomal RNA gene sequence analysis. – Syst. Appl. Microbiol. **17**: 206–210.

– VAN DE PEER, Y., WILMOTTE, A., DE WACHTER, R., 1995b: An early origin of plastids within the cyanobacterial divergence is suggested by evolutionary trees based on complete 16S rRNA sequences. – Molec. Biol. Evol. **12**: 1166–1173.

– DE BAERE, R., WILMOTTE, A., DE WACHTER, R., 1996: Phylogenetic relationships of nonaxenic filamentous cyanobacterial strains based on 16S rRNA sequence analysis. – J. Molec. Evol. **42**: 194–200.

OLSEN, G. J., MATSUKA, H., HAGSTROM, R., OVERBEEK, R., 1994: fastDNAml: A tool for construction of phylogenetic trees of DNA sequences using maximum likelihood. – Computer Applic. Biosci. **10**: 41–48.

PALINSKA, K. A., LIESACK, W., RHIEL, E., KRUMBEIN, W. E., 1996: Phenotype variability of identical genotypes: the need for a combined approach in cyanobacterial taxonomy demonstrated on *Merismopedia*-like isolates. – Arch. Microbiol. **166**: 224–233.

PARTENSKY, F., HOEPFFNER, N., LI, W. K. W., ULLOA, D., VAULOT, D., 1993: Photoacclimation of *Prochlorococcus* sp. (*Prochlorophyta*) strains isolated from the North Atlantic and the Mediterranean Sea. – Pl. Physiol. **101**: 285–296.

PORRA, R. J., SCHÄFER, W., GAD'ON, N., KATHEDER, I., DREWS, G., SCHEER, H., 1996: Origin of the two carbonyl oxygens of bacteriochlorophyll *a*. Demonstration of two different pathways for the formation of ring E in *Rhodobacter sphaeroides* and *Roseobacter denitrificans*, and a common hydratase mechanism for 3-acetyl group formation. – Eur. J. Biochem. **239**: 85–92.

POST, A. F., BULLERJAHN, G. S., 1994: The photosynthetic machinery in prochlorophytes: structural properties and ecological significance. – F.E.M.S Microbiol. Rev. **13**: 393–414.

PRUFERT-BEBOUT, L., GARCIA-PICHEL, F., 1994: Field and cultivated *Microcoleus chthonoplastes*: The search for clues to its prevalence in marine microbial mats. – In STAL, L. L., CAUMETTE, P., (Eds): Microbial mats. Structure development and environmental significance, pp. 111–116, – Berlin, Heidelberg New York: Springer.

RAVEN, P. H., 1970: A multiple origin for plastids and mitochondria. – Science **169**: 641–646.

RIPPKA, R., 1988: Recognition and identification of cyanobacteria. – Meth. Enzymol. **167**: 28–67.

– COHEN-BAZIRE, G., 1983: The *Cyanobacteriales*: a legitimate order based on the type strain *Cyanobacterium stanieri*? – Ann. Microbiol. **134B**: 21–36.

– HERDMAN, H., 1992: Pasteur Culture Collection of Cyanobacteria catalogue & taxonomic handbook. **1**. Catalogue of strains. – Paris: Institut Pasteur.

– WATERBURY, J., COHEN-BAZIRE, G., 1974: A cyanobacterium which lacks thylakoids. – Arch. Microbiol. **100**: 419–436.

– DERUELLES, J., WATERBURY, J. B., HERDMAN, M., STANIER, R. Y., 1979: Generic assignments, strain histories and properties of pure cultures of cyanobacteria. – J. Gen. Microbiol. **111**: 1–61.

RODRIGO, A. G., 1992: Two optimality criteria for selecting subsets of most parsimonious trees. – Syst. Biol. **41**: 33–40.

SAGAN, L., 1967: On the origin of mitosing cells. – J. Theoret. Biol. **14**: 225–274.

SAITOU, N., NEI, M., 1987: The neighbor-joining method: a new method for reconstructing phylogenetic trees. – Molec. Biol. Evol. **4**: 406–425.

SCHIMPER, A. F. W., 1883: Ueber die Entwickelung der Chlorophyllkörner und Farbkörper. – Bot. Zeitung (Leipzig) **41**: 105–112.

SCHOPF, J. W., 1994: Disparate rates, differing fates: tempo and mode of evolution changed from the Precambrian to the Phanerozoic. – Proc. Natl. Acad. Sci. USA **91**: 6735–6742.

SEEWALDT, E., STACKEBRANDT, E., 1982: Partial sequence of 16S ribosomal RNA and the phylogeny of *Prochloron*. – Nature **295**: 618–620.

SELSTAM, E., CAMPBELL, D., 1996: Membrane lipid composition of the unusual cyanobacterium *Gloeobacter violaceus* sp. PCC 7421, which lacks sulfoquinovosyl diacylglycerol. – Arch. Microbiol. **166**: 132–135.

STACKEBRANDT, E., GOEBEL, B. M., 1994: Taxonomic note: A place for DNA-DNA reassociation and 16S rRNA sequence analysis in the present species definition in bacteriology. – Int. J. Syst. Bacteriol. **44**: 846–849.

– SEEWALDT, E., FOWLER, V. J., SCHLEIFER, K. -H., 1982: The relatedness of *Prochloron* sp. isolated from different didemnid ascidian hosts. – Arch. Microbiol. **132**: 216–217.

STAM, W. T., 1980: Relationships between a number of filamentous blue-green algal strains (*Cyanophyceae*) revealed by DNA-DNA hybridization. – Arch. Hydrobiol. [Suppl.] **56**, Algol. Stud. **25**: 351–374.

– VENEMA, G., 1977: The use of DNA-DNA hybridization for determination of the relationship between some blue-green algae (*Cyanophyceae*). – Acta Bot. Neerl. **26**: 327–342.

– BOELE-BOS, S. A., STULP, B. K., 1985: Genotypic relationships between *Prochloron* samples from different localities and hosts as determined by DNA-DNA reassociations. – Arch. Microbiol. **142**: 340–341.

STANIER, R. Y., VAN NIEL, C. B., 1962: The concept of a bacterium. – Arch. Mikrobiol. **42**: 17–35.

– SISTROM, W. R., HANSEN, T. A., WHITTON, B. A., CASTENHOLZ, R. W., PFENNIG, N., GORLENKO, V. N., KONDRATIEVA, E. N., EIMHJELLEN, K. E., WHITTENBURY, R., GHERNA, R. L., TRUPER H. G., 1978: Proposal to place the nomenclature of the cyanobacteria (blue-green algae) under the rules of the International Code of Nomenclature of Bacteria. – Int. J. Syst. Bacteriol. **28**: 335–336.

STARMACH, K., 1996: *Cyanophyta*-sinice, *Glaucophyta*-glaukofity. – Flora Slodkowodna Polski, **2**. – Warszawa: Panstwowe wydawnietwo naukowe.

STRIMMER, K., VON HAESELER, A., 1996: Quartet puzzling: a quartet maximum-likelihood method for reconstructing tree topologies. – Molec. Biol. Evol. **13**: 964–969.

STULP, B. K., STAM, W. T., 1984: Genotypic relationships between strains of *Anabaena* (*Cyanophyceae*) and their correlation with morphological affinities. – Brit. Phycol. J. **19**: 287–301.

SWOFFORD, D. L., 1993: PAUP: Phylogenetic analysis using parsimony, version 3.1. – Computer program distributed by the Illinois Natural History Survey, Champaign, Illinois.

TAKAHASHI, Y., NAKANE, H., KOJIMA, H., SATOH, K., 1990: Chromatographic purification and determination of the carboxy-terminal sequences of photosystem II reaction center proteins, D1 and D2. – Pl. Cell Physiol. **31**: 273–280.

TAMURA, K., NEI, M., 1993: Estimation of the number of nucleotide substitutions in the control region of mitochondrial DNA in humans and chimpanzees. – Molec. Biol. Evol. **10**: 512–526.

THURET, G., 1875: Essai de classification des Nostochinees. – Ann. Sci. Nat., Bot. **6**, Ser. **1**: 372–382.

TOMIOKA, N., SHINOZAKI, K., SUGIURA, M., 1981: Molecular cloning and characterization of ribosomal RNA genes from a blue-green alga, *Anacystis nidulans*. – Molec. Gen. Genet. **184**: 359–363.

TURNER, S., DE LONG, E. F., GIOVANNONI, S. J., OLSEN, G. J., PACE, N. R., 1989a: Phylogenetic analysis of microorganisms and natural populations by using rRNA sequences. – In COHEN, Y., ROSENBERG, E., (Eds): Microbial mats. Physiological ecology of benthic microbial communities, pp. 390–401. – Washington D. C.: American Society for Microbiology.

– BURGER-WIERSMA, T., GIOVANNONI, S. J., MUR, L. R., PACE, N. R., 1989b: The relationship of a prochlorophyte, *Prochlorothrix hollandica*, to green chloroplasts. – Nature **337**: 380–382.

URBACH, E., ROBERTSON, D. L., CHISHOLM, S. W., 1992: Multiple evolutionary origins of prochlorophytes within the cyanobacterial radiation. – Nature **335**: 267–270.

VAN DER STAAY, G. W. M., DUCRET, A., AEBERSOLD, R., LI, R., GOLDEN, S. S., HILLER, R. G., WRENCH, P. M., LARKUM, A. W. D., GREEN, B. R., 1995: The chl *a/b* antenna from prochlorophytes is related to the iron stress-induced chl *a* antenna (*isi*A) from cyanobacteria. – In MATHIS, P., (Ed.): Photosynthesis: from light to biosphere, **1**, pp. 175–178. – Dordrecht: Kluwer.

VAN VALEN, L. M., 1982: Phylogenies in molecular evolution: *Prochloron*. – Nature **298**: 493–494.

WAKELEY, J., 1996: The excess of transitions among nucleotide substitutions: new methods of estimating transition bias underscore its significance. – Trends Ecol. Evol. **11**: 158–163.

WARD, D. M., FERRIS, M. J., NOLD, S. C., BATESON, M. M., KOPCZYNSKI, E. D., RUFF-ROBERTS, A. L., 1994: Species diversity in hot spring microbial mats as revealed by both molecular and enrichment culture approaches –relationship between biodiversity and community structure. – In STAL, L. L., CAUMETTE, P. (Eds): Microbial mats. Structure development and environmental significance, pp. 33–44. – Berlin, Heidelberg New York: Springer.

WATERBURY, J. B., 1989: Subsection II, order *Pleurocapsales* GEITLER 1925, emend. WATERBURY & STANIER 1978. – In STALEY, J. T., BRYANT, M. P., PFENNIG, N., HOLT, J. G., (Eds): BERGEY'S manual of systematic bacteriology. **3**, pp. 1746–1770. – Baltimore: Williams & Wilkins.

– RIPPKA, R., 1989: Subsection I, order *Chroococcales* WETTSTEIN 1924, emend. RIPPKA et al., 1979. – In STALEY, J. T., BRYANT, M. P., PFENNIG, N., HOLT, J. G., (Eds): BERGEY'S manual of systematic bacteriology. **3**, pp. 1728–1746. – Baltimore: Williams & Wilkins.

WELLER, R., WELLER, J. W., WARD, D. M., 1991: 16S rRNA sequences of uncultivated hot springs cyanobacterial mat inhabitants retrieved as randomly primed cDNA. – Appl. Environm. Microbiol. **57**: 1146–1151.

– BATESON, M. M., HEIMBUCH, B. K., KOPCZYNSKI, E. D., WARD, D. M., 1992: Uncultivated cyanobacteria, *Chlorflexus*-like inhabitants, and spirochete-like inhabitants of a hot spring microbial mat. – Appl. Environm. Microbiol. **58**: 3964–3969.

– WHITTON, B. A., 1989: Genus 1. *Calothrix* AGARDH 1824. – In STALEY, J. T., BRYANT, M. P., PFENNIG, N., HOLT, J. G., (Eds): BERGEY's manual of systematic bacteriology. **3** pp. 1791–1793. – Baltimore: Williams & Wilkins.

WILMOTTE, A., 1991: Taxonomic study of marine oscillatoriacean strains (*Cyanophyceae, Cyanobacteria*) with narrow trichomes. I. Morphological variability and autecological features. – Archiv. Hydrobiol. [Suppl.] **92**, Algol. Stud. **64**: 215–248.

– 1994: Molecular evolution and taxonomy of the cyanobacteria. – In BRYANT, D. A., (Ed.): The molecular biology of *Cyanobacteria*, pp. 1–25. – The Netherlands: Kluwer.

– GOLUBIC, S., 1991: Morphological and genetic criteria in the taxonomy of *Cyanophyta/ Cyanobacteria*. – Archiv. Hydrobiol. [Suppl.] **92**, Algol. Stud. **64**: 1–24.

– STAM, W. T., 1984: Genetic relationships among cyanobacterial strains originally designated as '*Anacystis nidulans*' and some other *Synechococcus* strains. – J. Gen. Microbiol. **130**: 2737–2740.

– URNER, S., VAN DE PEER, Y., PACE, N. R., 1992: Taxonomic study of marine *Oscillatoriacean* strains (*Cyanophyceae*, *Cyanobacteria*) with narrow trichomes. II. Nucleotide sequence analysis of the 16S ribosomal RNA. – J. Phycol. **28**: 828–838.

– VAN DER AUWERA, G., DE WACHTER, R., 1993: Structure of the 16S ribosomal RNA of the thermophilic cyanobacterium *Chlorogloeopsis* HTF ('*Mastigocladus laminosus* HTF') strain PCC7518, and phylogenetic analysis, – F.E.B.S Lett. **317**: 96–100.

– NEEFS, J.-M., DE WACHTER, R., 1994: Evolutionary affiliation of the marine nitrogen-fixing cyanobacterium *Trichodesmium* sp. strain NIBB 1067, derived by 16S ribosomal RNA sequence analysis. – Microbiology **140**: 2159–2164.

WOESE, C. R., 1987: Bacterial evolution. – Microbiol. Rev. **51**: 221–271.

– SOGIN, M. L., BONEN, L., STAHL, D., 1975: Sequence studies on 16S ribosomal RNA from a blue-green alga. – J. Molec. Evol. **4**: 307–315.

– KANDLER, O., WHEELIS, M. L., 1990: Towards a natural system of organisms: proposal for the domains *Archaea*, *Bacteria*, and *Eucarya*. – Proc. Natl. Acad. Sci. USA **87**: 4576–4579.

WOOD, A. M., TOWNSEND, D., 1990: DNA polymorphism within the WH7803 serogroup of marine *Synechococcus* spp. (*Cyanobacteria*). – J. Phycol. **26**: 576–585.

YAMADA, T., 1991: Repetitive sequence-mediated rearrangements in *Chlorella ellipsoidea* chloroplast DNA: completion of nucleotide sequence of the large inverted repeat. – Curr. Genet. **19**: 139–147.

YANG, Z., 1994: Statistical properties of the maximum likelihood method of phylogenetic estimation and comparison with distance matrix methods. – Syst. Biol. **43**: 329–342.

ZIMMERMAN, W. J., BERGMAN, B., 1990: The *Gunnera* symbiosis: DNA restriction fragment length polymorphism and protein comparisons of *Nostoc* symbionts. – Microbial Ecol. **19**: 291–302.

– ROSEN, B. H., 1992: Cyanobiont diversity within and among cycads of one field site. – Canad. J. Microbiol. **38**: 1324–1328.

The origin of plastids and their spread via secondary symbiosis

Charles F. Delwiche and Jeffrey D. Palmer

Key words: Plastid, chloroplast, endosymbiosis, cyanobacterium, evolution, phylogeny, eukaryote, organelle, red alga, green alga, glaucocystophyte, plastid genome, rubisco, photosynthetic pigment.

Abstract: The endosymbiotic, cyanobacterial nature of plastids is clearly established, but several fundamental issues concerning the origin and early evolution of plastids remain unresolved. One key question is whether plastids are monophyletic (derived from a single cyanobacterial ancestor) or polyphyletic (derived from more than one ancestor). This issue is complicated by the presence in many photosynthetic eukaryotes of secondary plastids, acquired by ingestion of a eukaryote, itself already equipped with plastids, rather than by direct ingestion of a free-living cyanobacterium. A review of the phylogenetic evidence from plastid genes indicates that the three major lineages of primary plastids (red, green, and glaucocystophyte) are probably monophyletic. Mitochondrial data further support this conclusion for red and green plastids (but are unavailable for glaucocystophytes), while nuclear data are largely unresolved. If plastids are monophyletic, then the pigment diversity of plastids must postdate their status as endosymbiotic organelles, but whether this diversity arose primarily by acquisition or loss is nuclear. Secondary endosymbiosis has greatly multiplied the variety of photosynthetic eukaryotes. A secondary origin of plastids is unequivocal for cryptomonads and chlorarachniophytes, is likely for heterokonts, haptophytes, and euglenophytes, and is suggested for the nonphotosynthetic parasites of phylum Apicomplexa. The remarkable plastid diversity of dinoflagellates appears to be the result of multiple secondary and tertiary endosymbiotic events. A consistent feature of all plastid genomes is extreme reduction relative to their cyanobacterial progenitors via outright gene loss, transfer of genes to the nuclear genome, and substitution by genes of nuclear ancestry. Most of this reduction seems to have occurred relatively soon after primary endosymbiosis, before the emergence of the major lineages of plastids, yet recent data also reveal surprising diversity of gene content among these lineages. The rubisco genes (*rbcLS*) of primary plastids on the red lineage are not related to those of cyanobacteria and seem to have been acquired via horizontal gene transfer.

The endosymbiotic origin of plastids, in which a previously free-living cyanobacterium took up permanent and heritable residence in a previously non-photosynthetic eukaryote, is now so well established as to be in essence "proven" (Gray 1992, Douglas 1994, Loiseaux-de Goër 1994, Bhattacharya & Medlin 1995). In this chapter, we consider the origin and early evolution of plastids, with emphasis on the various lines of evidence – plastid, mitochondrial, and nuclear –

that bear on the fundamental question of whether plastids are monophyletic or polyphyletic in origin: Did they arise from one or more cyanobacterial endosymbioses? Although on balance current evidence favors plastid monophyly, more evidence is needed, especially as it relates to host cell (nuclear) and mitochondrial phylogeny. In addition, better resolution of cyanobacterial phylogeny is required in order to identify the specific progenitor(s) of plastids and to determine the relative timing of plastid origin.

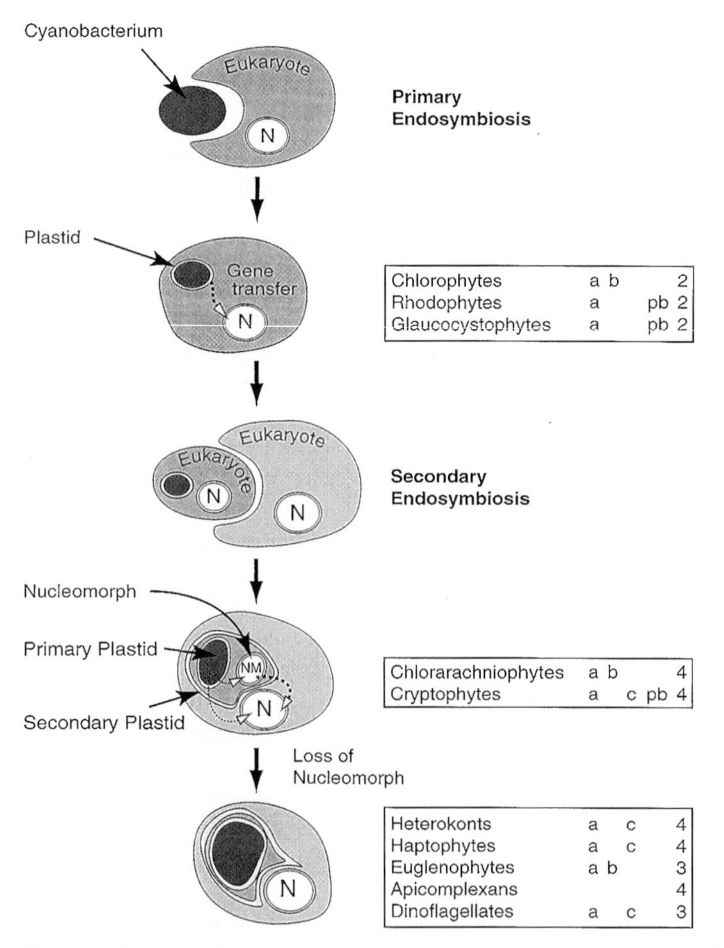

Fig. 1. Origin of plastids by primary endosymbiosis and their spread by secondary endosymbiosis. Groups of eukaryotes with primary or secondary plastids are listed to the right of the appropriate cell type. The "*a*", "*b*", "*c*", and /or "pb" shown to the right of each group name indicate the presence of chlorophyll *a*, chlorophyll *b*, chlorophyll *c*, and/ or phycobilins, respectively, in that group [note that chlorophyll *c* has also been found in at least one green alga (WILHELM 1987)]. The numbers to the far-right indicate the number of bounding membranes in each plastid type. Thus, while the bottom figure shows a secondary plastid with four membranes, some such plastids have only three membranes. Also, in some cases there is a direct continuity (not shown) between the outermost plastid membrane (which often contains ribosomes on its cytoplasmic side) and the outer membrane of the nuclear envelope (GIBBS 1993; WHATLEY 1993a, b)

While all plastids are ultimately derived from cyanobacteria, a number of algal lineages, and apparently even a nonphotosynthetic protistan lineage, acquired plastids indirectly, via *secondary* endosymbiosis (Fig. 1). In secondary endosymbiosis the host cell acquires an endosymbiont which is itself a eukaryote already equipped with plastids (GIBBS 1993, MCFADDEN & GILSON 1995, GILSON & MCFADDEN 1997). Three well-characterized algal lineages are thought to have primary plastids, the products of cyanobacterial/eukaryotic symbiosis only: the green algae, red algae, and glaucocystophytes (but see STILLER & HALL 1997, and below for an alternative view). All other algal lineages, plus the parasitic apicomplexans, have plastids that are believed, with varying levels of confidence, to be secondary in origin. A clear understanding of plastid phylogeny requires that the distinction between primary and secondary plastids be kept clear at all times, and that the history of plastid and nuclear lineages be clearly differentiated.

While much of this chapter is devoted to the issue of plastid phylogeny, from the perspective of both primary and secondary endosymbiosis, we also consider three other broad topics in plastid evolution. First, we review the surprisingly complicated phylogenetic history of the CO_2-fixing enzyme rubisco (ribulose-1,5-bisphosphate carboxylase/oxygenase), a history that is unusually rich in events of lateral gene transfer and gene duplication. Then, we discuss recent evidence that bears on the photopigment status of the ancestral plastid and cyanobacterium. Finally, we discuss the reductive evolution of plastid genomes, describing the processes that have led to the massive transfer, over and over again, of endosymbiotic genes to their various host nuclei, and reviewing the recent evidence that points to a surprising diversity and richness of genes still resident in certain plastid lineages.

Three lineages of primary plastids

Green algae, including their descendants the land plants, have stacked thylakoids, chlorophylls *a* and *b*, and carotenoids as accessory pigments. Essentially all available information suggests that green algal plastids are directly derived from cyanobacterial ancestors (but see STILLER & HALL 1997). Probable secondary plastids of green-algal origin are found in the chlorarachniophytes, the euglenophytes, some dinoflagellates, and apparently, the apicomplexans. Henceforth we shall use the terms "green algae" and "chlorophytes" interchangeably to refer to green algae plus land plants, and "chlorophytes s. l. " (*sensu lato*) to refer to these and their secondary derivatives (chlorarachniophytes and others).

The red algal plastid almost certainly represents a second lineage of primary plastids (but again, see STILLER & HALL 1997). Light capture in red plastids is performed by chlorophyll *a* and phycobilins. Taxa whose secondary plastids are probably of red-agal origin include the cryptophytes, the heterokonts (also known as chromophytes), the haptophytes, and stereotypical (peridinin-containing) dinoflagellates. All of these secondary plastids also contain chlorophyll *c*, but only the cryptophytes contain phycobilins as well. We shall use "rhodophytes" for red algae, and "rhodophytes s. l." for red algae and their secondary derivatives (cryptophytes and others).

The third lineage of probable primary plastids is found in a relatively small and poorly known group, the glaucocystophytes, which like the red algae contain chlorophyll *a* and phycobilins. The glaucocystophytes are noteworthy because their plastids retain a peptidoglycan cell wall between the two plastid envelope membranes (KIES & KREMER 1990). For many years this group was enigmatic, as it was unclear whether their photosynthetic organelles were more akin to free-living cyanobacteria or to true plastids (hence the name *Cyanophora paradoxa* and the term "cyanelle" for the organelle). Now, however, the peptidoglycan cell wall can be recognized as simply one ancestral feature that has been peculiarly retained in what is otherwise an authentic plastid, one that shows the marked reductions in genome size and gene content that are characteristic of plastids (LÖFFELHARDT & BOHNERT 1994, STIREWALT & al. 1995). The only alga which is even circumstantially suspected of having acquired its plastid via secondary symbiosis of a glaucocystophyte is the poorly understood *Paulinella chromatophora* (see below and BHATTACHARYA & al. 1995a, and BHATTACHARYA & MEDLIN 1995).

How many primary endosymbioses?

This question in effect reduces to "Were there one, two or three primary endosymbioses?". This is because essentially all available molecular and nonmolecular data strongly indicate that each of these three groups of primary plastid-containing organisms is itself monophyletic. On the molecular side, these data include trees based on plastid and nuclear SSU rRNAs, the only molecules sampled from multiple members of all three groups (CAVALIER-SMITH 1993a, BHATTACHARYA & MEDLIN 1995, BHATTACHARYA & al. 1995b, HELMCHEN & al. 1995, RAGAN & GUTELL 1995, VAN DER PEER & al. 1996); on a number of molecules multiply sampled from green and red algae but not glaucocystophytes [plastid *rbc*L] DELWICHE & PALMER 1996), nuclear actin (BHATTACHARYA & WEBER 1997); and a combination of multiple mitochondrial protein genes (LANG & al. 1997); and on additional molecules sampled among green algae only (e.g., β-tubulin, KEELING & DOOLITTLE 1996).

Phylogenetic analysis of the six plastid genes that have been examined for multiple cyanobacteria and for all three primary lineages of plastids indicates, with varying degrees of confidence, a monophyletic origin of plastids. The strongest evidence comes from 16s rRNA, which is well sampled among both plastids and cyanobacteria, and for which plastids emerge as a monophyletic group with moderate (NELISSEN & al. 1995) to strong (BHATTACHARYA & MEDLIN 1995; see also fig. 3 of BHATTACHARYA & SCHMIDT 1997, this volume, and fig. 7 of MEDLIN & al. 1997, this volume) bootstrap support. The *tuf*A gene is well sampled among plastids, but only modestly so among cyanobacteria (seven), and provides weak to moderate evidence for plastid monophyly (Fig. 2; DELWICHE & al. 1995, KOHLER & al. 1997). The *atpB* gene provides moderate support for plastid monophyly in an analysis (DOUGLAS & MURPHY 1994) with nine diverse plastids but only four cyanobacteria. The *rpo*Cl gene is moderately well sampled among cyanobacteria (16 taxa) and gives moderate to strong bootstrap support for plastid monophyly in analyses that are problematic with respect to plastid sampling. In one case, only *Cyanophora* and four greens were included (PALENIK & HASELKORN 1992), in the other only *Cyanophora* and two reds were analyzed (Palenik & Swift 1996). The *psb*A gene is moderately well sampled among cyanobacteria and also gives moderate to strong support for monophyly of the eight diverse plastids examined (Hess & al. 1995). Finally, the *rbc*L gene strongly supports monophyly of greens and *Cyanophora*, but with two caveats: cyanobacterial sampling is weak (6 taxa), and reds are not open to scrutiny because their *rbc*L is of proteobacterial origin, probably due to lateral gene transfer (Delwiche & Palmer 1996; also see section on "Rubisco phylogeny and horizontal gene transfer", including Fig. 3).

Viale & Arakaki (1994) found 100% bootstrap support, and a signature amino acid deletion, for the grouping of the red alga *Cyanidium* with two cyanobacteria to the exclusion of several green HSP60s. This would appear, as the authors conclude, to constitute impressive evidence against a monophyletic origin of all plastids. However, in a contemporary paper (Viale & al. 1994), the same authors showed, but failed to comment on, an analysis that contained one additional, critically important HSP60 sequence from cyanobacteria. This second HSP60 gene from *Synechocystis* 6803 groups with the green sequences, whereas the other HSP60 gene 6803 groups with *Cyanidium* (and with *Synechococcus*). Thus, the best explanation from this very limited sampling is that the green and red HSP60s are paralogs derived from an ancient gene duplication present in the cyanobacterial ancestor of plastids, in which case HSP60 is at present simply neutral on the question of plastid monophyly.

On balance, then, there is an emerging consensus from six plastid genes for a monophyletic, cyanobacterial origin of all plastids, with no dissenting plastid genes if one accepts the above interpretation of the HSP60 data. At the same time, we must reemphasize two limitations of the current data: bootstrap support for plastid monophyly is often modest, and sampling among cyanobacteria is usually poor. The latter means that two or more independent endosymbiotic events involving (closely) related cyanobacteria could not be distinguished with the present data.

In addition to sequence-based phylogenies, several other plastid traits also suggest a common endosymbiotic origin of all primary plastids. First, Wolfe & al. (1994) showed that the light-harvesting antenna complex proteins of photosystem I are immunologically related between green and red algae, but not with cyanobacteria; comparable data for glaucocystophytes are clearly needed here. Second, the genomes of all three primary plastids possess two gene clusters (potential operons) – *psbB/N/H* and *atp/rps/rpo* – that are not found in cyanobacteria (Reith & Munholland 1993), although interpretation here is confounded by the highly reductive, perhaps convergently so, evolution of chloroplast gene content. Third, and somewhat similarly, genomes of all three lineages of primary plastids usually contain a large, rRNA-encoding, inverted repeat; however, because these repeats otherwise are highly variable (i.e., in size, gene content, spacing, and orientation), and because rRNA-gene duplications are commonly observed in other genomes as well, the phylogenetic significance of this pattern is unclear at best (Palmer 1985, 1991; Loiseaux-de Goër 1994; Reith 1995). Fourth, the transit peptides of nuclear genes encoding plastid glyceraldehyde-3-phosphate dehydrogenase (GAPDH) from red algae and land plants have an intron of similar position and therefore potential homology; however, the tenuous alignment of these very divergent transit peptides precludes any strong conclusions (Liaud & al. 1993,

ZHOU & RAGAN 1994). Fifth, and more impressively, transit peptides from *Cyanophora* direct efficient import of cytoplasmically synthesized proteins into plastids of land plants, and vice-versa (JAKOWITCH & al. 1996), while a red algal transit peptide can also direct import into plant plastids (APT & al. 1993). Finally, KOWALLIK (1994) has argued that the extensive overlap in gene content among plastid genomes, viewed in the context of their universally extreme reduction from cyanobacterial genomes, supports a monophyletic origin of all plastids. This argument seems rather compelling for glaucocystophytes and red algae, which are relatively similar in plastid gene content and diversity, and for which all but eight of the 191 plastid genes found in *Cyanophora paradoxa* are also present among the 251 genes of *Porphyra purpurea* (REITH 1995). For green algae (here represented by the land plant *Marchantia polymorpha*), the case is somewhat less compelling, as the *Marchantia* genome is significantly more reduced in both gene content (120 genes) and, especially, gene diversity, yet actually contains more unique genes (~20) (REITH 1995).

Contrary to early expectations, the chlorophyll *a*/*b*-containing "prochlorophytes" are not the specific progenitors of green (or any other) plastids in phylogenetic analyses of all six of the plastid genes that indicate plastid monophyly [see above references; also see TURNER (1997) for the latest discussion of the 7 amino acid *psb*A deletion originally (MORDEN & GOLDEN 1989), now infamously, thought to link prochlorophytes and green plastids]. Moreover, for all four of those genes, 16s, *rpo*C, *psb*A, and *rbc*L, that have been multiply sampled among the prochlorophytes, *Prochloron*, *Prochlorococcus*, and *Prochlorothrix* do not even form a monophyletic group, but instead are scattered widely across the cyanobacterial radiation (PALENIK & HASELKORN 1992, URBACH & al. 1992, PALENIK & SWIFT 1996, DELWICHE & PALMER 1996, TURNER 1997). It is thus clear that the prochlorophytes are a polyphyletic group; they are fundamentally cyanobacteria which have thrice independently derived an analogous photosynthetic apparatus (PALENIK & HASELKORN 1992, URBACH & al. 1992). The frequency with which the prochlorophyte syndrome has arisen suggests that the number of changes involved in the cyanobacterium-to-prochlorophyte transition may be small.

If a single endosymbiotic event gave rise to all primary plastids, as suggested by essentially all the available plastid molecular evidence, then the three groups with primary plastids should, in the simplest scenario, cluster together in all phylogenies, plastid, mitochondrial (mt), and nuclear. Although mt data are unavailable for any glaucocystophytes, a sister-group relationship of red and green mitochondria is indeed strongly supported. The first evidence for this came from the mitochondrial *cox3* gene, which showed strong clustering of reds and greens in the context of limited sampling of eukaryotes overall (BOYEN & al. 1994). [We are excluding from our purview as grossly unreliable for phylogeny reconstruction the mitochondrial rRNA genes, whose highly uneven, often very rapid, "clock-speeds" produce highly artefactual trees, e.g., in which even land plants and green algae fail to cluster properly (GRAY 1995, GRAY & SPENCER 1996)]. Very recently, however, the combined analysis of five "universal" or nearly so mt protein genes has revealed high bootstrap support (98%) for a clade comprising three diverse red algae and four diverse green algae and land plants in the context of reasonably good sampling of eukaryotes (LANG & al. 1997, also see PAQUIN & al. 1997).

If both plastid and mitochondrial data point, rather strongly, to a monophyletic origin of primary plastids, then what does the nucleus say? Although often viewed as incompatible with plastid monophyly, the available nuclear data are in our view largely inconclusive. Phylogenetic analyses which include red and green algae are, to our knowledge, available for nine different nuclear genes (six encoding protein and three rRNA), only two of which (18s rRNA and actin) include sequences from glaucocystophytes. First, the proteins: β-tubulin from the red alga *Chondrus crispus* is extraordinarily divergent, and its phylogenetic placement (with the slime mold *Physarum*) is almost certainly an artefact of long-branch-attraction (LIAUD & al. 1995). Analyses of triosephosphate isomerase (TPI) are very poorly resolved with respect to the major groups of eukaryotes, reds and greens included (KEELING & DOOLITTLE 1997), although RAGAN & GUTELL (1995) cite a subset of their unpublished analyses in which red and green TPI's group together. The red alga *Porphyra purpurea* contains two genes for elongation factor-1α (EF-1α), a highly divergent gene whose deep position in global eukaryotic trees is probably a long-branch-attraction artefact, and a more conserved sequence which groups with green algae in distance but not parsimony analyses (LIU & al. 1996). Phylogenetic studies of cytosolic GAPDH, carried out by two different groups, place reds and greens together with weak support (LIAUD & al. 1994, ZHOU & RAGAN 1995). Actin has evolved rapidly in red algae, and two early analyses with relatively poor overall sampling each placed a different red sequence in a fundamentally unresolved position among the "crown-group" of eukaryotes (BOUGET & al. 1995, TAKAHASHI & al. 1995). However, an expanded set of analyses that included many more eukaryotic sequences, each of the above red sequences, and the first glaucocystophyte sequence (from *Cyanophora paradoxa*) usually recovered a red/green grouping and sometimes even a red/green/glaucocystophyte grouping (BHATTACHARYA & WEBER 1997).

Finally, in analyses of the largest subunit of RNA polymerase II (RPBI), green algae group with *Acanthamoeba*, fungi, and animals, and red algae form the sister group to this entire clade with high (88–93%) bootstrap support (STILLER & HALL 1997). This study also presents formal statistical tests of alternate hypotheses and finds that trees that place rhodophytes and chlorophytes as sister taxa are significantly worse than those that separate them.

Despite the value of this study, it should be noted that its taxonomic sampling is relatively poor (e.g., only 1–2 green and 2 red sequences are used in the various analyses), which is a weakness that holds for many of the above studies as well. Also, the red algal RPB1 sequences are relatively divergent compared to the green algal sequence and those they cluster with, raising the possibility that the former have "gone deep" by way of long-branch-attraction to the even more divergent protistan sequences included in these analyses.

All three cytoplasmic rRNAs, 18s, 28s, and 5s, have been applied to the issue of red/green relationships. The 5s rRNA tree of HORI & OSAWA (1987), which placed red algae at the base of eukaryotes, has been criticized by RAGAN & GUTELL (1995) as "seriously misleading" because of their use of the UPGMA method, whose ultrametric assumption is grossly violated by the rapidly evolving red sequences. RAGAN & GUTELL's (1995) reanalysis of 5s rRNA sequences using neighbor-joining placed the reds instead in an unresolved "crown-group" position together with green algae and many other eukaryotes. Interestingly, KRISHNAN & al. (1990) report a 5s analysis in which reds and greens cluster together, along with a mesozoan. The only relevant 28s rRNA study (PERASSO & al. 1989) does not group the one red sequence included with the green clade, but we feel these results should be disregarded because (1) only a short tract of sequence (450 nucleotides) was analyzed; (2) the resultant tree was not subjected to any type of branch-support assessment, such as bootstrapping; and (3) one of the key results obtained (a specific grouping of animals and plants to the exclusion of fungi) has since been shown to be almost certainly incorrect (BALDAUF & PALMER 1993, WAINRIGHT & al. 1993, NIKOH & al. 1994). Literally dozens of 18s rRNA analyses have been published that include both red and green algal sequences (e.g., CAVALIER-SMITH 1993a; BHATTACHARYA & MEDLIN 1995, BHATTACHARYA & al. 1995a, NELISSEN & al. 1995; RAGAN & GUTELL 1995; KUMAR & RZHETSKY 1996; PAWLOWSKI & al. 1996; SOGIN 1994, 1996; VAN DE PEER & al. 1996). Although these two groups only rarely cluster as sisters (e.g., PAWLOWSKI & al. 1996), they virtually always form part of a large nexus, termed the "crown-group" radiation, consisting of a tight knot of branches separated by tiny internodes and representing many of the major groups of eukaryotes (e.g., fig. 1 of MEDLIN & al. 1997, this volume). In essence, 18s sequences from red algae, green algae, and many other eukaryotes form an unresolved star phylogeny. While glaucocystophyte 18s sequences also conjoin with this crown-group nexus, they specifically and consistently group, albeit with modest bootstrap support, with the host nuclear lineage from cryptophytes (BHATTACHARYA & MEDLIN 1995; BHATTACHARYA & al. 1995b; fig. 1 of MEDLIN & al. 1997, this volume).

On balance, we regard the evidence from these nine different nuclear genes as inconclusive on the question of monophyly/polyphyly of the host cell lineages of primary plastids. On the protein side, we regard one molecule (β-tubulin) as completely inconclusive, two (TPI and EF-1α) as largely inconclusive, albeit providing occasional weak support for red/green monophyly, two (GAPDH and actin) as providing more consistent but still weak support for monophyly, and only *RPB1* as providing significant evidence against red/green monophyly. The available 5s and 28s analyses involve only short sequence lengths, are predictably unresolved, and should be disregarded on other grounds as well. The 18s relationships of reds and green are also, in our view, unresolved, but there is weak evidence that the sister-group of glaucocystophytes is *not* a primary plastid lineage. We thus agree with the introductory statement by STILLER & HALL (1997) that "Other nuclear genes (in addition to 18s rDNA) examined thus far also fail to provide robust support either for or against a monophyletic relationship of plants with red algae". Accordingly, we disagree with their subsequent discussionary statement that "...the vast majority of evidence from nuclear genes suggests that the host cell lines are not related", when essentially the only additional evidence supporting the latter assertion is their own *RPB1* data set.

In summary, plastid data – both sequence-based and idiosyncratic – overwhelmingly support a monophyletic origin of the plastids of red algae, green algae, and glaucocystophytes; mitochondrial data strongly support monophyly of reds and greens but are unavailable for glaucocystophytes; and nuclear data are largely inconclusive. Overall, then, the total weight of evidence clearly supports a single primary, cyanobacterial origin of plastids, but equally clearly, substantial more data are needed to achieve a robust, unequivocal, and complete picture of the cyanobacterial ancestry of plastids.

With respect to plastid genome phylogeny, little more needs to be done on the plastid side, as whole-genome sequences are already available for a diverse variety of plastid types (see below), with more genome sequences on the way. Instead, much work is needed on cyanobacteria, for which only one whole-genome sequence is available (KANEKO & al. 1996), and for which only one gene (16s rRNA) has been widely sequenced (TURNER 1997, this volume). Broad sampling, among dozens of cyanobacteria, should be performed for the most promising plastid genes, such as *tuf*A, *atp*B, *rpo*C1, 23s rRNA, and perhaps, *psb*A (see above). When combined with the 16s data, these data sets should considerably improve the current poorly resolved picture of cyanobacterial phylogeny (TURNER 1997, this volume), and in so doing, should allow an even firmer conclusion as to whether plastid genomes arose from one or more than one lineage of cyanobacteria. In particular, the better the sampling of cyanobacteria, the better our chances of distinguishing true monophyly from cryptic polyphyly, i.e. endosymbioses involving two or more closely related cyanobacterial lineages (see below). These data should also allow, for the first time, identification of the specific cyanobacterial lineage(s) that gave rise to plastids. This, in turn, should stimulate further biochemical and molecular (e.g., complete genome sequencing) study of appropriate extant members of that cyanobacterial clade, in order to improve our understanding of the properties of the immediate ancestor(s) of plastids and of the ways in which plastids have changed as a consequence of their highly integrated genetic, cellular, and biochemical partnership with their host cell.

Finally, a vastly improved understanding of cyanobacterial phylogeny and of the place(s) of plastids among cyanobacteria should clarify the current "molecules vs. morphology" paradox regarding the age of cyanobacteria and plastids. The fossil and biogeochemical records tell us that cyanobacteria are an ancient group of at least 2.8 and quite likely 3.5 billion years of age, which was well diversified 2 billion years ago, whereas plastids are thought to be much younger, only about 1 billion years old (KNOLL 1992, 1994; KNOLL & GOLUBIC 1992; SCHOPF 1993, 1994, 1996). Yet 16s rRNA phylogenies generally place plastids close to the base of cyanobacteria (NELISSEN & al. 1995; BHATTACHARYA & MEDLIN 1995; TURNER 1997 this volume; see fig. 3 of BHATTACHARYA & SCHMIDT 1997, this volume), implying that the two groups are not so disparate in age. While this paradox can, of course, be resolved by reinterpreting the fossil record (e.g., RUNNEGAR 1992), an alternative must also be considered: that 16s rRNA places plastids misleadingly deeply within cyanobacteria, most likely because of high rates of change in the former compared to the latter (GIOVANNONI & al. 1988). Again, information from several other molecules will be very instructive here.

Although sufficient data on plastid genomes are already, or will soon be, available, the comparative study of other aspects of plastid molecular and cell biology should also illuminate our understanding of primary plastid origins. For instance, as mentioned above, initial experiments support plastid monophyly by demonstrating the functional interchangeability of transit peptides among greens, reds, and glaucocystophytes (APT & al. 1993, JAKOWITCH & al. 1996). This type of experiment needs to be pursued further, for example, to evaluate the specificity of these plastid interchanges relative to other organelles in these cells (especially the mitochondrion). Of equal importance, the constituents of the protein import apparati of these three plastid types need to be scrutinized for evidence of homology.

On the mitochondrial side, genome sequence data are sorely needed (and should soon be available, thanks to the efforts of F. LANG and collaborators (in the Organelle Genome Megasequencing Program) for *Cyanophora paradoxa* and a few other glaucocystophytes. This program also promises to enlarge the general sampling of protistan mt genomes. Although extremely valuable and promising, global mt phylogenies do have their limits, both because a number of major lineages of eukaryotes no longer have mt genomes (including, apparently, the poorly characterized microaerophilic amoebomastigote *Psalteriomonas lanterna*, which is claimed to possess photosynthetic organelles termed "thylakosomes" (HACKSTEIN 1995), and because some eukaryotes have extremely rapidly evolving and/or highly edited mt genomes, which are quite problematic for phylogeny reconstruction.

On the nuclear side, far better taxonomic sampling is clearly needed, both among eukaryotes in general and also within the three groups in question, especially the reds and glaucocystophytes. This is true for the most promising of the proteins cited above (i.e. actin, GAPDH, EF-1α, and *RPB1*), as well as for molecules such as 28s rRNA and various other proteins that appear to have promise for unraveling deep eukaryotic phylogeny (e.g., HSP60, HSP70, *recA/rad51*, α-tubulin, *atp*B, other RNA polymerase subunits). For all of these molecules (especially 18s rRNA because we already so have much data for it), we are in need of better methods for constructing and evaluating phylogenetic trees, especially in light of the often severe rate heterogeneity that unpredictably plagues all molecules.

While we think it likely that more and better data, especially on the nuclear side, will ultimately build an unassailable case for plastid monophyly, what if they don't? What if instead, they support the *RPB1* evidence against the sisterhood of green algae and red algae (STILLER & HALL 1997), and the 18s evidence that the cryptophyte host nucleus is the sister-group of the glaucocystophyte nucleus (BHATTACHARYA & al. 1995b; fig. 1 of MEDLIN & al. 1997, this volume)? or, to put it another way, what if a fundamental incongruity emerges, between plastid gene phylogenies supporting monophyly of primary plastids and strongly supported nuclear gene phylogenies that do not? Assuming that these gene trees are reliable, three main hypotheses have been advanced to reconcile such a conflict, as recently reviewed by STILLER & HALL (1997). First, as suggested above and by many previous authors, it is entirely possible that the different groups of primary plastids arose from independent endosymbioses involving closely related cyanobacteria, such that the current monophyly in plastid gene trees is an artefact of inadequate sampling of this progenitor group, perhaps due to its extinction (STILLER & HALL 1997). In support of this notion, STILLER & HALL (1997) cite "Numerous examples of modern symbiotic relationships...(in which)...certain lineages are adopted preferentially as endosymbionts by widely divergent hosts."

A second hypothesis, again advanced by many (see especially CAVALIER-SMITH 1993b), is that primary plastids did arise only once, relatively early in eukaryotic evolution, but were subsequently lost in many descendant lineages. We agree with STILLER & HALL (1997) that this hypothesis is unlikely, not only because it posits plastid loss in so many lineages, but for two other reasons as well. While many instances of the permanent loss of photosynthesis are known among algae and plants, in all cases, the plastid (and even its genome) persists, indicating that it provides some role that is essential even in the context of parasitism, including intracellular parasites such as the apicomplexans (WILSON & al. 1996; MCFADDEN & al. 1997a, this volume). Why would plastids be entirely lost from so many other eukaryotes in the face of their retention in so many long-term nonphotosynthetic organisms? Furthermore, this hypothesis would posit plastid loss from the ancestors of animals and fungi, and it seems highly unlikely that all traces of cyanobacterial genes would have vanished from such well-studied, even completely sequenced, nuclear genomes.

STILLER & HALL (1997) advance a third, apparently novel hypothesis to reconcile what they perceive as a conflict between plastid and nuclear phylogenies. They postulate that perhaps the plastids of red algae and/or green algae are in fact "secondary plastids that have masqueraded as primary", because they have lost the two extra bounding membranes regarded as signatures of secondary symbiosis. This idea is attractive because of the known variation in membrane number among secondary plastids (three in some, four in others; Fig. 1), which suggests, but does not prove, loss of at least one secondary membrane in euglenophytes and dinoflagellates. Thus, according to STILLER & HALL (1997), reds and/or greens might have gone one step further, i.e., lost both secondary membranes. Furthermore, this is the only one of these three hypotheses that attempts to reconcile the nuclear evidence for apparent non-sisterhood of reds and greens not only with the strong evidence for monophyly of their plastids, but also the strong evidence for sisterhood of their mitochondria (LANG & al. 1997, PAQUIN & al. 1997). That is, STILLER & HALL (1997) imply that reds might have acquired not only their plastids secondarily from a green alga, or vice-versa, but also their mitochondria.

Although superficially attractive, this scenario has several problems. First, it posits that the endocytosed cell's mitochondrion must have contained homologs of all the genes of the host cell's mitochondrion at the time of the latter's loss via secondary replacement, which seems somewhat unlikely in the face of the huge differences in mitochondrial gene content known among the major groups of eukaryotes (GRAY 1992, GRAY & SPENCER 1996, LANG & al. 1997). Second, it also requires that the six major multisubunit complexes of the mitochondrion (i.e. the ribosome, ATP synthase, and complexes I–IV of the electron transport chain) whose genetic specification is split between the nucleus and mitochondrion must be able to interact and function together seamlessly when combined between highly disparate evolutionary lineages. This seems highly unlikely, i.e. that the nuclear- and mitochondrially-encoded partners would show no significant incompatibilities in such a "wide cross". Alternatively, to prevent such an incompatibility, virtually all of the many nuclear genes for these proteins would have to have been transferred from the endosymbiont's nucleus to the host nucleus prior to loss of the host mitochondrion, yet it seems hard to believe that a cell would maintain both types of mitochondria and their genomes for the many millions of years such massive transfer would require. Third, the secondary-mitochondrion corollary of the membrane-loss hypothesis of Stiller & Hall (1997) implies that, in both plastid and mitochondrial phylogenies, the group that arose by secondary endosymbiosis (either reds or greens) should emerge as a monophyletic group from within a paraphyletic group of primary-plastid-containing organisms (either greens or reds). There is simply no evidence for this in any plastid or mitochondrial gene phylogenies (see above). The only way around this conundrum is to postulate that this secondary endosymbiosis occurred before the diversification of the host cell lineage into extant groups of algae, in which case this hypothetical endosymbiosis is both quantitatively (more ancient) and qualitatively (involving secondary mitochondria too) different than any of the known cases of secondary plastid endosymbiosis. Finally, all three of these problems only multiply when glaucocystophytes are brought into the picture. In summary, then, we believe that the evidence for a true nuclear/plastid phylogenetic conflict is more apparent than real, and if real, then none of the extant hypotheses to reconcile the conflict suffices to account for the mitochondrial evidence as well.

Finally, we must stress the importance of gathering extensive molecular data from all relevant genomes in order to understand the phylogenetic history of the enigmatic photosynthetic "organelles" present in such poorly characterized creatures as *Paulinella chromatophora* (Bhattacharya & al. 1995a, Bhattacharya & Medlin 1995), *Psalteriomonas lanterna* (see above, and Hackstein 1995), and *Dinophysis* spp. (Schnepf 1993). The only molecular data available for these are a nuclear 18s sequence from *P. chromatophora*, which establishes a very strong, nested grouping of it with two nonphotosynthetic lineages (Bhattacharya & al. 1995a, Bhattacharya & Medlin 1995). Furthermore, *P. chromatophora* has a close relative, *P. ovalis*, which is largely identical morphologically excect that it lacks any vestige of a photosynthetic organelle. Current thinking is that *P. chromatophora* acquired its plastid (or "cyanelle", because it is bounded by a peptidoglycan cell wall) either by an independent primary endosymbiosis from that which gave rise to the green, red, and glaucocystophyte plastids, or else by secondary endosymbiosis of a glaucocystophyte (Bhattacharya & Medlin 1995, Bhattacharya & al. 1995).

Taxa with secondary plastids

Although the number of primary endosymbiotic events is very small, and there is even reason to think that all primary plastids are monophyletic secondary endosymbioses greatly complicate plastid phylogeny. The plastids of cryptophytes and chlorarachinophytes are unquestionably secondary plastids, derived from eukaryotic endosymbionts (reviewed in Whatley 1993a, b; McFadden & Gilson 1995; Gilson & McFadden 1997; McFadden & al. 1997b, this volume). Ultrastructural studies have shown that in these taxa a total of four membranes surround the plastid. Two are thought to be homologous to the dual membranes surrounding the primary plastids of green algae, red algae, and glaucocystophytes, while two additional membranes apparently correspond to the eukaryotic endosymbiont's cell membrane and to the food vacuole of the host cell. These outer two membranes (sometimes called the chloroplast endoplasmic reticulum) surround a periplastidal space that includes not only the plastid, but also the 'nucleomorph', a vestigial eukaryotic nucleus with a highly reduced genome (380–660 kb) comprising but three chromosomes (Gibbs 1981; Whatley 1993a, b; Gilson & McFadden 1996, 1997; McFadden & al. 1997b).

Phylogenetic analyses place the nucleus and nucleomorph of both cryptophytes and chlorarachniophytes in clearly separate clades, strongly supporting the view that both groups acquired their plastids via secondary endosymbiosis and, therefore, that the nucleomorph is the remnant of the nucleus of the engulfed algal cell (Van de Peer & al. 1996, Cavalier-Smith & al. 1997). Although the arrangement and genome sizes of both the plastid and nucleomorph are strikingly similar in chlorarachniophytes and cryptophytes, pigment composition, ultrastructure, and molecular phylogenies indicate that these taxa independently acquired their secondary endosymbionts, the chlorarachniophytes from a green alga, and the cryptophytes from a red alga (Van de Peer & al. 1996, Cavalier-Smith & al. 1997). Although until recently a relatively obscure and poorly understood group, the chlorarachniophytes are now, thanks to the efforts of G. McFadden and collaborators, perhaps the best documented case of secondary endosymbiosis (McFadden & Gilson 1995; Gilson & McFadden 1996, 1997; McFadden & al. 1997b).

Several other algal groups seem to have secondary plastids, but with further reduction of the endosymbiont to the extent that the nucleomorph has been entirely lost (Gibbs 1993, Palmer & Delwiche 1996). As normally interpreted, the primary plastid is surrounded by two membranes, and additional membranes are taken to be evidence of secondary endosymbiosis (Cavalier-Smith 1995). The plastids of heterokonts (which include brown algae, diatoms, chrysophytes, xanthophytes, and other algae) and haptophytes (also known as prymnesiophytes) are surrounded by four membranes in a structure reminiscent of the plastid of cryptophytes, but do not retain a nucleomorph. Phylogenetic analyses of plastid rRNAs interleave heterokont and haptophyte plastids among the red algae and cryptophytes, consistent with the belief that both groups acquired their plastids via secondary endosymbiosis from red algae (see Fig. 7 of Medlin & al. 1997, this volume; Fig. 3 of Bhattacharya & Schmidt 1997, this volume; Bhattacharya & Medlin 1995; Medlin & al. 1995). Ironically, the strongest evidence that all three of the above groups acquired their plastids from red algae (or at least their ancestors) comes from rubisco analyses, which show that this entire group has a rubisco operon of proteobacterial origin, as opposed to the cyanobacterial rubisco present in glaucocystophytes and chlorophytes s. l. (Fig. 3 and following section on rubisco evolution; also see Fig. 8 of Medlin & al. 1997, this volume; Fujiwara & al. 1993; Delwiche & Palmer 1996). The unrelated positions of heterokonts and haptophytes in analyses of both plastid (16s rRNA and *rbc*L) and nuclear (18s rRNA and actin) sequences indicate that the two groups probably acquired their plastids via independent secondary events (see Figs. 1, 7, and 8 of Medlin & al. 1997, this volume; Leipe & al. 1994; Medlin & al. 1995; Bhattacharya & Weber 1997).

The euglenophytes have green-alga-like pigmentation, but differ from green algae in many characteristics of the host cell. The euglenophyte plastid is surrounded by three membranes, and this was interpreted early on as evidence of a secondary plastid acquired from a green alga (Gibbs 1978). Subsequent molecular phylogenetic and ultrastructural studies have led to the present view that the host cell in this secondary endosymbiosis was a relatively derived, nonphotosynthetic member of a clade that includes the kinetoplastids and euglenoids (TRIEMER & FARMER 1991; SOGIN 1994; R. TRIEMER, pers. comm.). The peculiar mix of characters seen in the euglenophytes could be attributed to the chimeric nature of the cell resulting from secondary endosymbiosis. Molecular studies of the GAPDH gene support this conclusion, although the frequency of this gene's duplication and loss complicates interpretation of the data (HENZE & al. 1995).

Finally, two groups of protists – the dinoflagellates and apicomplexans-have plastids that are probably of secondary origin (in some cases even tertiary) and whose enigmatic features warrant more extended discussion, as presented in the next two sections.

The diverse plastids of dinoflagellates

The most familiar plastids in the dinoflagellates are the peridinin-containing plastids (DODGE 1989, SCHNEPF 1993, WHATLEY 1993). These plastids are surrounded by three membranes and have distinctive pigmentation (chlorophylls *a* and c2, with peridinin, diadinoxanthin, and β-carotene as major secondary pigments). The peridinin plastids are found only within the orders *Gonyaulacales* (*Peridiniales*) and *Prorocentrales*, but they account for the majority of photosynthetic dinoflagellates, about half of which are heterotrophic (DODGE 1989).

Are peridinin plastids secondary plastids? There is indirect evidence that this may be the case. Although they are not obviously similar to the plastids of any other algal group, they resemble those of heterokonts and haptophytes in the presence of chlorophylls *a* and *c*, and in the stacking of thylakoids into groups of three. However, the dinoflagellate host cell bears little resemblance to the host cells of other taxa with similar plastids (rather, it is related to ciliates and apicomplexans; e.g., see CAVALIER-SMITH 1993a; SOGIN 1994, 1996; VAN DER PEER & al. 1996), suggesting that they may have acquired their plastids independently. In addition, like the euglenophytes, peridinin plastids have three bounding membranes. Finally, many dinoflagellates are phagotrophic (in fact only about half are photosynthetic; DODGE 1989), providing a ready mechanism for secondary endosymbiosis. Unfortunately no molecular information is yet available for the plastid genome of any dinoflagellate, and consequently all phylogenetic inferences must be based on ultrastructure and biochemistry alone.

The *rbc*L gene, encoding the large subunit of rubisco, has been isolated from the peridinin-containing dinoflagellates *Gonyaulax* and *Symbiodinium*, but remarkably, it is not located on the plastid genome. Rather, apparently all peridinin-containing dinoflagellates rely on a nuclear-encoded, form II rubisco (MORSE & al. 1995, WHITNEY & al. 1995, ROWAN & al. 1996). Form II rubisco is composed of a dimer (L_2) of large subunits, and is otherwise known only among the proteobacteria, whereas all cyanobacteria and all other plastids contain form I rubisco (L_8S_8). How a form II rubisco came to replace the putatively native rubisco of cyanobacterial form I origin is completely conjectural, but because the mitochondrion is of proteobacterial descent, one of several potential mechanisms would involve horizontal gene transfer of the proteobacterial/mitochondrial form II gene to the nuclear genome, followed by targeting of the form II rubisco to the plastid and eventual loss of the original plastid form I gene (MORSE & al. 1995, PALMER 1995, ROWAN & al. 1996).

VAN DEN HOEK & al. (1995) argue that because the two dinoflagellate orders (the *Gonyaulacales* and *Prorocentrales*) with peridinin plastids are quite distinct and do not seem to be sister taxa, these plastids are the ancestral plastid type for the dinoflagellates and many non-photosynthetic dinoflagellates are descendants of photosynthetic forms that have lost their plastids. According to their interpretation, some of those taxa which have lost their original plastids have subsequently reacquired new plastids via further secondary endosymbioses. An alternative explanation would involve independent initial acquisitions, via secondary endosymbiosis, in the two orders. Because of the great diversity of plastids found among the dinoflagellates it is clear that plastid evolution within the group has been complex, and there is no compelling reason to favor either hypothesis.

In addition to the peridinin-containing plastids, there is a bewildering array of plastids among the dinoflagellates, all of which are of apparent secondary origin. Dinoflagellate plastids of putative green algal, haptophytic, and cryptophytic origin have all been documented, the latter two representing apparent cases of *tertiary* endosymbiosis. In addition, some dinoflagellates have long-term associations with cyanobacteria that do not have the reduced genomes characteristic of plastids. A particularly interesting case is *Lepidodinium* (WATANABE & al. 1990), which contains secondary plastids that resemble those found among the prasinophyte green algae, but which also displays complex scales on the surface of the dinoflagellate host cell that closely resemble scales found among the prasinophyceae. Other scale-forming dinoflagellates are known, but the complexity of these scales and their striking similarity to those found in the prasinophyceae suggests that the genes responsible for scale formation in the prasinophyte endosymbiont (which does not possess a nucleomorph) were transferred to the host nucleus, where they are successfully expressed. Verification of this hypothesis will require identification and study of the genes responsible for scale formation in prasinophytes and in *Lepidodinium*.

Apicomplexan plastids

The most recent addition to the ranks of taxa known to have plastids is the phylum *Apicomplexa*. These organisms are obligate intracellular parasites of animals and include such important pathogens as *Plasmodium*, the cause of malaria, and *Toxoplasma*, the cause of toxoplasmosis. All known apicomplexans are parasitic, and none are known to be photosynthetic (VIVIER & DESPORTES 1989). Phylogenetic analyses of nuclear genes agree with structural evidence that places the apicomplexans in a group (the *Alveolata*) with the ciliates and the dinoflagellates (CAVALIER-SMITH 1993a; SOGIN 1994, 1996; VAN DE PEER & al. 1996).

The presence of a plastid in apicomplexans was originally recognized through the identification of two extrachromosomal genomes, a 35 kb circular genome, and a linear genome of tandemly repeated 6 kb elements. The 35 kb genome of *Plasmodium* has been completely sequenced and resembles those of plastids in containing a rRNA-encoding inverted repeat, genes encoding a eubacterial RNA polymerase, elongation factor Tu (*tuf*A), and ORF470, and a group I intron in the

UAA anticodon of the gene for tRNA-Leu (PREISER & al. 1995; WILSON & al. 1996; MCFADDEN & al. 1997b, this volume). All of these features are widespread in plastid genomes, but rare or absent in mitochondria. By contrast, the 6 kb genome of *Plasmodium* resembles those of mitochondria, encoding three proteins from the mitochondrial electron system and fragmented rRNA genes. Thus, gene content and organization indicate that the 35 kb genome is plastidic and the 6 kb genome is mitochondrial (PALMER 1992, FEAGIN 1994, WILSON & al. 1996). Both apicomplexan extrachromosomal genomes are uniparentally inherited, as are organellar genomes in most taxa (VAIDYA & al. 1993).

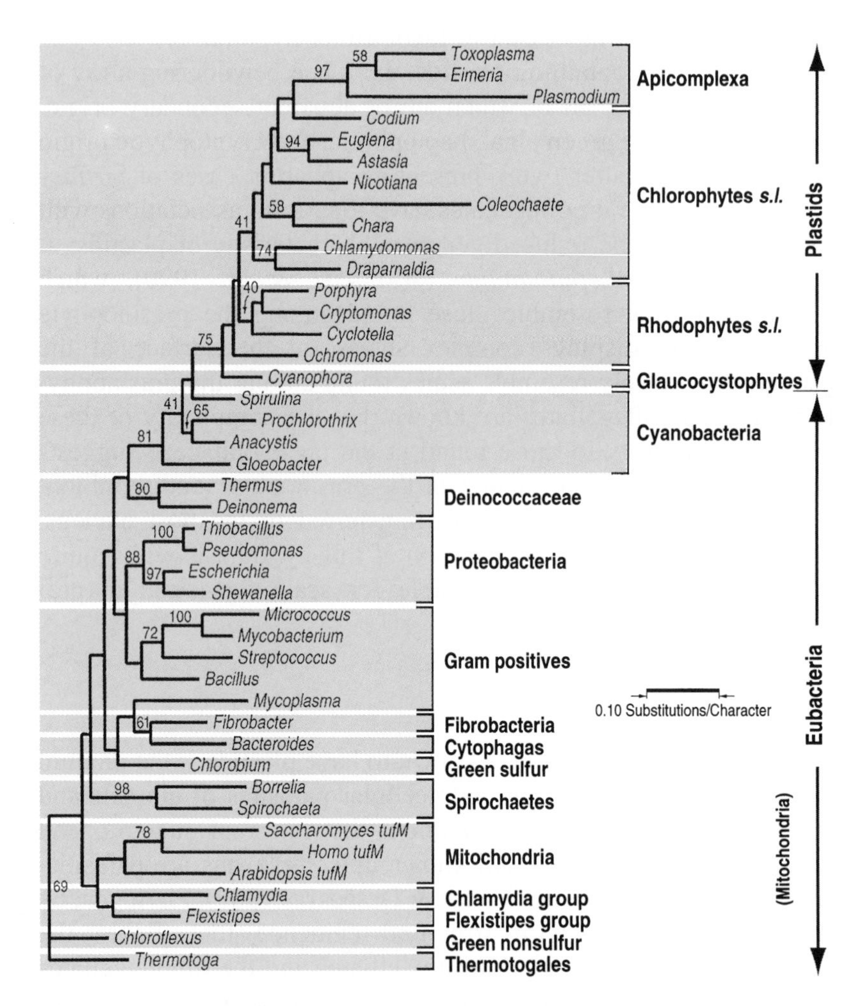

Fig. 2. Phylogeny of *tuf*A sequences from eubacteria and their endosymbiotic descendants. Shown is the highest likelihood tree found in a maximum likelihood analysis of first and second nucleotide positions. Branch lengths are proportional to the number of inferred substitutions (see scale bar). Bootstrap values 40% are given above the corresponding branch. This figure is modified from KÖHLER & al. (1997: fig. 3), which should be consulted for more details on methodology. Also see KÖHLER & al. (1997) for an example of a tree showing monophyly of plastids relative to seven cyanobacterial sequences

The highly divergent and AT-rich nature of the 35 kb genome has confounded most attempts at molecular phylogenetic identification of their origin (e.g., EGEA & LANG-UNNASCH 1995). Recently, however, analyses of *tuf*A from the 35 kb genomes of three diverse apicomplexans, *Plasmodium*, *Toxoplasma*, and *Eimeria*, have clearly placed these taxa among the plastids (Fig. 2; KÖHLER & al. 1997). Furthermore, these analyses consistently place the apicomplexa within the clade of green plastids (KÖHLER & al. 1997). Thus, these data strongly support the view that the 35 kb genome is indeed plastidic and provisionally suggest that this plastid is of green algal origin. Because the apicomplexan nucleus is clearly *not* of green algal origin, but instead related to those of dinoflagellates and ciliates (see above), these data also imply that the apicomplexan plastid was acquired via secondary endosymbiotic uptake of a green alga.

Ultrastructural evidence, although controversial (McFADDEN & al. 1997b, this volume), supports the notion that the apicomplexan plastid arose via secondary endosymbiosis. The 35 kb genome has recently been localized via in situ hybridization in *Toxoplasma* to a spherical, membrane-bound organelle (now identifiable as a degenerate plastid) (McFADDEN & al. 1996, KÖHLER & al. 1997). McFADDEN & al. (1996) report having found sections in which the *Toxoplasma* plastid is surrounded by only two membranes, and suggest that additional membranes may be closely appressed portions of the golgi apparatus. However, KÖHLER & al. (1997) present electron micrographs from *Toxoplasma* that clearly show the presence of four bounding membranes, and report having followed these membranes around the plastid in serial sections of several individual plastids. The number of membranes surrounding the plastid is important because, as described earlier, membrane number is a widely accepted indicator of primary (2 membranes) vs. secondary origin (>2) of the plastid. Because there may be pores or evaginations of the plastid envelope that would lead to the appearance of a different number of membranes than are actually present, studies with full three-dimensional reconstruction of the plastid envelope will probably be necessary to fully resolve this question. Study of the apicomplexan nuclear genome may also reveal genes of (green) algal ancestry that would help clarify the origin of this plastid.

Rubisco phylogeny and horizontal gene transfer

Overwhelming biochemical, structural, and molecular evidence indicates that all plastids are derived from cyanobacteria, and substantial evidence suggests that they may be monophyletic. Yet, in plastid genomes of red algae all of their secondary symbiotic derivatives (cryptophytes, heterokonts, and haptophytes), with the prominent exception of dinoflagellates (see previous section), the genes encoding the key photosynthetic enzyme rubisco are clearly not cyanobacterial in origin but are instead of proteobacterial provenance (BOCZAR & al. 1989, VALENTIN & ZETSCHE 1989, DOUGLAS 1994, DELWICHE & PALMER 1996). Because the presence of a proteobacterial rubisco in reds and their derivatives is contrary to such a massive monolith of evidence favoring a cyanobacterial origin of all plastids, most authors have assumed that this reflects a single ancient horizontal transfer of the rubisco operon from a proteobacterium to the common ancestor of red-like plastids, after their divergence from the green and glaucocystophyte lineages (see above references and MORDEN & al. 1992). An alternative explanation, first presented by MARTIN & al. (1992), is an ancient duplication of the rubisco operon followed by differential loss of one of the two copies from various plastid (and bacterial) lineages.

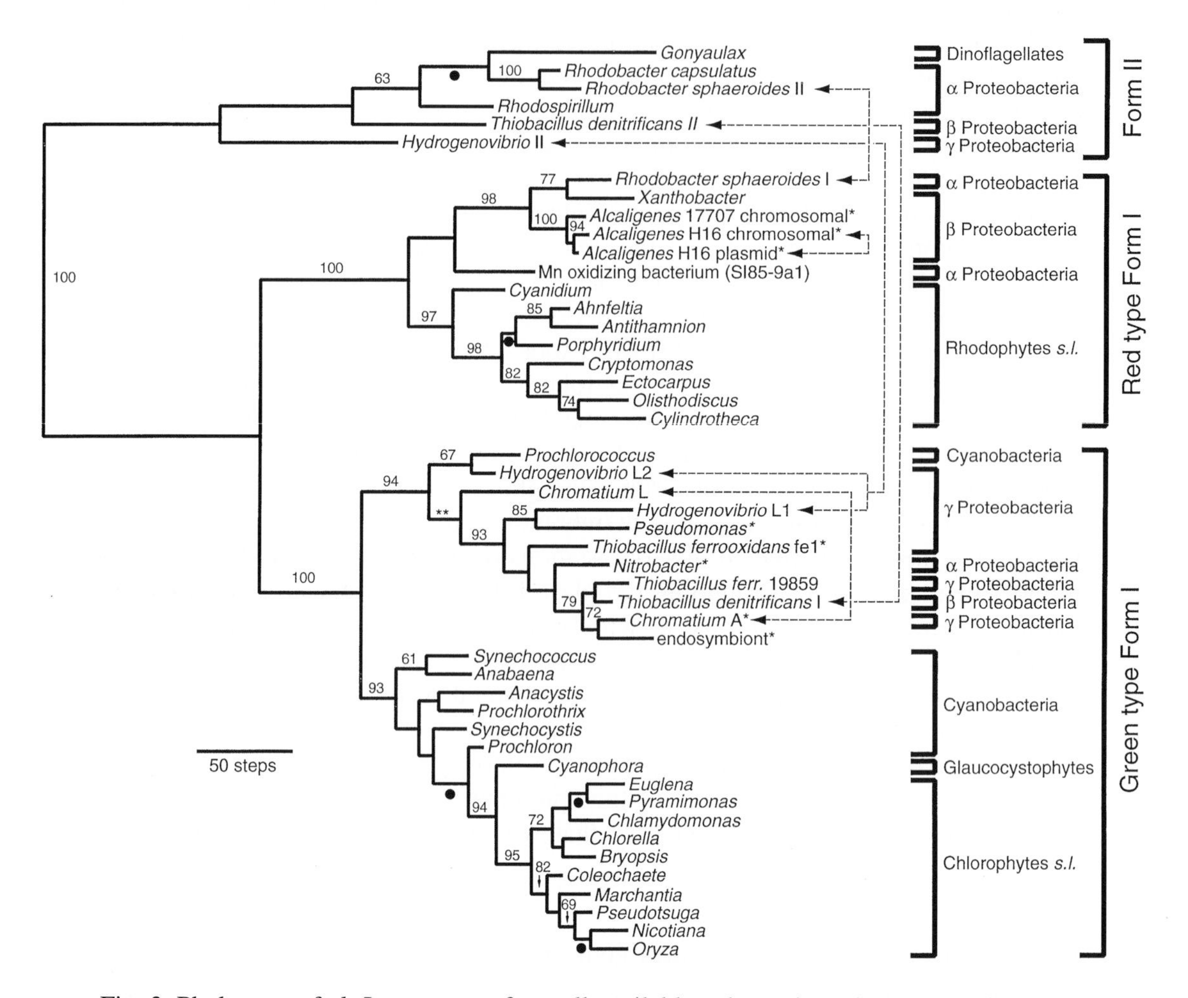

Fig. 3. Phylogeny of *rbc*L sequences from all available eubacteria and representative algae and plants (this figure is modified from DELWICHE & PALMER 1996). The tree shown is one of 24 shortest trees, selected arbitrarily from a maximum parsimony analysis of *rbc*L amino acid sequences. Classification of taxa based on 16s rRNA and other evidence is indicated to the immediate right of the taxon names. Bootstrap values above 60% are indicated above the branches, and branches which were not present in the strict consensus of the 24 shortest trees are indicated by a bullet (•) below the branch. Branch lengths correspond to the inferred number of character state changes on each branch (see scale bar). Dashed, arrowheaded lines connect multiple *rbc*L sequences determined from the same taxon. *Ribosomal RNA sequences were not available for this strain, and the proteobacterial subgroup may be suspect. ** Bootstrap analysis placed *Chromatium* L. in a clade with *Hydrogenovibrio* L2 and *Prochlorococcus* with 60% bootstrap support, but this topology did not occur among the shortest trees

We recently performed phylogenetic analyses of the rubisco *rbc*L gene in order to further evaluate the complex phylogenetic history of this key enzyme from the standpoint of horizontal gene transfer and gene duplication and loss (Delwiche & Palmer 1996, also see Watson & Tabita 1996). When rooted by form II *rbc*L (present in certain proteobacteria and the dinoflagellate nucleus), the *rbc*L tree (Fig. 3) identifies two groups of form I rubisco, the green type found in plants, green algae, glaucocystophytes, cyanobacteria, and some proteobacteria (termed types IA and IB by Tabita 1995), and a red type seen in red algae, their secondary derivatives, and some proteobacteria (types IC and ID). Within both of these groups, *rbc*L sequences are relatively similar (70% amino acid identity or greater), while between the groups there is substantially more divergence (53–60% identity; Delwiche & Palmer 1996). The *rbc*L phylogeny (Fig. 3) reveals a surprisingly large number of gene relationships which are fundamentally incongruent with organismal relationships as inferred from multiple lines of other molecular evidence. Roughly six different horizontal gene transfers are implied by the form I rubisco phylogeny alone, five involving strictly proteobacteria or proteobacteria and cyanobacteria, and the sixth being the putative transfer of proteobacterial rubisco to a red algal ancestor. Alternatively, much of the incongruence could be accounted for by an ancient duplication of the form I rubisco operon, followed by repeated and pervasive loss of one operon or the other. Although on balance, horizontal transfer seems the best explanation for the presence of proteobacterial rubisco genes in rhodophytic s.1. plastid genomes, the rubisco operon has most likely undergone multiple events of both horizontal gene transfer and gene duplication/loss across bacteria.

Pigmentation in the ancestral chloroplast

Traditionally, the great diversity in photopigment composition of the major lineages of algae has been viewed in two phylogenetic contexts, polyphyletic and acquisitive. The first of these postulates that whereas red algae acquired their plastids from cyanobacteria (both have chlorophyll *a* and phycobilins), green algae acquired theirs from hypothetical "green" bacteria possessing chlorophylls *a* and *b*, and heterokonts and haptophytes theirs from hypothetical "yellow" bacteria with chlorophylls *a* and *c* (Raven 1970). The subsequent discovery of the "prochlorophytes", oxygenic photobacteria with green alga-like pigmentation, led to widespread acceptance of this hypothesis (Bullerjahn & Post 1993). However, as discussed earlier, plastids are probably monophyletic in origin, and it is now clear that green algae are not derived from (at least the known) prochlorophytes. Instead the three characterized lineages of prochlorophytes are best regarded as independently derived types of cyanobacteria (Palenik & Haselkorn 1992; Urbach & al. 1992; Turner 1997, this volume). Intriguingly, however, the three prochlorophyte lineages appear to have either converged on, or retained, a similar photosynthetic apparatus, and despite superficial similarity, "this" photosynthetic apparatus of prochlorophytes is rather different from that of green plastids (Bullerjahn & Post 1993). Most notably, it has recently been shown that the chlorophyll *a*/*b* light-harvesting complex (LHC) proteins of the three prochlorophyte lineages are related to one another and to two other classes of cyanobacterial proteins, but are unrelated to the functionally analogous light-harvesting proteins of plastids (LA ROCHE & al. 1996).

The acquisitive theory of photopigment evolution in plastids fits much better with the current evidence that primary plastids probably arose through a single endosymbiotic event. This theory proposes that the ancestral plastid was red algal- (and glaucocystophyte-) like in containing the conventional cyanobacterial photopigments chlorophyll *a* and phycobilins. Subsequently, chlorophyll *c* was acquired by heterokonts, haptophytes, stereotypical dinoflagellates, and cryptophytes, with all but the latter losing phycobilins, while green algae acquired chlorophyll *b* and also lost phycobilins. While this theory may be largely correct, proving or disproving it is likely to be difficult, and recent findings on photopigment diversity in *Cyanobacteria* s. str. make equally tenable a more reductive theory of photopigment evolution. This theory, first developed by BRYANT (1992), proposes that ancestral, and perhaps most extant, cyanobacteria, including the progenitor(s) of plastids, had many or all of the photopigments currently found in various combinations in the different types of plastids, with plastid diversity largely reflecting differential loss of these pigments. This reductive theory also accounts for the seemingly parallel evolution of photopigments, and associated LHC proteins (LA ROCHE & al. 1996), in the three unrelated lineages of prochlorophytes, by proposing instead that these are retained primitive features, with these lineages having each lost phycobilisomes.

A number of recent observations are consistent with the theory that the cyanobacterial ancestors of plastids and prochlorophytes possessed essentially the entire spectrum of photopigments and associated LHC proteins found variously among extant plastids. This evidence includes the following: First, WILHELM (1987) reported the existence of a chlorophyll *c*-like pigment in prasinophytic green algae such as *Mantoniella squamata*. Second, LARKUM & al. (1994) recapitulated this finding by discovering a chlorophyll *c*-like pigment, in addition to chlorophylls *a* and *b*, in the prochlorophyte *Prochloron* sp. Third, in the same year, WOLFE & al. (1994) identified serological similarities between one of two sets of LHC proteins in red algae and the LHC proteins of plants (but not of cyanobacteria). This finding suggests that ancestral plastids may have had two functional LHCs, one using phycobilins as antenna pigments, and the other using a LHC more similar to those of green algae and heterokonts. Fourth, MIROSHNICHENKO DOLGANOV & al. (1995) identified proteins in cyanobacteria that are similar to the chlorophyll *a/b*-binding proteins of plastids. Finally, HESS & al. (1996) discovered that the marine prochlorophyte *Prochlorococcus marinus* contains phycobilins in addition to the usual chlorophylls *a* and *b*. None of these observations is convincing by itself, but together they suggest that light harvesting systems exist in nature that have more complex combinations of pigments and associated proteins than has been traditionally recognized. Ultimately, in order to critically evaluate the extent to which photopigment evolution has been primarily reductive or acquisitive, we need to know the identity of the complete set of photopigments and associated light-harvesting proteins present in a broad spectrum of oxygenic phototrophs, as well as garner information on the homology relationships of these pigments and proteins.

Evolution of plastid gene content

Plastids are greatly modified from their cyanobacterial ancestors. They have become highly integrated with the host cell over the course of their 1–2 billion year permanent symbiotic relationship. The most dramatic manifestation of this intimate and highly interdependent relationship between plastids and their host cells is the tremendous reduction of plastid gene content and concomitant transfer of most plastid gene function to the nucleus (PALMER 1991, LOISEAUX-DE GOËR 1994, LÖFFELHARDT & BOHNERT 1994, REITH 1995). Whereas many cyanobacterial/ plastid genes have presumably been lost because their products are no longer needed in the endosymbiotic environment, many others have been replaced by genes in the nucleus. Two processes have been identified that lead to the replacement of a plastid gene by a nuclear gene. In the case of gene transfer, a plastid gene is directly moved to the nucleus and establishes functionality there. This process is thought to always involve a transition or intermediate stage in which copies of the same gene are present in both compartments, the gene in the plastid being lost only after the duplicated and transferred nuclear copy becomes properly expressed and has its product correctly targeted back to the plastid (BALDAUF & al. 1990, GANTT & al. 1991, NUGENT & PALMER 1991). Alternatively, in gene *substitution,* a nuclear gene of strictly eukaryotic history, whose product is normally expressed in the cytosol, has its product retargeted to the plastid (PALMER 1991). Usually this involves a gene duplication event within the nucleus, with one gene still supplying the cytosol and the other newly supplying the plastid (e.g., TINGEY & al. 1988, HENZE & al. 1994, SCHMIDT & al. 1995). A number of more complicated situations have recently been identified, which involve both transfer and substitution. For example, the gene encoding 3-phosphoglycerate kinase appears to have been transferred from the plastid to the nucleus early in plastid evolution; later, in land plant or green algal evolution, this now-nuclear gene was duplicated, with one copy replacing the preexisiting nuclear gene of eukaryotic descent in supplying this enzyme to the cytosol (BRINKMAN & MARTIN 1996). Thus, substitutional events can work both ways, leading to loss of an ancestrally cyanobacterial/plastid gene via retargeting of an ancestrally nuclear gene's product to the plastid, as well as the reverse, loss of an ancestrally nuclear gene due to retargeting of a transferred (and duplicated) plastid gene's product to the cytosol.

All plastid genomes are substantially reduced in size and gene content from the genome of their cyanobacterial ancestor. The only sequenced cyanobacterial genome, from *Synechocystis* PCC 6803, is 3573 kb in size and contains about 3000 genes (KANEKO & al. 1996). This is one of the smaller cyanobacterial genomes; many are 2–4 times larger (HERDMAN & al. 1979). Table 1 summarizes the size and gene content of the 11 plastid genomes that have been completely sequenced to date. Even the largest and most gene-rich of these, from the red alga *Porphyra purpurea* (REITH & MUNHOLLAND 1993, 1995), has at most 10% of the genes found in the free-living *Synechocystis.* [A few poorly characterized plastid genomes from green algae are, at 300–400 kb and perhaps larger, significantly larger than even the *Porphyra* genome (PALMER 1991). However, this extra DNA more likely consists of extra repeats and introns rather than many additional genes (e.g., TURMEL & al. 1987).] Although considerably reduced compared to their cyanobacterial

Table 1. Size, gene content, and intron content of sequenced plastid genomes[1].

Group	Organism	Genome size (bp)	Number of genes[2]					Number of introns
			Total	Genetic	Photo-synthetic	Misc-ellaneous	ORF	
Glaucocystophyte	*Cyanophora paradoxa*	135599	191	87	48	20	36	1
Rhodophyte	*Porphyra purpurea*	191028	251	103	53	30	65	0
Heterokont	*Odontella sinesis*	119704	165	77	41	10	37	0
Euglenophyte	*Euglena gracilis*	143170	97	56	27	1	13	>149
Land plant	*Marchantia polymorpha*	121024	120	62	42	7	9	19
Land plant	*Pinus thunbergii*	119707	108	62	30	5	11	16
Land plant	*Oryza sativa*	134525	110	61	41	1	7	18
Land plant	*Zea mays*	140386	110	61	41	1	7	18
Land plant	*Nicotiana tabacum*	155884	113	61	41	2	9	21
Land plant	*Epifagus virginiana*[3]	70028	42	38	0	2	2	6
Apicomplexan	*Plasmodium falciparum*[3]	34682	57	48	0	2	7	1

[1] References for the complete sequences are: *Cyanophora*, STIREWALT & al. (1995); *Porphyra*, REITH & MUNHOLLAND (1995); *Odontella*, KOWALLIK & al. (1995); *Euglena*, HALLICK & al. (1993); *Marchantia*, OHYAMA & al. (1986); *Pinus*, WAKASUGI & al. (1994); *Oryza*, HIRATSUKA & al. (1989); *Zea*, MAIER & al. (1995); *Nicotiana*, SHINOZAKI & al. (1986); *Epifagus*, WOLFE & al. (1992); *Plasmodium*, WILSON & al. (1996).
[2] Duplicated genes are counted only once. "Genetic" genes are involved in translation, transcription, RNA processing, and DNA replication. "Photosynthetic" genes includes standard photosynthetic genes and the chlororespiratory NADH dehydrogenase genes. "Miscellaneous" genes are involved in the biosynthesis of chlorophyll, carotenoids, phycobilins, amino acids, fatty acids, NAD^+, pyrimidines, thiamine, and the peptidoglycan wall of the *Cyanophora* plastid; the assimilation of nitrogen and sulfur; protein transport, processing, and assembly/chaperoning; and the specification of thioredoxin and pyruvate dehydrogenase. "ORF" indicates highly conserved open reading frames of unknown function, but which are almost certainly functional genes.
[3] Nonphotosynthetic organisms.

progenitors, plastids are nonetheless very diverse organelles metabolically, and one can roughly estimate that they contain about 1000 different proteins (EMES & TOBIN 1993). Thus, about 80% of plastid proteins are encoded by nuclear genes even in *Porphyra*, and 90% or more in land plants, *Euglena*, and the apicomplexan parasite *Plasmodium* (Table 1).

Assuming a monophyletic origin of plastids, two observations – that all plastids are considerably reduced in gene content and that the smaller genomes largely overlap the larger ones in gene content (e.g., all but eight of the 191 plastid genes in the glaucocystophyte *Cyanophora* are present among the 251 plastid genes of the rhodophyte *Porphyra*, as are all but 20 of the 120 genes in the chlorophyte *Marchantia*) – lead to an important conclusion. Namely, the last common ancestor of all extant plastids had an already highly reduced genome, containing on the order of 300 genes, again, only about 10% as many as its cyanobacterial progenitor. To restate this conclusion: Most gene transfer and substitutional relocations of gene function to the nucleus occurred relatively soon after cyanobacterial/plastid endosymbiosis ensued, before any of the organismal divergences leading to the extant groups of plastid-containing eukaryotes. Although we don't know how long this early period of massive transfer and substitution was, it seems safe to conclude that the pace of functional relocation was much higher early on than it has been during the subsequent and relatively long period leading to the establishment and diversification of the many groups of extant algae.

Nonetheless, the range in gene content (97–251 genes) of the small number (9) of sequenced plastid genomes from photosynthetic eukaryotes reveals that there has still been considerable loss, transfer, and substitution of plastid gene function over the course of algal and plant evolution. A core set of ~80 genes is found in all of these photosynthetic plastid genomes. These genes function virtually exclusively in but two general processes – gene expression (primarily translation) and photosynthesis. Indeed, this statement is largely true for the entire plastid genomes of all examined photosynthetic chlorophytes and their secondary derivatives (i.e. euglenophytes; a chlorarachniophyte plastid genome has not yet been sequenced, Table 1). As a consequence, the permanently nonphotosynthetic derivatives of these chlorophytes s. l., organisms such as the beechtree parasite *Epifagus virginiana* (WOLFE & al. 1992), the malarial parasite *Plasmodium falciparum* (WILSON & al. 1996), and the colorless euglenophyte *Astasia longa* (GOCKEL & al. 1994), have extremely small plastid genomes (Table 1) geared virtually entirely towards the mere act of gene expression itself. Each of these three organisms appears to contain but one or a very few "raison d'être" genes, i.e. a gene whose product plays some essential metabolic role even in a parasite or colorless plastid. Given this, one might well expect to find that some of the many other, as yet molecularly uncharacterized, independent lineages of nonphotosynthetic chlorophytes might contain a plastid entirely lacking a genome, just as we now know that a surprisingly large number of lineages of mitochondria (usually called "hydrogenosomes") have independently dispensed with their genomes (PALMER 1997).

The extra genes found in the gene-rich photosynthetic genomes, of *Porphyra, Cyanophora*, and *Odontella*, fall into several categories. First, these genomes contain substantially more genes involved in gene expression, chiefly owing to a larger complement of ribosomal protein genes (47 in *Porphyra* compared to 21 in *Euglena* and most land plants). Second, they contain more photosynthetic genes, although the difference here is obscured by the "special retention" in most land plants of 11 NADH dehydrogenase genes (see below). These 20-odd extra photosynthetic genes include both additional genes for components of the photosynthetic apparatus already represented in other plastid genomes, and also 7–10 genes for phycobilisomes, which are simply absent from chloroplasts and many other plastids (including that of *Odontella*) (Fig. 1). Third, and most notably, the *Porphyra* and *cyanophora* plastid genomes contain 20–30 genes (included in the "Miscellaneous" column of Table 1) which greatly enlarge the contribution of the plastid genome to the diverse metabolic economy of plastids. On the one hand, the depauperate rice and maize plastid genomes encode as few as a single nonphotosynthetic, nongenetic protein (a subunit of the Clp protease). On the other hand, the *Porphyra* genome encodes at least 30 such proteins, including enzymes involved in the biosynthesis of amino acids, fatty acids, various types of pigments, pyrimidine, and thiamine; pyruvate dehydrogenase subunits and thioredoxin; and factors involved in nitrogen assimilation and in protein transport, processing, and assembly (REITH & MUNHOLLAND 1993, 1995). Furthermore, even the genetic category of plastid genes in *Porphyra* (and also *Cyanophora*) is far more diverse than in chlorophytes, as it includes such otherwise "novel" genes as those encoding transcriptional regulators, a DNA replication helicase, tRNA synthetases, novel translation factors, the RNA component of RNAse P, and an RNAse E homolog. Finally, these gene-rich genomes also contain substantially more genes of unknown function (ORFs), and it seems a safe bet that many of these will further widen the metabolic diversity of the rhodophyte s. l. and glaucocystophyte plastid genomes relative to their depauperate (in both number and kind) chlorophytic brethren. Why the non-green plastids have retained so many more genes and such a greater diversity of genes is not at all clear.

Present data thus allow us to classify plastid genomes into three categories with respect to the kinds of genes they contain: (1) plastid genomes of glaucocystophytes and perhaps all rhodophytes s. l. (e.g., including *Cryptomonas*; DOUGLAS, 1992, REITH 1995) contain a significant number of genes belonging to each of three classes – genes involved in gene expression, genes involved in photosynthesis, and genes involved in a wide array of other kinds of plastid metabolism and biochemistry. (2) plastid genomes of all examined chlorophytes s. l. contain many belonging to these first two classes, but very few of the third category. (3) plastid genomes of nonphotosynthetic chlorophytes s. l. contain a significant number of genes for gene expression only, along with as few as a single "raison d'être" metabolic gene. We have already predicated the likely existence of a fourth type of plastid, one with no genome at all in some nonphotosynthetic chlorophyte, and can also predict a fifth type: permanently nonphotosynthetic rhodophytes s. l. should contain many genes for both gene expression and plastid metabolism, but none for photosynthesis.

The pattern of gene loss across these sequenced plastid genomes is noteworthy only in the case of the "specially retained" NADH dehydrogenase (*ndh*) genes. That is , excluding of course the 80 or so core "universal" genes noted above, most genes have been lost only once or at most a few times among these and other examined photosynthetic plastid genomes. The entire set of 11 *ndh* genes, however, has been lost repeatedly and almost pervasively, such that none of these genes is present in any examined algal genome, including not only the ones listed in Table 1, but also the green alga *Chlamydomonas* (BOUDREAU & al. 1994), leaving them "specially retained" in land plants only. (The charophytic ancestors of land plants will be very interesting to examine in this regard.) Furthermore, the entire set of *ndh* genes is missing or defunct in the plastid genomes of the black pine *Pinus thunbergii* (WAKASUGI & al. 1994) and the flowering plant *Cuscuta reflexa*, a holoparasite which otherwise retains a normal set of photosynthetic genes (HABERHAUSEN & ZETSCHE 1994). An analogous, seemingly illuminating situation is observed in mitochondrial genomes of ascomycetes: *Saccharyomyces* and *Schizosaccharomyces* have each independently lost the entire set of 7 mt *ndh* genes that are retained in other ascomycetes, including other yeasts (GRAY 1992). It is thought that *Saccharomyces* and *Schizosaccharomyces* have entirely dispensed with complex I of the mitochondrial respiratory chain, and hence with all of their *ndh* genes too. The extreme lability of the *entire* set of plastid *ndh* genes may reflect a marginal selective value of this complex in plastids, perhaps owing to the existence of some nearly redundant or readily substitutable complex.

Although all examined plastid genomes are highly reduced in size relative to cyanobacteria and also compactly organized, with short intergenic spacers and retention of many cyanobacterial-like operons, there is nonetheless significant (2.7-fold) variation in gene density among the 11 sequenced genomes. *Plasmodium* (1.64 genes per kb) and *Cyanophora* (1.41) have the highest density of genes, while *Epifagus* (0.60) and *Euglena* (0.68) have the lowest. These density differences can be accounted for by four parameters – amount of repeated DNA, amount of intronic DNA, average length of intergenic spacers, and average gene length. The huge number of introns (>149, comprising >38% of the genome) alone accounts for the low gene density of the *Euglena* plastid genome (Hallick & al. 1993). The relatively low densities of the land plant genomes, especially those of flowering plants, reflect all four parameters: These genomes usually have a large inverted repeat (20–25 kb in size in most flowering plants), on average about 20 introns, have somewhat larger spacers than the non-chlorophytes, and contain two land-plant-specific genes that are the largest known plastid genes (Palmer 1991). The especially low density in *Epifagus* reflects the disproportionately large size of its inverted repeat (nearly half its sequence complexity is duplicated) and its disproportionate retention of large genes. The record high density in *Plasmodium* is paradoxical, for it too contains a relatively large inverted repeat (about a fifth of its complexity is duplicated), but this is more than countered by very short spacers, a lack of introns, and especially, an extremely short average gene size (many little tRNA, ribosomal protein, and ORF genes) (Wilson & al. 1996).

Functional gene relocation (via transfer or substitution) has clearly played a major role in the evolution of primary plastids alone, having occurred hundreds of times in the rapid early phase leading to the last common ancestor of extant plastids, and hundreds more since then (some genes probably having been repeatedly relocated), during the diversification of algae and plants. These numbers are dwarfed, however, by what has happened over the repeated history of secondary plastid endosymbiosis. Excluding the potentially many cases involving dinoflagellates, for which molecular data are lacking, there are six independent cases of secondary symbiosis where we can examine the gene content of the secondary plastid and its endosymbiosed cognate nucleus (i.e. nucleomorph). In all six cases, plastid gene content is conventionally small (Table 1; Douglas 1992, Reith 1995), and in four (heterokonts, haptophytes, euglenophytes, apicomplexans) there is no detectable trace of a nucleomorph genome. Thus, in each of these four lineages, the entire set of many hundreds of nucleomorph genes necessary for plastid function, which were already transferred once – from plastid to primary nucleus (=nucleomorph), have been transferred again – from nucleomorph to secondary (host) nucleus. Moreover, in the two groups (chlorarachniophytes and cryptophytes) with residual nucleomorphs, these genomes are very highly reduced in size and function chiefly to maintain and express the nucleomorph's own genetic apparatus (Gilson & McFadden 1996, 1997; Palmer & Delwiche 1996; McFadden & al. 1997a, this volume). In other words, even in organisms with a nucleomorph, most plastid proteins are encoded by the host nucleus. Thus, in at least six independent cases, and probably many more if one considers dinoflagellates, genes for several hundred plastid proteins have been functionally relocated a second time, from nucleomorph to host nucleus. This repeated pattern of massive gene flow implies strong pressures to reduce plastid and nucleomorph genomes and relocate their genes to the main nucleus, but the nature of these pressures, including whether they are primarily selective or mechanistic, is unclear. In any event, secondary plastid endosymbiosis, both the primary phenomenon itself and the underlying massive flow of genes between nuclear compartments, represents a spectacular case of parallel evolution writ large.

We thank DEBASHISH BHATTACHARYA, GERTRAUD BURGER, MIKE GRAY, BEN HALL, FRANZ LANG, CHARLEY O'KELLY, JOHN STILLER, and RICHARD TRIEMER for generously supplying unpublished results, SEÁN TURNER for reading the manuscript, and the NSF (grant DEB-9318594 to J. D. P.) for financial support.

Note added in proof

The first complete sequence of a green algal plastid genome was reported while this chapter was being edited and typeset (see WAKASUGI, T., NAGAI, T., KAPOOR, M., SUGITA, M., ITO, M., ITO, S., TSUDZUKI, J., NAKASHIMA, K., TSUDZUKI, T., SUZUKI, Y., HAMADA, A., OHTA, T., INAMURA, A., YOSHINAGA, K., SUGIURA, M. 1997: Complete nucleotide sequence of the chloroplast genome from the green alga *Chlorella vulgaris*: the existence of genes possibly involved in chloroplast division. – Proc. Natl. Acad. Sci. USA **94**: 5967–5972). The *Chlorella vulgaris* plastid genome is 150,613 bp in size, lacks a large inverted repeat typical of most ptDNAs, and contains three introns. It contains at least 111 genes, similar to land plant genomes (Table 1), although it lacks the 11 NADH dehydrogenase genes found in most land plant genomes (but absent from all examined algal genomes; see text) and instead contains 10 genes absent from land plant genomes. However, most of these extra genes are present in one or more non-chlorophyte genomes. Two notable exceptions are *minC* and *minD*, which are homologous to bacterial genes involved in cell division, and which are unique to the *Chlorella* genome among all examined plastid genomes.

References

APT, K. E., HOFFMAN, N. E., GROSSMAN, A. R., 1993: The gamma subunit of R-phycoerythrin and its possible mode of transport into the plastids of red algae. – J. Biol. Chem. **268**: 16206–16215.

BALDAUF, S., MANHART, J., PALMER, J. D., 1990: Different fates of the chloroplast *tuf*A gene following its transfer to the nucleus in green algae. – Proc. Natl. Acad. Sci. USA **87**: 5317–5321.

– PALMER, J. D., 1993: Animals and fungi are each other's closest relatives: congruent evidence from multiple proteins. – Proc. Natl. Acad. Sci. USA **90**: 11558–11562.

BHATTACHARYA, D., MEDLIN, L., 1995: The phylogeny of plastids: A review based on comparisons of small-subunit ribosomal RNA coding regions. – J. Phycol. **31**: 489–498.

– SCHMIDT, H. A., 1997: Division Glaucocystophyta. – In BHATTACHARYA, D., (Ed): The origins of algae and their plastids. – Pl. Syst. Evol. [Suppl.] **11**: 139–148.

– WEBER, K., 1997: The actin gene of the glaucocystophyte *Cyanophora paradoxa*: Analysis of the coding region and introns and an actin phylogeny of eukaryotes. – Curr. Genet. **31**: 439–446.

– HELMCHEN, T., MELKONIAN, M., 1995a: Molecular evolutionary analyses of nuclear-encoded small subunit ribosomal RNA identify an independent rhizopod lineage containing the *Euglyphina* and the *Chlorarachniophyta*. – J. Eukaryote. Microbiol. **42**: 65–69.

– – Bibeau, C., Melkonian, M., 1995b: Comparisons of nuclear-encoded small-subunit ribosomal RNAs reveal the evolutionary position of the *Glaucocystophyta.* – Molec. Biol. Evol. **12**: 415-420.

Boczar, B. A., Delaney, T. P., Cattolico, R. A., 1989: Gene for the ribulose-1,5-bisphosphate carboxylase small subunit protein is similar to that of a chemoautotrophic bacterium. – Proc. Natl. Acad. Sci. USA **86**: 4996–4999.

Boudreau, E., Otis, C., Turmel, M., 1994: Conserved gene clusters in the highly rearranged chloroplast genomes of *Chlamydomonas moewusii* and *Chlamydomonas reinhardtii.* – Pl. Molec. Biol. **24**: 585–602.

Bouget, F.-Y., Kerbourc'h, C., Liaud, M.-F., Loiseaux de Goër, S., Quatrano, R. S., Cerff, R., Kloareg, B., 1995: Structural features and phylogeny of the actin gene of *Chondrus crispus* (*Gigartinales*, *Rhodophyta*). – Curr. Genet. **28**: 164–172.

Boyen, C., Leblanc, C., Bonnard, G., Grienenberger, J. M., Kloareg, B., 1994: Nucleotide sequence of the *cox3* gene from *Chondrus crispus*: evidence that UGA encodes tryptophan and evolutionary implications. – Nucl. Acids Res. **22**: 1400–1403.

Brinkmann, H., Martin, W., 1996: Higher-plant chloroplast and cytosolic 3-phosphoglycerate kinases: a case of endosymbiotic gene replacement. – Pl. Molec. Biol. **30**: 65–75.

Bryant, D. A., 1992: Puzzles of chloroplast ancestry. – Curr. Biol. **2**: 240–242.

Bullerjahn, G. S., Post, A. F., 1993: The prochlorophytes: are they more than just chlorophyll *a*/*b*-containing cyanobacteria? – Crit. Rev. Microbiol. **19**: 43–59.

Cavalier-Smith, T., 1993a: Kingdom *Protozoa* and its 18 phyla. – Microbiol. Rev. **47**: 953–994.

– 1993b: The origin, losses and gains of chloroplasts. – In Lewin, R., (Ed.): Origins of plastids, pp. 291–348. – New York, London: Chapman & Hall.

– 1995: Membrane heredity, symbiogenesis, and the multiple origins of algae. – In Arai, R., Kato, M., Doi, Y., (Eds): Biodiversity and evolution, pp. 75–114. – Tokyo: National Science Museum.

– Couch, J. A., Thorsteinsen, K. E., Gilson, P., Deane, J. A., Hill, D. R. A., McFadden, G. I., 1997: Cryptomonad nuclear and nucleomorph 18s rRNA phylogeny – Eur. J. Phycol. **31** (in press).

Delwiche, C. F., Kuhsel, M., Palmer, J. D., 1995: Phylogenetic analysis of *tuf*A sequences indicates a cyanobacterial origin of all plastids. – Molec. Phylogenet. Evol. **4**: 110–128.

– Palmer, J. D., 1996: Rampant horizontal gene transfer and duplication of rubisco genes in eubacteria and plastids. – Molec. Biol. Evol. **13**: 873–882.

Dodge, J. D., 1989: Phylogenetic relationships of dinoflagellates and their plastids. – In Green, J. C., Leadbeater, B. S. C., Diver, W. L., (Eds): The chromophyte algae: problems and perspectives, pp. 207–227. – Oxford: Clarendon Press.

Douglas, S. E., 1992: Eukaryote-eukaryote endosymbioses: insights from studies of a cryptomonad alga. – BioSystems **8**: 57–68.

– 1994: Chloroplast origins and evolution. – In Bryant, D. A., (Ed.): The molecular biology of cyanobacteria, pp. 91–118. – Amsterdam: Kluwer.

– Murphy, C. A., 1994: Structural, transcriptional, and phylogenetic analyses of the *atp*B gene cluster from the plastid of *Cryptomonas* sp. (*Cryptophyaceae*). – J. Phycol. **30**: 329–340.

Egea, N., Lang-Unnasch, N., 1995: Phylogeny of the large extrachromosomal DNA of organisms in the Phylum *Apicomplexa.* – J. Eukaryote Microbiol. **42**: 679–684.

Emes, M. J., Tobin, A. K., 1993: Control of metabolism and development in higher plant plastids. – Int. Rev. Cytol. **145**: 149–215. – Orlando: Academic Press.

Feagin, J. E., 1994: The extrachromosomal DNAs of apicomplexan parasites. – Annual Rev. Microbiol. **48**: 81–104.

FUJIWARA, S., IWAHASHI, H., SOMEYA, J., NISHIKAWA, S., MINAKA, N., 1993: Structure and cotranscription of the plastid-encoded *rbc*L and *rbc*S genes of *Pleurochrysis carterae* (*Prymnesiophyta*). – J. Phycol. **29**: 347–355.

GANTT, J. S., BALDAUF, S., CALIE, P., WEEDEN, N., PALMER, J. D., 1991: Transfer of *rpl22* to the nucleus greatly preceded its loss from the chloroplast and involved the gain of an intron. – EMBO J. **10**: 3073–3078.

GIBBS, S. P., 1978: The chloroplasts of *Euglena* may have evolved from symbiotic green algae. – Canad. J. Bot. **56**: 2883–2889.

– 1981: The chloroplast endoplasmic reticulum, structure, function, and evolutionary significance. – Int. Rev. Cytol. **72**: 49–99.

– 1993: The evolution of algal chloroplasts. – In LEWIN, R. A., (Ed): Origins of plastids. pp. 107–121. – New York: Chapman and Hall.

GILSON, P. R., MCFADDEN, G. I., 1996: The miniaturized nuclear genome of a eukaryotic endosymbiont contains genes that overlap, genes that are cotranscribed, and the smallest known spliceosomal introns. – Proc. Natl. Acad. Sci. USA **93**: 7737–7742.

– 1997: Good things in small packages: the tiny genomes of chlorarachniophyte endosymbionts. – Bioessays **19**: 167–173.

GIOVANNONI, S., TURNER, S., OLSEN, G., BARNS, S., LANE, D., PACE, N., 1988: Evolutionary relationships among cyanobacteria and green chloroplasts. – J. Bacteriol. **170**: 3584–3592.

GOCKEL, G., HACHTEL, W., BAIER, S., FLISS, C., HENKE, M., 1994: Genes for components of the chloroplast translational apparatus are conserved in the reduced 73-kb plastid DNA of the nonphotosynthetic euglenoid flagellate *Astasia longa*. – Curr. Genet. **26**: 256–262.

GRAY, M. W., 1992: The endosymbiont hypothesis revisited. – Int. Rev. Cytol. **141**: 233–357.

– 1995: Mitochondrial evolution. – In LEVINGS, C. S. III, VASIL, I. K., (Eds): The molecular biology of plant mitochondria, pp. 635–659. – Dordrecht: Kluwer.

SPENCER, D. F., 1996: Organellar evolution. – In ROBERTS, D. M., SHARP, P., ALDERSON, G., COLLINS, M., (Eds): Evolution of microbial life, pp. 109–126. – Cambridge: Cambridge University Press.

HABERHAUSEN, G., ZETSCHE, K., 1994: Functional loss of all *ndh* genes in an otherwise relatively unaltered plastid genome of the holoparasitic flowering plant *Cuscuta reflexa*. – Pl. Molec. Biol. **24**: 217–222.

HACKSTEIN, J. H. P., 1995: A photosynthetic ancestry for all eukaryotes? – Trends Ecol. Evol. **10**: 247.

HALLICK, R., HONG, L., DRAGER, R., FAVREAU, M., MONFOR, A., ORSAT, B., SPIELMANN, A., STUTZ, E., 1993: Complete sequence of *Euglena gracilis* chloroplast DNA. – Nucl. Acids Res. **21**: 3537–3544.

HELMCHEN, T. A., BHATTACHARYA, D., MELKONIAN, M., 1995: Analyses of ribosomal RNA sequences from glaucocystophyte cyanelles provide new insights into the evolutionary relationships of plastids. – J. Molec. Evol. **41**: 203–210.

HENZE, K., SCHNARRENBERGER, C., KELLERMANN, J., MARTIN, W., 1994: Chloroplast and cytosolic triosephosphate isomerase from spinach: Purification, microsequencing and cDNA sequence of the chloroplast enzyme. – Pl. Molec. Biol. **26**: 1961–1973.

– BADR, A., WETTERN, M., CERFF, R., MARTIN, W., 1995: A nuclear gene of eubacterial origin in *Euglena gracilis* reflects cryptic endosymbioses during protist evolution. – Proc. Natl. Acad. Sci. USA **92**: 9122–9126.

HERDMAN, M., JANVIER, M., RIPPKA, R., STANIER, R., 1979: Genome size of cyanobacteria. – J. Gen. Microbiol. **111**: 73–85.

HESS, W. R., WEIHE, A., LOISEAUX-DE GOËR, S., PARTENSKY, F., VAULOT, D., 1995: Characterization of the single *psb*A gene of *Prochlorococcus marinus* CCMP 1375 (*Prochlorophyta*). – Pl. Molec. Biol. **27**: 1189–1196.

– PARTENSKY, F., VAN DER STAAY, G. W. M., GARCIA-FERNANDEZ, J. M., BÖRNER, T., VAULOT, D., 1996: Coexistence of phycoerythrin and a chlorophyll *a/b* antenna in a marine prokaryote. – Proc. Natl. Acad. Sci. USA **93**: 1126–1130.

HIRATSUKA, J., SHIMADA, H., WHITTIER, R., ISHIBASHI, T., SAKAMOTO, M., MORI, M., KONDO, C., HONJI, Y., SUN, C.-R., MENG, B.-Y., LI, Y.-Q., KANNO, A., NISHIZAWA, Y., HIRAI, A., SHINOZAKI, K., SUGIURA, M., 1989: The complete sequence of the rice (*Oryza sativa*) chloroplast genome: Intermolecular recombination between distinct tRNA genes accounts for a major plastid DNA inversion during the evolution of the cereals. – Molec. Gen. Genet. **217**: 185–194.

HORI, H., OSAWA, S., 1987: Origin and evolution of organisms as deduced from 5s ribosomal RNA sequences. – Molec. Biol. Evol. **4**: 445–472.

JAKOWITCH, J., NEUMANN-SPALLART, C., MA, Y., STEINER, J., SCHENK, H. E. A., BOHNERT, H. J., LÖFFELHARDT, W., 1996: In vitro import of pre-ferredoxin-NADP$^+$-oxidoreductase from *Cyanophora paradoxa* into cyanelles and into pea chloroplasts. – FEBS Lett. **381**: 153–155.

KANEKO, T., SATO, S., KOTANI, H., TANAKA, A., ASAMIZU, E., NAKAMURA, Y., MIYAJIMA, N., HIROSAWA, M., SUGIURA, M., SASAMOTO, S., KIMURA, T., HOSOUCHI, T., MATSUNO, A., MURAKI, A., KAKAZAKI, N., NARUO, K., OKUMURA, S., SHIMPO, S., TAKEUCHI, C., WADA, T., WATANABE, A., YAMADA, M., YASUDA, M., TABATA, S., 1996: Sequence analysis of the genome of the unicellular cyanobacterium *Synechocystis* sp. strain PCC6803. II. sequence determination of the entire genome and assignment of potential protein-coding regions. – DNA Res. **3**: 109–136.

KEELING, P. J., DOOLITTLE, W. F., 1996: Alpha-tubulin from early-diverging eukaryotic lineages and the evolution of the tubulin family. – Molec. Biol. Evol. **13**: 1297–1305.

– DOOLITTLE, F., 1997: Evidence that eukaryotic triosephosphate isomerase is of apha-proteobacterial origin. – Proc. Natl. Acad. Sci. USA **94**: 1270–1275.

KIES, L., KREMER, B. P., 1990: Phylum *Glaucocystophyta*. – In MARGULIS, L., CORLISS, J. O., MELKONIAN, M., CHAPMAN, D. J., (Eds): Handbook of *Protoctista*, pp. 152–166. – Boston: Jones and Bartlett.

KNOLL, A., 1992: The early evolution of eukaryotes: a genological perspective. – Science **256**: 622–627.

– 1994: Proterozoic and early cambrian protists: evidence for accelerating evolutionary tempo. – Proc. Natl. Acad. Sci. USA **91**: 6743–6750.

– GOLUBIC, S., 1992: Proterozoic and living cyanobacteria. – In SCHIDLOWSKI, M., & al. (Eds): Early organic evolution: Implications for mineral and energy resources, pp. 455 457–458. – Berlin, Heidelberg, New York: Springer.

KÖHLER, S., DELWICHE, C. F., DENNY, P. W., TILNEY, L. G., WEBSTER, P., WILSON, R. J. M., PALMER, J. D., ROOS, D. S., 1997: A plastid of probable green algal origin in apicomplexan parasites. – Science **275**: 1485–1489.

KOWALLIK, K. V., 1994: From endosymbionts to chloroplasts: evidence for a single prokaryotic/eukaryotic endocytobiosis. – Endocytobiosis Cell Res. **10**: 137–149.

– STOEBE, B., SCHAFFRAN, I., KROTH-PANIC, P., FREIER, U., 1995: The chloroplast genome of a chlorophyll *a+c*-containing alga, *Odontella sinensis*. – Pl. Molec. Biol. Reporter **13**: 336–342.

KRISHNAN, S., BARNABAS, S., BARNABAS, J., 1990: Interrelationships among major protistan groups based on a parsimony network of 5s rRNA sequences. – BioSystems **24**: 135–144.

Lang, B. F., Burger, G., O'Kelly, C. J., Gray, M. W., 1997: Organelle genome megasequencing program. – http://megasun.bch. umontreal.ca/maps/globaltree.gif

La Roche, J., van der Staay, G. W. M., Partensky, F., Ducret, A., Aebersold, A., Li, R., Golden, S. S., Hiller, R. G., Wrench, P. M., Larkum, A. W. D., Green, B. R., 1996: Independent evolution of the prochlorophyte and green plant chlorophyll *a/b* light-harvesting proteins. – Proc. Natl. Acad. Sci. USA **93**: 15244–15248.

Larkum, A. W. D., Scaramuzzi, C., Cox, G. C., Hiller, R. G., Turner, A. G., 1994: Light-harvesting chlorophyll *c*-like pigment in *Prochloron*. – Proc. Natl. Acad. Sci. USA **91**: 679–683.

Leipe, D. D., Wainwright, P. O., Gunderson, J. H., Porter, D., Patterson, D. J., Valois, F., Himmerich, S., Sogin, M. L., 1994: The stramenophiles from a molecular perspective: 16s-like rRNA sequences from *Labyrinthuloides minuta* and *Cafeteria roenbergensis*. – Phycologia **33**: 369–377.

Liaud, M.-F., Valentin, C., Brandt, U., Bouget, F.-Y., Kloareg, B., Cerff, B., Cerff, R., 1993: The GAPDH gene system of the red alga *Chondrus crispus*: promotor structures, intron/exon organization, genomic complexity and differential expression of genes. – Pl. Molec. Biol. **23**: 981–994.

– Valentin, C., Martin, W., Bouget, F.-Y., Kloareg, B., Cerff, R., 1994: The evolutionary origin of red algae as deduced fom the nuclear genes encoding cytosolic and chloroplast glyceraldehyde-3-phosphate dehydrogenases from *Chondrus Crispus*. – J. Molec. Evol. **38**: 319–327.

– Brandt, U., Cerff, R., 1995: The marine red alga *Chondrus crispus* has a highly divergent β-tubulin gene with a characteristic $5'$ intron: functional and evolutionary implications. – Pl. Molec. Biol. **28**: 313–325.

Liu, Q., Baldauf, S., Reith, M., 1996: Elongation factor 1α genes of the red alga *Porphyra purpurea* include a novel, developmentally specialized variant. – Pl. Molec. Biol. **31**: 77–85.

Löffelhardt, W., Bohnert, H. J., 1994: Molecular biology of cyanelles. – In Bryant, D. A., (Ed.): The molecular biology of the *Cyanobacteria*, pp. 65–89. – Netherlands: Kluwer.

Loiseaux-de Göer, S., 1994: Plastid lineages. – Phycol. Res. **10**: 137–177.

Maier, R. M., Neckermann, K., Igloi, G. L., Kössel, H., 1995: Complete sequence of the maize chloroplast genome: gene content, hotspots of divergence and fine tuning of genetic information by transcript editing. – J. Molec. Biol. **251**: 614–628.

Martin, W. S., Somerville, C. C., Loiseaux-de Göer, S., 1992: Molecular phylogenies of plastid origins and algal evolution. – J. Molec. Evol. **35**: 385–404.

McFadden, G. I., Gilson, P., 1995: Something borrowed, something green: lateral transfer of chloroplasts by secondary endosymbiosis. – Trends Ecol. Evol. **10**: 12–17.

– Reith, M. E., Munholland, J., Lang-Unnasch, N., 1996: Plastid in human parasites. – Nature **381**: 482.

– Waller, R. F., Reith, M. E., Lang-Unnasch, N., 1997a: Plastids in apicomplexan parasites. – In Bhattacharya, D., (Ed.): The origins of algae and their plastids. – Pl. Syst. Evol., [Suppl.] **11**: 261–287.

– Gilson, P. R., Hofmann, C. J. B., 1997b: Division *Chlorarachniophyta*. – In Bhattacharya, D., (Ed.): The origins of algae and their plastids. – Pl. Syst. Evol., [Suppl.] **11**: 175–185.

Medlin, L. K., Cooper, A., Hill, C., Wrieden, S., Wellbrock, U., 1995: Phylogenetic position of the *Chromista* plastids based on small subunit rRNA coding regions. – Curr. Genet. **28**: 560–565.

– Kooistra, W. H. C. F., Potter, D., Saunders, G. W., Andersen, R. A., 1997: Phylogenetic relationships of the 'golden algae' (haptophytes, heterokont chromophytes) – In

BHATTACHARYA, D., (Ed.): The origins of algae and their plastids. – Pl. Syst. Evol., [Suppl.] **11**: 187–219.

MIDROSHNICHENKO DOLGANOV, N. A., BHAYA, D., GROSSMAN, A. R., 1995: Cyanobacterial protein with similarity to the chlorophyll *a/b* binding proteins of higher plants: evolution and regulation. – Proc. Natl. Acad. Sci. USA **92**: 636–640.

MORDEN, C. W., GOLDEN, S. S., 1989: *psb*A genes indicate common ancestry of prochlorophytes and chloroplasts. – Nature **337**: 382–385.

– DELWICHE, C. F., KUHSEL, M., PALMER, J. D., 1992: Gene phylogenies and the endosymbiotic origin of plastids. – Biosystems **28**: 75–90.

MORSE, D., SALOIS, P., MARKOVIC, P., HASTINGS, J. W., 1995: A nuclear-encoded form II RuBisCo in dinoflagellates. – Science **268**: 1622–1624.

NELISSEN, B., VAN DE PEER, Y., WILMOTTE, A., DE WACHTER, R., 1995: An early origin of plastids within the cyanobacterial divergence is suggested by evolutionary trees based on complete 16s rRNA sequences. – Molec. Biol. Evol. **6**: 1166–1173.

NIKOH, N., HAYASE, H., IWABE, H., KUMA, K.-I., MIYATA, T., 1994: Phylogenetic relationship of the kingdoms *Animalia*, *Plantae*, and *Fungi*, inferred from 23 different protein species. – Molec. Biol. Evol. **11**: 762–768.

NUGENT, J., PALMER, J. D., 1991: RNA-mediated transfer of the gene *coxII* from the mitochondrion to the nucleus during flowering plant evolution. – Cell **66**: 473–481.

OHYAMA, K., FUKUZAWA, H., KOHCHI, T., SHIRAI, H., SANO, T., SANO, S., UMESONO, K., SHIKI, Y., TAKEUCHI, M., CHANG, Z., AOTA, S.-I., INOKUCHI, H., OZEKI, H., 1986: Chloroplast gene organization deduced from complete sequence of liverwort *Marchantia polymorpha* chloroplast DNA. – Nature **322**: 572–574.

PALENIK, B., HASELKORN, R., 1992: Multiple evolutionary origins of prochlorophytes, the chlorophyll *b*-containing prokaryotes. – Nature **355**: 265–267.

– SWIFT, H., 1996: Cyanobacterial evolution and prochlorophyte diversity as seen in DNA-dependent RNA polymerase gene sequences. – J. Phycol. **32**: 638–646.

PALMER, J. D., 1985: Comparative organization of chloroplast genomes. – Annual Rev. Genet. **19**: 325–354.

– 1991: Plastid chromosomes: structure and evolution. – In BOGORAD, L., VASIL, I. K., (Eds): The molecular biology of plastids, pp. 5–53. – San Diego: Academic Press.

– 1992: Green ancestry of malarial parasites. – Curr. Biol. **2**: 318–320.

– 1995: Rubisco rules fall; gene transfer triumphs. – Bioessays **17**: 1005–1008.

– 1997: Organelle genomes: going, going, gone ! – Science **275**: 790–791.

– DELWICHE, C., 1996: Second-hand chloroplasts and the case of the disappearing nucleus. – Proc. Natl. Acad. Sci. USA **93**: 7432–7435.

PAQUIN, B., LAFOREST, M. J., FORGET, L., ROEWER, I., WANG, A., LONGCORE, J., LANG, B. F., 1997: The fungal mitochondrial genome project: evolution of fungal mitochondrial genomes and their gene expression. – Curr. Genet. **31**: 380–395.

PAWLOWSKI, J., BOLIVAR, I., FAHRNI, J., CAVALIER-SMITH, T., GOUY, M., 1996: Early origin of *Foraminifera* suggested by SSU rRNA gene sequences. – Molec. Biol. Evol. **13**: 445–450.

PERASSO, R., BAROIN, A., QU, L., BACHELLERIE, J. P., ADOUTTE, A., 1989: Origin of the algae. – Nature **339**: 142–144.

PREISER, P., WILLIAMSON, D. H., WILSON, R. J. M., 1995: tRNA genes transcribed from the plastid-like DNA of *Plasmodium falciparum*. – Nucl. Acids Res. **23**: 4329–4336.

RAGAN, M. A., GUTELL, R. R., 1995: Are red algae plants? – Bot. L. Linn. Soc. **118**: 81–105.

RAVEN, P. H., 1970: A multiple origin for plastids and mitochondria. – Science **169**: 641–646.

Reith, M., 1995: Molecular biology of rhodophyte and chromophyte plastids. – Annual. Rev. Pl. Physiol. Pl. Molec. Biol. **46**: 549–575.

– Munholland, J., 1993: A high-resolution gene map of the chloroplast genome of the red alga *Porphyra purpurea*. – Pl. Cell. **5**: 465–475.

– – 1995: Complete nucleotide sequence of the *Porphyra purpurea* chloroplast genome. – Pl. Molec. Biol. Reporter **13**: 333–335.

Rowan, R., Whitney, S. M., Fowler, A., Yellowlees, D., 1996: Rubisco in marine symbiotic dinoflagellates: form II enzymes in eukaryotic oxygenic phototrophs, encoded by a nuclear multi-gene family. – Pl. Cell **8**: 539–553.

Runnegar, B., 1992: The tree of life. – In Schopf, J. W., Klein, C., (Eds): The proterozoic biosphere. A multidisciplinary study, pp. 471–475. – New York: Cambridge University Press.

Schmidt, M., Dvensen, I., Feierabend, J., 1995: Analysis of the primary structure of the chloroplast isozyme of triosephosphate isomerase from rye leaves by protein and cDNA sequencing indicates a eukaryotic origin of its gene. – Biochim. Biophys. Acta **1261**: 257–264.

Schnepf, E., 1993: From prey via endosymbiont to plastids: comparative studies in dinoflagellates. – In Lewin, R. A., (Ed.): Origins of plastids, pp. 53–76. – New York: Chapman & Hall.

Schopf, J. W., 1993: Microfossils of the early archean apex chert: new evidence of the antiquity of life. – Science **260** : 640–646.

– 1994: Disparate rates, differing fates: Tempo and mode of evolution changed from the Precambrian to the Phanerozoic. – Proc. Natl. Acad. Sci. USA **91**: 6735–6742.

– 1996: Are the oldest fossils cyanobacteria? – In Roberts, D. M., Sharp, P., Alderson, G., Collins, M. A., (Eds): Evolution of microbial life, pp. 23–61. – Cambridge: Cambridge University Press.

Shinozaki, K., Ohme, M., Tanaka, M., Wakasugi, T., Hayashida, N., Matsubayashi, T., Zaita, N., Chunwongse, J., Obokata, J., Yamaguchi-Shinozaki, K., Ohto, C., Torazawa, K., Meng, B. Y., Kusuda, J., Takaiwa, F., Kato, A., Tohdoh, N., Shimada, H., Sugiura, M., 1986: The complete nucleotide sequence of the tobacco chloroplast genome: its gene organization and expression. – EMBO J. **5**: 2043–2049.

Sogin, M., 1994: The origin of eukaryotes and evolution into major kingdoms. – In Bengston, S., (Ed.): Early life on earth. Nobel Symposium No. 84, pp. 181–194. – New York: Columbia University Press.

– 1996: Problems with molecular diversity in the eukarya. – In Roberts, D. M., Sharp, P., Alderson, G., Collins, M. A., (Eds): Evolution of microbial life, pp. 167–184. – Cambridge: Cambridge University Press.

Stiller, J. W., Hall, B. D., 1997: The origin of red algae: implications for plastid evolution. – Proc. Natl. Acad. Sci. USA

Stirewalt, V., Michalowski, C., Löffelhardt, W., Bohnert, H., Bryant, D., 1995: Nucleotide sequence of the cyanelle genome from *Cyanophora paradoxa*. Pl. Molec. Biol. Reporter **13**: 327–332.

Tabita, F. R., 1995: The biochemistry and metabolic regulation of carbon metabolism and CO_2 fixation in purple bacteria. – In Blankenship, R. E., Madigan, M. T., Bauer, C. E., (Eds): Anoxygenic photosynthetic bacteria, pp. 885–914. – Amsterdam: Kluwer.

Takahashi, H., Takano, H., Yokoyama, A., Hara, Y., Kawano, S., Toh-e, A., Kuroiwa, T., 1995: Isolation, characterization and chromosomal mapping of an actin gene from the primitive red alga *Cyanidioschyzon merolae*. – Curr. Genet. **28**: 484–490.

Tingey, S. V., Tsai, F.-Y., Edwards, J.W., Walker, E. L., Coruzzi, G. M., 1988: Chloroplast and cytoplasmic glutamine synthetase are encoded by homologous nuclear genes which are differentially expressed in vivo. – J. Biol. Chem. **263**: 9651–9657.

TRIEMER, R. E., FARMER, M. A., 1991: An ultrastructural comparison of the mitotic apparatus, feeding apparatus, flagellar apparatus and cytoskeleton in euglenoids and kinetoplastids. – Protoplasma **164**: 91–104.

TURMEL, M., BELLEMARE, G., LEMIEUX, C., 1987: Physical mapping of differences between the chloroplast DNAs of the interfertile algae *Chlamydomonas eugametos* and *Chlamydomonas moewusii*. – Curr. Genet. **11**: 543–552.

TURNER, S., 1997: Molecular systematics of oxygenic photosynthetic bacteria. – In BHATTACHARYA, D., (Ed.): The origins of algae and their plastids. – Pl. Syst. Evol., [Suppl.] **11**: 13–52.

URBACH, E., ROBERTSON, D. L., CHISHOLM, S. W., 1992: Multiple evolutionary origins of prochlorophytes within the cyanobacterial radiation. – Nature **355**: 267–269.

VAIDYA, A. B., MORRISEY, M., PLOWE, C. V, KASLOW, D. C., WELLEMS, T. E., 1993: Unidirectional dominance of cytoplasmic inheritance in two genetic crosses of *Plasmodium falciparum*. – Molec. Cell. Biol. **13**: 7349–7357.

VALENTIN, K., ZETSCHE, K., 1989: The genes of both subunits of ribulose-1,5-bisphosphate carboxylase constitute an operonon the plastome of a red alga. – Curr. Genet. **16**: 203–209.

VAN DEN HOEK, C., MANN, D. G., JAHNS, H. M., 1995: *Algae*: an introduction to phycology. – Cambridge: Cambridge University Press.

VER DE PEER, Y., RENSING, S. A., MAIER, U.-G., DE WACHTER, R., 1996: Substitution rate calibration of small subunit ribosomal RNA identifies chlorarachniophyte endosymbionts as remnants of green algae. – Proc. Natl. Acad. Sci. USA **93**: 7732–7736.

VIALE, A. M., Arakaki, A. K., 1994: The chaperone connection to the origins of the eukaryotic organelles. – FEBS Lett. **341**: 146–151.

– ARAKAKI, A. K., SONCINI, F. C., FERREYRA, R. G., 1994: Evolutionary relationships among eubacterial groups as inferred from GroEL (chaperonin) sequence comparisons. – Int. J. Syst. Bacteriol. **44**: 527–533.

VIVIER, E., DESPORTES, I., 1989: Phylum *Apicomplexa*. – In MARGULIS, L., CORLISS, J. O., MELKONIAN, M., CHAPMAN, D. J., (Eds): Handbook of *Protoctista*, pp. 549–573. – Boston: Jones & Bartlett.

WAINWRIGHT, P. O., HINKLE, G., SOGIN, M. L., STICKEL, S. K., 1993: Monophyletic origins of the metazoa: An evolutionary link with fungi. – Science **260**: 340–342.

WAKASUGI, T., TSUDZUKI, J., ITO, S., NAKASHIMA, K., TSUDZUKI, T., SUGIURA, M., 1994: Loss of all *ndh* genes as determined by sequencing the entire chloroplast genome of the black pine *Pinus thunbergii*. – Proc. Natl. Acad. Sci. USA **91**: 9794–9798.

WATANABE, M. M., SUDA, S., INOUYE, I., SAWAGUCHI, T., CHIHARA, M., 1990: *Lepidodinium viride* gen. et sp. nov. (*Gymnodiniales*, *Dinophyta*), a green dinoflagellate with a chlorophyll *a*- and *b*-containing endosymbiont. – J. Phycol. **26**: 741–751.

WATSON, G. M. F., TABITA, F. R., 1996: Regulation, unique gene organization, and unusual primary structure of carbon fixation genes from a marine phycoerythrin-containing cyanobacterium. – Pl. Molec. Biol. **32**: 1103–1115.

WHATLEY, J. M., 1993a: Chloroplast ultrastructure. – In BERNER, T., (Ed.): Ultrastructure of microalgae, pp. 135–204. – Boca Raton, FL: CRC Press.

– 1993b: Membranes and plastid origins. – In LEWIN, R. A., (Ed.): Origins of plastids, pp. 77–106. – New York: Chapman & Hall.

WHITNEY, S. M., SHAW, D. C., YELLOWLEES, D., 1995: Evidence that some dinoflagellates contain a ribulose-1,5-bisphosphate carboxylase/oxygenase related to that of the a-proteobacteria. – Proc. Roy. Soc. London, Ser. B, Biol. Sci. **259**: 271–275.

WILHELM, C., 1987: The existence of chlorophyll *c* in the Chl *b*-containing light-harvesting complex of the green alga *Mantoniella squamata* (*Prasinophyceae*). – Bot. Acta **101**: 7–10.

Wilson, R. J. M., Denny, P. W., Preiser, P. R., Rangachari, K., Roberts, K., Roy, A., Whyte, A., Strath, M., Moore, D. J., Moore, P. W., Williamson, D. H., 1996: Complete gene map of the plastid-like DNA of the malaria parasite *Plasmodium falciparum*. – J. Molec. Biol. **261**: 155–172.

Wolfe, G. R., Cunningham, F. X., Durnford, D., Green, B. R., Gantt, E., 1994: Evidence for a common origin of chloroplasts with light-harvesting complexes of different pigmentation. – Nature **367**: 566–568.

Wolfe, K., Morden, C., Palmer, J. D., 1992: Function and evolution of a minimal plastid genome from a nonphotosynthetic parasitic plant. – Proc. Natl. Acad. Sci. USA **89**: 10648–10652.

Zhou, Y. H., Ragan, M. A., 1994: Cloning and characterization of the nuclear gene encoding plastid glyceraldehyde-3-phosphate dehydrogenase from the marine red alga *Gracilaria verrucosa*. – Curr. Genet. **26**: 79–86.

– – 1995: The nuclear gene and cDNAs encoding cytosolic glyceraldehyde-3-phosphate dehydrogenase from the marine red alga *Gracilaria verrucosa*: cloning, characterization and phylogenetic analysis. – Curr. Genet. **28**: 324–332.

The evolution of the Green Algae

THOMAS FRIEDL

Key words: *Chlorophyta, Chlorophyceae, Prasinophyceae, Streptophyta, Trebouxiophyceae, Ulvophyceae*, 18S rDNA, coccoid green algae, lichen algae, molecular evolution.

Abstract: Phylogenetic analyses of complete 18S rDNAs from the green algae reveal the monophyletic origin of three independent evolutionary lineages, the classes *Chlorophyceae, Trebouxiophyceae* and *Ulvophyceae* with a basal divergence of a heterogenous assemblage of scaled flagellates, the prasinophytes. The *Trebouxiophyceae* comprise mainly coccoid green algae from terrestrial habitats or lichen symbioses. The *Chlorophyceae* unite green algae that have flagellated cells with basal bodies oriented in the clockwise direction or with directly opposed basal bodies. 18S rDNA sequence comparisons substantiate the importance of flagellated cell morphology over vegetative cell features for tracing evolutionary relationships among the green algae. However, the counterclockwise basal body orientation which is shared among the *Trebouxiophyceae* and *Ulvophyceae* is likely a symplesiomorphy. Coccoid green algae, formerly placed in a single order (*Chlorococcales*), are distributed over several lineages.

The Green Algae are a eukaryotic lineage that is characterized by the presence of chloroplasts with two envelope membranes, stacked thylakoids and the chlorophylls *a* and *b*. Another important marker is the presence of starch as a reserve polysaccharide that is deposited inside the plastids. These synapomorphies clearly distinguish the Green Algae and their sister-group, the embryophytes, from other eukaryotic lineages. The monophyletic origin of all green plants is supported in phylogenetic analyses based on 18S rDNA sequence comparisons of various lineages of eukaryotes (e.g., in BHATTACHARYA & MEDLIN 1995; Fig. 1); these analyses also suggest the origin of land plants (embryophytes) from green algal ancestors (see chapter 4). Intensive ultrastructural work on the Green Algae during the past 25 years has lead to fundamental changes in green algal taxonomy compared to traditional concepts (for review: PICKETT-HEAPS 1975; MATTOX & STEWART 1984; VAN DEN HOEK & al. 1988, 1992). An exciting period in phycology started in the 1990s as molecular techniques, i.e. sequence comparisons mainly of 18S ribosomal DNAs, were introduced into the investigation of the evolution of Green Algae. The aim of the present article is to give an overview of the current status of our knowledge of green algal phylogeny as it relates to recent analyses of 18S rDNA sequences.

Two datasets have been prepared from complete 18S rDNA sequences that are available from the Genbank/EBI/DDBJ databases. The sequences may be directly accessed from the databases using the species names. The alignments used for phylogenetic analyses in this chapter are available from the author. Dataset 1 comprises 53 sequences representing all major lineages of the *Chlorophyta* sensu SLUIMAN (1985), *Charophyceae* sensu STEWART & MATTOX (1975) and six taxa from the embryophytes. For this dataset (1719 unambiguously aligned sequence positions, 781/587 variable/informative positions for parsimony analyses), the glaucocystophytes *Cyanophora paradoxa* KORSHIKOV and *Glaucocystis nostochinearum* ITZIGSOHN were used as outgroup taxa (STEINKÖTTER & al. 1994). Dataset 2 comprises 56 sequences representing major lineages of the *Chlorophyta* only (*Trebouxiophyceae*, *Chlorophyceae*, *Ulvophyceae*, and *Prasinophyceae*) that are distantly related to the *Charophyceae* and embryophytes (see chapter 4) with 1719 unambigously aligned sequence positions (665 variable, 454 informative for parsimony analyses). The prasinophytes *Pseudoscourfieldia marina* (THRONDSEN) MANTON and *Nephroselmis olivacea* STEIN (*Pseudoscourfieldiales*) were used as the outgroup taxa for this dataset. Gene phylogenies were constructed from both datasets using neighbor-joining (KIMURA 1980, SAITOU & NEI 1987) of PHYLIP 3.5 (FELSENSTEIN 1993), maximum parsimony of PAUP 3.1.1 (SWOFFORD 1993) and PAUP* 4.0 (with permission of the author of the program), with sequence positions weighted according to their rescaled consistency index over an interval of 1–1000 (HILLIS & al. 1994), and maximum likelihood of fastDNAml (G. J. OLSEN & al. 1994). Robustness of groups was tested in bootstrap analyses (FELSENSTEIN 1985) using neighbor-joining (500 replications) and weighted maximum parsimony (100 replications).

Early divergence of the Green Algae

Phylogenetic relationships between the 53 18S rDNA sequences representing all major lineages of green plants are shown in Fig. 1. The Green Algae are divided into two major clades: one clade contains the major green algal diversification representing the division *Chlorophyta* s. str. (SLUIMAN 1985), the other unites the *Charophyceae* (represented by the *Klebsormidiales, Coleochaetales, Zygnematales*, and *Charales* in Fig. 1) with the embryophytes representing the *Streptophyta* (BREMER 1985). The monophyletic origin of both major clades is well resolved (see WILCOX & al. 1993; BHATTACHARYA & al. 1994, 1996; SUREK & al. 1994; and chapter 4). This result substantiates earlier assumptions and taxonomic proposals based on ultrastructure, e.g., those of STEWART & MATTOX (1975) who divided the Green Algae into two classes, the "*Chlorophyceae*" (all green algae that are distantly related to the embryophytes) and the *Charophyceae* (which is closely linked with the embryophytes). The two clades may be most conveniently ranked at the level of division, the *Streptophyta* sensu BREMER (1985) and *Chlorophyta* s.str. sensu SLUIMAN (1985), respectively.

The distinction of these two principal clades of green plants in the rDNA phylogeny is congruent with fundamental differences in the architecture of flagellated cells of the *Chlorophyta* and *Streptophyta*; these may display the only derived morphological characters (apomorphies) that define both divisions. The flagellar roots in zoids of the *Chlorophyta* s.str. are in a cruciate condition while those of the *Streptophyta* possess an unilateral flagellar root which is associated

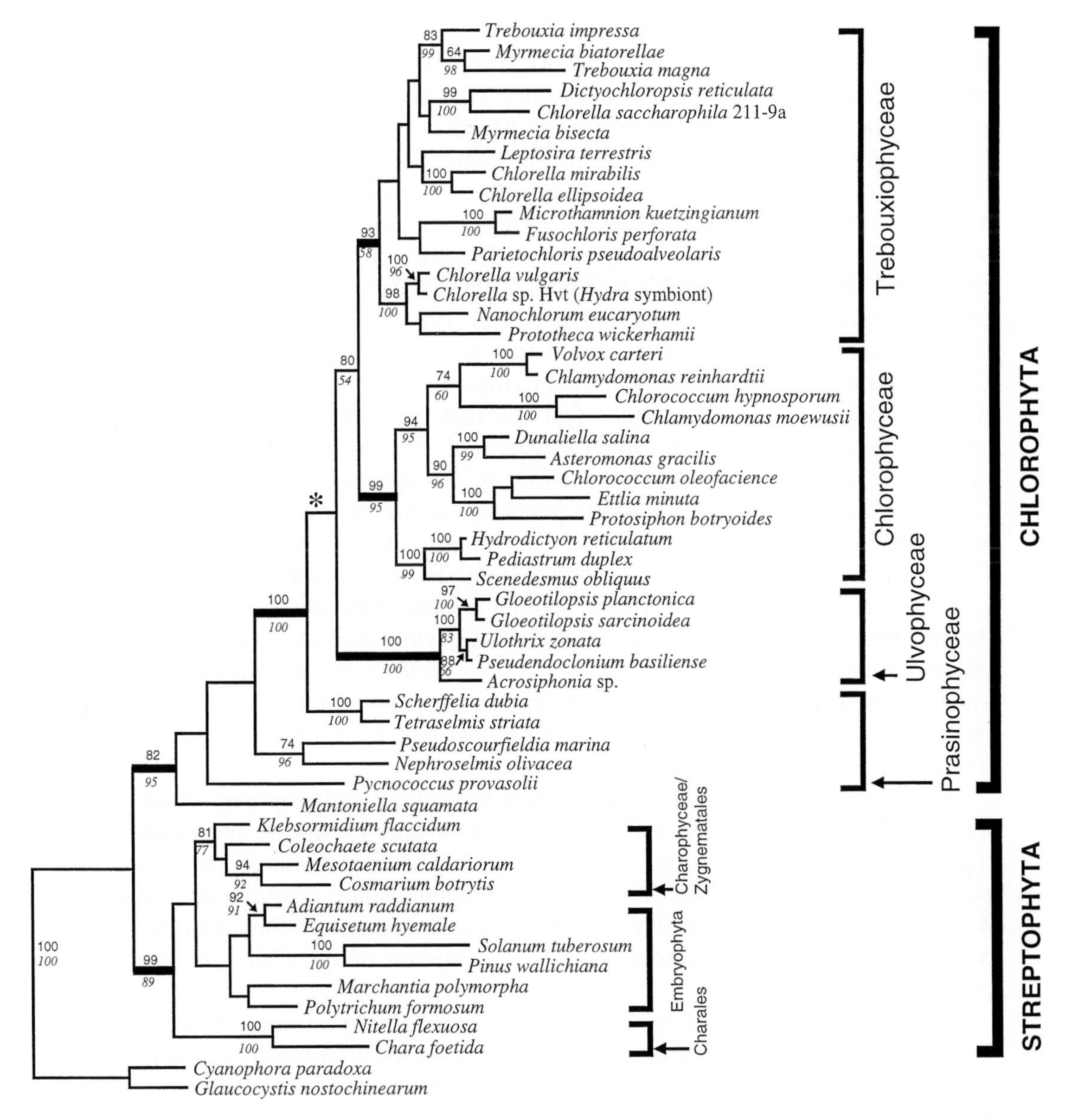

Fig. 1. Phylogenetic analysis of 18S rDNA sequences from all major lineages of the green plants (*Chlorophyta* and *Streptophyta*) using the weighted maximum parsimony method (CI=0.6650). The internal nodes shown with thick lines indicate the major divergences within the green plants. Bootstrap values were computed independently for 500 resamplings using the neighbor-joining method (above lines) and for 100 resamplings using the weighted maximum parsimony method (below the lines). Only bootstrap values that define nodes shared by three independent methods (maximum likelihood, neighbor-joining, weighted parsimony) are recorded. The asterisk indicates the branching point of rDNAs from the Dasycladales (sequences not included in the analysis). Arrows are used to indicate bootstrap values at nodes where these numbers do not fit on the branches

anteriorly with a multilayered structure (MLS) (e.g., PICKETT-HEAPS 1975; STEWART & MATTOX 1975, 1978; MOESTRUP 1978). The MLS, however, may be a shared primitive character (symplesiomorphy) since MLS-like structures are also found in zoids of the *Trentepohliales* (GRAHAM & MCBRIDE 1975; CHAPMAN 1980, 1981; GRAHAM 1984; ROBERTS 1984), and in the trebouxiophyte *Myrmecia israeliensis* (CHANTANACHAT & BOLD) FRIEDL (described as "terminal cap-like structures," MELKONIAN 1982, MELKONIAN & BERNS 1983). *Trentepohlia* MARTIUS is associated with the *Ulvophyceae* in rDNA phylogenies (ZECHMAN & al. 1990; FRIEDL, unpubl.; see below). A cruciate flagellar root system that is associated with MLSs is also found in prasinophyte green algae (ROGERS & al. 1981, MELKONIAN 1989) and in the *Glaucocystophyta* (KIES 1979); this latter group is distantly related to the *Chlorophyta/Streptophyta* in rDNA phylogenies (STEINKÖTTER & al. 1994, BHATTACHARYA & al. 1995). Other features that are frequently assumed to be apomorphic of the "land plant line of evolution" are the phragmoplast/cell plate cytokinetic system (PICKETT-HEAPS 1975) in charophytes (*Coleochaetales* and *Charales*) and embryophytes, and the persistent nature of the mitotic spindle that keeps the nuclei distantly separated at telophase in the *Klebsormidiales* and *Chlorokybales* (*Charophyceae*) (PICKETT-HEAPS 1975, LOKHORST & STAM 1985, LOKHORST & al. 1988). However, a parallel evolutionary development of these features may be envisioned (GRAHAM 1993) since cytokinesis in the *Trentepohliales* (which belongs to the *Chlorophyta* s. str., see below) involves a microtubular system somewhat similar to charophyte/embryophyte phragmoplasts (CHAPMAN & HENK 1986). Persistent spindles also occur in the mitosis of the *Ulvophyceae* (e.g., SLUIMAN & al. 1983, SLUIMAN 1991).

Resolution of the most ancient divergences within the charophyte – land plant lineage (the *Streptophyta*) remains elusive which may be due to the rapidity of these evolutionary radiations (MCCOURT 1995). For instance, the precise phylogenetic position of the *Charales* is ambiguous; some analyses (e.g., KRANZ & al. 1995, and Fig. 1) suggest a para- or polyphyletic origin of the *Charophyceae*. See chapter 4 for further discussion of the evolution of the *Charophyceae*/land plant lineages.

The monophyletic *Chlorophyta* are subdivided into at least four distinct lineages, the *Trebouxiophyceae*, the *Chlorophyceae*, the *Ulvophyceae* and a group of scaly flagellates, the *Prasinophyceae* that divergence at the base of the *Chlorophyta* (Fig. 1).

Trebouxiophyceae

The class *Trebouxiophyceae* has been described only recently on the basis of rDNA sequence comparisons (FRIEDL 1995). Phylogenetic inferences from 18S rDNA sequence data demonstrate a single origin of many coccoid green algae that completely lack motile stage (autosporic coccoids, e.g., *Chlorella* spp.) With those that are defined by the ultrastructure of their zoospores as members of the *Microthamniales* sensu MELKONIAN (1982, 1990) (*Pleurastrales* sensu MATTOX & STEWART 1984) (FRIEDL & ZELTNER 1994, FRIEDL 1995; Figs. 1,2). This is not only in perfect concordance with a description of that order based on motile cell ultrastructure (for review see MELKONIAN 1990) but expands this previous concept

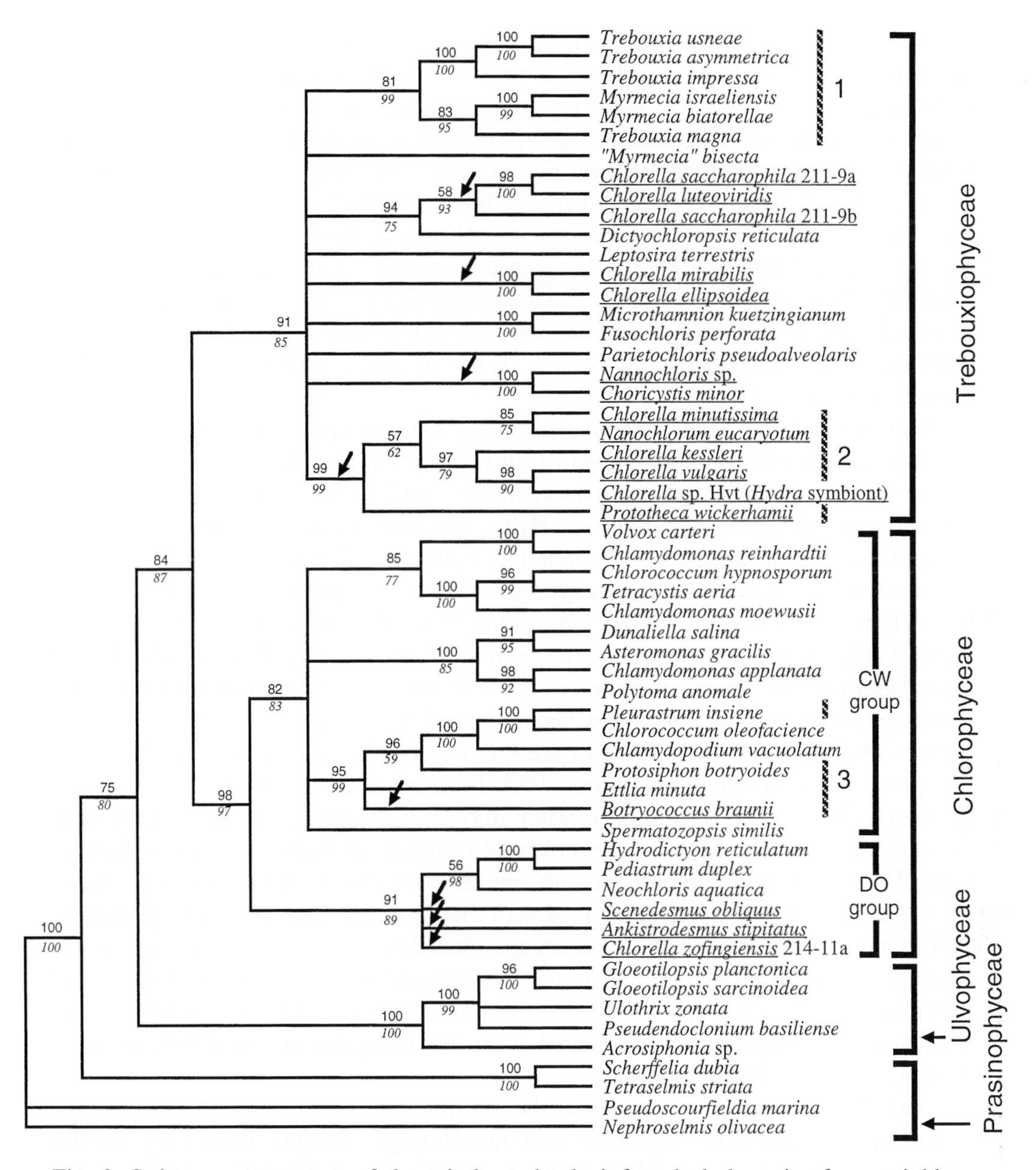

Fig. 2. Strict consensus tree of three independently inferred phylogenies from neighbor-joining, weighted maximum parsimony and maximum likelihood methods. Bootstrap values were computed independently for 500 resamplings using the neighbor-joining method (above line) and for 100 resamplings using the weighted parsimony analysis (below line). Parsimony analysis of the weighted data resulted in a single most parsimonious tree with a consistency index (Cl) of 0.6970, the maximum likelihood phylogeny had a Ln likelihood of – 15860.23. Nodes with less than 50% bootstrap support or nodes that are not shared by the maximum-likelihood, neighbor-joining and weighted parsimony trees are drawn as collapsed branch lengths. Arrows on branch lengths indicate putative independent losses of flagellated stages. 1 *Trebouxiales*; 2 *Chlorellales*; 3 *Chlorococcales* s. str. (see text)

to the inclusion of many autosporic coccoids. There are no morphological characters available that would reveal these relationships independent from the rDNA data. The specific tree topology suggests a zoosporic ancestry of autosporic coccoids. The loss of flagellated stages may have occurred independently several times within the *Trebouxiophyceae*. For example, the zoosporic *Dictyochloropsis reticulata* (TSCHERMAK-WOESS) TSCHERMAR-WOESS shares a sister group relationship with two strains of *Chlorella saccharophila* (KRÜGER) MIGULA and *C. luteoviridis* CHODAT (FRIEDL 1995; Fig. 2). Autosporic coccoids define additional lineages within the *Trebouxiophyceae* [e.g., *C. ellipsoidea* GERNECK/*C. mirabilis* ANDREYEVA, *Nannochloris* sp./*Choricystis minor* (SKUJA) FOTT; and *Chlorella* spp., *Nanochlorum eucaryotum* WILHELM, EISBEIS, WILD & ZAHN, *Prototheca wickerhamii* TUBAKI & SONEDA; Fig. 2].

Members of the *Trebouxiophyceae* share various common features such as ecology and growth morphology. Most known trebouxiophytes are coccoids, they live in terrestrial habitats or occur in symbioses with lichen fungi (e.g., *Trebouxia* spp.) or invertrebrates (*Chlorella* sp. HvT). The rDNA analyses indicate that the capacity to exist in lichen associations has multiple independent origins and that lichen algae are derived from non-symbiotic terrestrial green algae. For example, the lichen symbionts *Myrmecia biatorellae* BOYE-PETERSEN and *Trebouxia* spp. share a common origin with the terrestrial *Myrmecia israeliensis*. The lichen photobiont *Dicytochloropsis reticulata* has a sister group relationship with species of *Chlorella* BEIJERINCK that have been isolated from tree barks (FRIEDL 1995). Species of *Dictyochloropsis* GEITLER are known as lichen phycobionts and as free-living soil algae (TSCHERMAK-WOESS 1984, NAKANO & ISAGI 1987), as well, while *Trebouxia* spp. and *Myrmecia biatorellae* may be restricted to the association with lichen-forming fungi. Most of the eukaryotic lichen photobionts studied thus far with rDNA sequence comparisons are members of the *Trebouxiophyceae*. However, there are several other lichenized taxa that have unclear taxonomic positions. Preliminary sequence analyses also suggest that lichenized species of *Coccomyxa* SCHMIDLE, *Diplosphaera* BIALOSUKINA, *Elliptochloris* TSCHERMAK-WOESS and *Stichococcus* NÄGELI are members of the *Trebouxiophyceae*, whereas the filamentous *Trentepohlia* is associated with the *Ulvophyceae* in rDNA analyses (FRIEDL & NAKAYAMA, unpubl.).

The rDNA analyses substantiate the significance of motile cell features for delineating genera over vegetative cell morphology. Despite their very similar vegetative morphology, the rDNA analyses resolve a monophyletic origin only for those species of *Myrmecia* PRINTZ that share identical zoospore characters (*M. biatorellae* and *M. israeliensis*), whereas *M. bisecta* REISIGL which has zoospore features different from the other *Myrmecia* species forms an independent lineage in the rDNA phylogeny (Fig. 2) (FRIEDL 1995). A grouping of *Trebouxia* spp. with most species of *Myrmecia* (but excluding *M. bisecta*) is well resolved, the order *Trebouxiales* has been accorded to this particular clade (1 in Fig. 2). This is in concordance with the presence of unique motile cell characters that separate these taxa from other zoosporic members of the *Trebouxiophyceae* (FRIEDL & ZELTNER 1994); these include the pronounced flattening of the cell, a reduced transition region, and a reduced system II fibre in the flagellar apparatus (MELKONIAN & BERNS 1983, MELKONIAN & PEVELING 1988). The rDNA analyses provide an even finer resolution of evolutionary relationships than motile cell ultrastructure. A paraphyletic origin of the genus *Trebouxia* DE PUYMALY is suggested in the rDNA phylogenies, i.e. *T. magna* ARCHIBALD groups with *Myrmecia* spp. and not with other *Trebouxia* spp. (Figs. 1,2), although all members of the *Trebouxiales* share identical zoospore characters.

The specific topology of the rDNA phylogeny (Fig. 2) suggests that a filamentous organization has several independent origins within the *Trebouxiophyceae* which is in sharp contrast to traditional taxonomic schemes of the Green Algae (e.g., FRITSCH 1935, FOTT 1971). *Leptosira* spp. and *Microthamnion kuetzingianum* NÄGELI are filamentous algae; *M. kuetzingianum* is evolutionarily very closely related to the coccoid *Fusochloris perforata* (LEE & BOLD) FLOYD, WATANABE & DEASON (FRIEDL & ZELTNER 1994; Fig. 2). *Leptosira terrestris* (FRITSCH & JOHN) PRINTZ defines a single line whose position among other trebouxiophycean lineages is unresolved (FRIEDL 1996). 18S rDNA sequence comparison also suggest that the multiseriate filaments-forming *Prasiola* spp. are members of *Trebouxiophyceae* (NAKAYAMA & FRIEDL, unpubl.).

The monophyletic origin of *Chlorella vulgaris* BEIJERINCK, the type species of the genus *Chlorella*, together with other *Chlorella* species, *Nanochlorum eucaryotum* and *Prototheca wickerhamii* is clearly resolved in the 18S rDNA analyses (Fig. 2). It is suggested to regard this grouping as the *Chlorellales* (2 in Fig. 2) which is an order of the *Trebouxiophyceae*. This is different from a previous taxonomic scheme (ETTL & KOMÁREK 1982) in which all coccoid green algae that do not have flagellated stages with a cell wall were accommodated into an order "*Chlorellales*" under the class *Chlorophyceae* (see also discussion below). *Chlorella* is clearly a polyphyletic genus in rDNA analyses. Most species of *Chlorella* investigated by rDNA analyses so far lie within the *Trebouxiophyceae*, but *Chlorella* species are also found within the *Chlorophyceae* (Fig. 2; HUSS & KESSLER, unpubl.). Only those species closely related to *C. vulgaris* (e.g., *C. kessleri* FOTT & NOVÁKOVA and *C.* sp. HvT) within the *Trebouxiophyceae* may represent the genus *Chlorella* (BEIJERINCK 1890), while those that have their origin in other lineages of the *Trebouxiophyceae* (e.g., *C. saccharophila*, *C. mirabilis*) or the *Chlorophyceae* (e.g., *C. zofingiensis* DÖNZ) need to be included in other genera.

The monophyletic origin of the class *Trebouxiophyceae* is well resolved in most analyses (e.g., 91% and 85% in Fig. 2). That the topologies within the maximum parsimony trees are less robust (e.g., 58% in Fig. 1) than those in the neighbor-joining analyses may reflect the greater sensitivity of parsimony methods to unequal rates of changes within lineages (FELSENSTEIN 1978, 1981) (note the different lengths of internal branches leading to the *Chlorellales* and the *Microthamnion/Fusochloris* group in Fig. 1). The *Trebouxiophyceae* appear as an array of several independent lineages whose interrelationships are not resolved in the rDNA analyses. The relative branching order for lineages within the *Trebouxiophyceae* obtained from three independent phylogeny reconstruction methods are different; these divergences are also not resolved in bootstrap resamplings using neighbor-joining and parsimony methods (FRIEDL 1995).

Zoosporic members of the *Trebouxiophyceae* have earlier been grouped with the class *Pleurastrophyceae* (order *Pleurastrales*, MATTOX & STEWART 1984). This class also included *Tetraselmis/Scherffelia*, a clade that diverges prior to the radiation of the *Trebouxiophyceae/Chlorophyceae/Ulvophyceae* (Figs. 1,2). The *Pleurastrophyceae* is, therefore, polyphyletic in the rDNA analyses (STEINKÖTTER & al. 1994). The "metacentric" type of phycoplast (MOLNAR & al. 1975) shared by the *Trebouxiophyceae* and *Scherffelia/Tetraselmis* is a symplesiomorphy (FRIEDL 1996). The type species of *Pleurastrum* (CHODAT 1894), *P. insigne* CHODAT (SLUIMAN & GÄRTNER 1990), lies outside the monophyletic clade that contains the *Pleurastrales* sensu MATTOX & STEWART. Hence, "*Pleurastrophyceae*", permanently tied to *P. insigne*, becomes a later synonym of the *Chlorophyceae*.

Chlorophyceae

The class *Chlorophyceae* is represented as a clade whose monophyletic origin is well supported in bootstrap analyses (Figs. 1, 2). It is divided into two monophyletic lineages that are defined by green algae with a clockwise basal body configuration in flagellated cells (CW group; Fig. 2) and with directly opposed basal bodies (DO group; Fig. 2), respectively (LEWIS & al. 1992, NAKAYAMA & al. 1996). Presence of a cell wall (like in *Chlamydomonas* EHRENBERG) occurs only in members of the CW group. However, this feature does not delimit a monophyletic group. Within the CW group cell walls on motile cells were putatively lost independently several times [*Dunaliella* spp., *Protosiphon botryoides* (KÜTZING) KLEBS, *Spermatozopsis similis* PREISIG & MELKONIAN] or modified and thinned [e.g., *Pleurastrum insigne*, *Ettlia minuta* (ARCE & BOLD) KOMÁREK]. Previous taxonomic schemes (ETTL 1981, ETTL & KOMÁREK 1982) that emphasized the presence of cell walls on motile cells as a feature to distinguish orders (e.g., *Chlamydomonadales, Chlorococcales*) or even a separate class, the "*Chlamydophyceae*", are not supported by the rDNA analyses. *Chlamydomonas* and *Chlorococcum* MENEGHINI as well are seen as polyphyletic taxa within the CW group (BUCHHEIM & al. 1996, NAKAYAMA & al. 1996; Fig. 2). This group is divided into at least four independent monophyletic lineages whose interrelationships are not resolved (Fig. 2). These lines represent the *Volvox*, *Tetracystis* and *Dunaliella* clades of NAKAYAMA & al. (1996) and *Spermatozopsis similis*. Monophyly of the *Dunaliella* clade sensu NAKAYAMA & al., however, is not supported in Fig. 2, it is split into two independent lineages comprising members with only flagellated and coccoid vegetative cells, respectively. The rDNA data are consistent with a multiple origin of the coccoid organization in different lineages of the *Chlorophyceae* (e.g., *Tetracystis aeria* BROWN & BOLD and *Pleurastrum insigne* in the CW group, and all members of the DO group). This again demonstrates polyphyly of all previous concepts of the *Chlorococcales* (e.g., ETTL 1981, ETTL & KOMÁREK 1982, DEASON & al. 1991). There is one particular lineage within the CW group which contains only coccoid members and thus may represent the order *Chlorococcales* s.str. (3 in Fig. 2). However, it is not possible at the moment to distinguish this lineage from other lines of the CW group using morphology; there are also no morphological characters known that would support the monophyly of the other lineages within the CW group (NAKAYAMA & al. 1996). The multinucleate condition may have been independently developed [*Protosiphon botryoides* and *Hydrodictyon reticulatum* (L.) LAGERHEIM; Fig. 2] and flagellated stages may have been independently lost [*Botryococcus braunii* KÜTZING, *Scenedesmus obliquus* (TURP.) KÜTZING, *Chlorella zofingiensis*, Fig. 2] in both CW and DO groups.

The taxonomic status of the lineages within the *Chlorophyceae* resolved by the rDNA analyses is still unknown (NAKAYAMA & al. 1996). Additional sequence data from coccoid and filamentous members of the DO group (LEWIS & FUERST, unpubl.; BOOTON & al. 1996) resolve better the phylogenetic relationships of the DO-group and these data suggest that directly opposed basal bodies may be the ancestral condition within the *Chlorophyceae*.

Ulvophyceae

The class *Ulvophyceae* (represented by members of the order *Ulotrichales*; Figs. 1,2) defines a clade in the rDNA phylogenies that diverges with a long branch length prior to the nearly simultaneous radiation of the green algal lineages included in the analyses (FRIEDL 1996; Fig. 1). The *Ulvophyceae/Ulotrichales* occupy a basal position within the *Ulvophyceae/Chlorophyceae/Trebouxiophyceae* clade; the *Trebouxiophyceae* and *Chlorophyceae* share a sister group relationship. As zoosporic taxa within the *Ulvophyceae/Ulotrichales* and the *Trebouxiophyceae* share the counterclockwise basal body orientation, they have been regarded as closely related orders within one class, the *Ulvophyceae* sensu SLUIMAN (1989) (LOKHORST & RONGEN 1994). However, from the rDNA phylogenies there is strong evidence that the *Ulotrichales* and *Trebouxiophyceae* are not evolutionarily close to one another. Instead, they represent two independent lineages of green algae. The rDNA analyses (Figs. 1, 2) show that the counterclockwise basal body orientation is a symplesiomorphic character as well as the "phycoplast" type mitosis/cytokinesis pattern that is present in the *Ulvophyceae/Ulotrichales* and the *Trebouxiophyceae* (STEINKÖTTER & al. 1994, FRIEDL 1996). However, a number of ultrastructural features distinguish flagellated cells of the *Ulvophyceae/Ulotrichales* from the *Trebouxiophyceae* (summarized in FLOYD & O'KELLY 1990, MELKONIAN 1990) and their phylogenetic significance is substantiated by the rDNA analyses. Positioning of the *Ulvophyceae/Ulotrichales* at the base of the radiation of the *Chlorophyceae* and *Trebouxiophyceae* is not unequivocally resolved in the rDNA phylogenies (Figs. 1, 2). Analyses of user-defined trees show that a forced sister-group relationship of the *Ulvophyceae* with the *Chlorophyceae* (while the *Trebouxiophyceae* is basal) is not significantly worse. These tests provide additional support for the hypothesis that the *Ulvophyceae/Ulotrichales* are evolutionarily removed from the *Trebouxiophyceae* (FRIEDL 1996). It is plausible, on the basis of morphology, that the *Ulvophyceae/Ulotrichales* assume a basal position relative to the *Trebouxiophyceae* and *Chlorophyceae* (MATTOX & STEWART 1984). Members of the *Ulotrichales* (sensu FLOYD & O'KELLY 1990) have a scaly covering on their flagellated cells; this character is shared with green flagellates that clearly belong to early diverging lineages (STEINKÖTTER & al. 1994, MELKONIAN & SUREK 1995). Body scales have, however, not been found on motile cells of the *Trebouxiophyceae* or the *Chlorophyceae*.

Additional sequence analyses for members of the *Ulotrichales* and *Ulvales* show the *Ulotrichales* as defined by zoospore characters (FLOYD & O'KELLY 1990) as a monophyletic group (NAKAYAMA & FRIEDL, unpubl.). A few marine coccoid taxa [*Halochlorococcum* sp. and *Chlorocystis cohnii* (WRIGHT) REINHARD] form a clade that is specifically related to the *Ulotrichales* and *Ulvales* within the *Ulvophyceae* (NAKAYAMA & FRIEDL, unpubl.). Although not included in the analyses presented here, the 18S rDNA sequences from the *Cladophorales* (J. L. OLSEN & al. 1994) also cluster within the ulvophyte lineages. The exact position of the *Cladophorales* and also that of the *Dasycladales* is difficult to analyse with 18S rDNA sequence data due to homoplesious branch attraction in the phylogenetic analysis of these data (SWOFFORD & al. 1996). However, it is almost certain from rDNA sequence comparisons that the *Dasycladales* represent a lineage outside the radiation of the *Ulvophyceae/Chlorophyceae/Trebouxiophyceae* ("*" in Fig. 1) which supports findings from previous rRNA analyses (ZECHMAN & al. 1990).

Prasinophyceae

The *Prasinophyceae* are represented as a heterogenous assemblage of lineages that arise at the base of the radiation of the *Chlorophyta* (Fig. 1) and thus were used as outgroup taxa for sequence comparsions that resolve closer relationships among the radiation of the *Trebouxiophyceae/Chlorophyceae/Ulvophyceae* (Fig. 2) (STEINKÖTTER & al. 1994). The rDNA analyses clearly demonstrate that the *Prasinophyceae* (CHRISTENSEN 1962, MOESTRUP & THRONDSEN 1988, MOESTRUP 1991) is not a monophyletic group of green algae. There is also no unique morphological character that unites the *Prasinophyceae* to the exclusion of other green algae (for review see STEINKÖTTER & al. 1994). The exact position of the prasinophyte lineages shown in Fig. 1 is not unequivocally resolved. However, the *Tetraselmis/Scherffelia* clade represents a monophyletic lineage that is later-diverging compared to other prasinophytes; it immediately precedes the radiation of the "more advanced" green algal lineages. The basal position of a group of scaly flagellates in the rDNA phylogenies substantiates hypotheses already made early in this century (BLACKMAN & TANSLEY 1902, PASCHER 1918, 1924, 1931) that the flagellated condition is ancestral and that it gave rise to other more complex organizational levels of green algae. Some prasinophyte lineages branch near the divergence point of the *Chlorophyta* and the *Streptophyta*. Therefore, it seems likely that the common ancestor of both green plant divisions was among prasinophyte-like scaly flagellates. A prasinophyte that shares certain ultrastructural features with charophyte embryophyte flagellated cells, *Mesostigma viride* MELKONIAN (1989), was found being associated with one charophyte lineage in rDNA analyses (MELKONIAN & al. 1995). 18S rDNA sequence data have been obtained from almost all genera previously accomodated in the *Prasinophyceae* (MARIN & MELKONIAN, unpubl.) and these data will soon become the basis for a new taxonomic description of the scaly green flagellates.

Conclusions

18S rDNA sequence comparisons resolve nearly congruent relationships as motile cell ultrastructure, they substantiate the phylogenetic significance of these morphological characters for the distinction of classes of green algae (LEWIS & al. 1992, STEINKÖTTER & al. 1994). These characters are shown to be important even for the delineation of genera (FRIEDL 1995). However, the rDNA analyses have not only substantiated previous assumptions for the evolution of the Green Algae that were based on ultrastructural evidence, but have expanded taxonomic concepts to the inclusion of green algal taxa for which no zoospore characters are available (as in the case of autosporic coccoid green algae, WILCOX & al. 1992, FRIEDL 1995). The rDNA sequences provide an independent data set that allows the testing of hypotheses based on morphology and to reveal cases of parallel evolution (i.e. certain mitosis/cytokinesis patterns, STEINKÖTTER & al. 1994, FRIEDL 1996, and the coccoid organization, HUSS & SOGIN 1990, LEWIS & al. 1992, FRIEDL 1995). More sequence data are needed to evaluate the taxonomic position of certain filamentous green algae, such as the *Chaetophorales* and the *Oedogoniales*, and also for taxa that were regarded as members of the *Ulvophyceae* (see VAN DEN HOEK & al. 1992). Also, the significance of mitosis/cytokinesis patterns (certainly of secondary importance in comparisons to motile cell features) for resolving evolutionary relationships at lower taxonomic levels needs to be examined by molecular analyses; these characters may further substantiate the close relationships of autosporic and zoosporic taxa. For certain relationships where only evidence from rDNA sequence comparisons (i.e. also not from morphology) is available, additional testing with independent molecular markers is desirable. Preliminary analyses of actin coding regions, for example, support the *Trebouxiophyceae* as an independent evolutionary lineage and the position of autosporic coccoids as seen from rDNA analyses (NICKLES & al., unpubl.). The rapidly increasing 18S rDNA sequence data will provide a phylogenetic basis for a refined taxonomy of the green algae.

References

BEIJERINCK, M. W., 1890: Culturversuche mit Zoochlorellen, Lichenengonidien und anderen niederen Algen, – Z. Bot. **45–48**: 725–785.

BHATTACHARYA, D., MEDLIN, L., 1995: The phylogeny of plastids: a review based on comparisons of small-subunit ribosomal RNA coding regions. – J. Phycol. **31**: 489–498.

– FRIEDL, T., DAMBERGER, S., 1996: Nuclear-encoded rDNA group I introns: origin and phylogenetic relationships of insertion site lineages in the green algae. – Mol. Biol. Evol. **13**: 978–989.

– HELMCHEN, T., BIBEAU, C., MELKONIAN, M., 1995: Comparisons of nuclear-encoded small-subunit ribosomal RNAs reveal the evolutionary position of the *Glaucocystophyta*. – Mol. Biol. Evol. **12**: 415-420.

– SUREK, B., RÜSING, M., DAMBERGER, S., MELKONIAN, M., 1994: Group I introns are inherited through common ancestry in the nuclear-encoded rDNA of *Zygnematales* (*Charophyceae*). – Proc. Natl. Acad. Sci. USA **91**: 9916–9920.

BLACKMAN, F. E., TANSLEY, A. G., 1902: A revision of the classification of the green algae. – New Phytol **1**: 17–24, 67–77, 89–96, 114–120, 133–144, 163–168, 189–192, 213–220, 258–264.

BOOTON, G. C., FLOYD, G. L., FUERST, P. A., 1996: Polyphyletic relationships of the *Tetrasporales* supported by 18S ribosomal RNA gene sequence analysis. – J. Phycol. **32** [Suppl.]: 8.

BREMER, K., 1985: Summary of green plant phylogeny and classification. – Cladistics **1**: 369–385.

BUCHHEIM, M. A., LEMIEUX, C., OTIS, C., GUTELL, CHAPMAN, R. L., TURMEL, M., 1996: Phylogeny of the *Chlamydomonadales* (*Chlorophyceae*): a comparison of ribosomal RNA gene sequence from the nucleus and the chloroplast. – Mol. Phylogen. Evol. **5**: 391–402.

CHAPMAN, R. L., 1980: Ultrastructure of *Cephaleuros virescens* (*Chroolepidaceae, Chlorophyta*). II. Gametes. – Amer. J. Bot. **67**: 10–17.

– 1981: Ultrastructure of *Cephaleuros virescens* (*Chroolepidaceae: Chlorophyta*). III. Zoospores. – Amer. J. Bot. **68**: 544–556.

– HENK, M. C., 1986: Phragmoplasts in cytokinesis of *Cephaleuros parasiticus* (*Chlorophyta*) vegetative cells. – J. Phycol. **22**: 83–88.

CHODAT, R., 1894: Matériaux pur servir à l'histoire des Protococcoidées. – Bull. Herb. Boissier **2**: 585–616.

CHRISTENSEN, T., 1962: Systematik Botanik, Alger, – In BÖCHER, T. W., LANGE, M., SORENSEN, T., (Eds): Botanik, pp. 1–178. – København: Munksgaard.

DEASON, T. R., SILVA, P. C., WATANABE, S., FLOYD, G. L., 1991: Taxonomic status of the green algal genus *Neochloris*. – Pl. Syst. Evol. **177**: 213–219.

ETTL, H., 1981: Die neue Klasse *Chlamydophyceae*, eine natürliche Gruppe der Grünalgen (*Chlorophyta*) (Systematische Bemerkungen zu den Grünalgen I). – Pl. Syst. Evol. **137**: 107–126.

– KOMÁREK, J., 1982: Was versteht man unter dem Begriff "kokkale Grünalgen"? (Systematische Bemerkungen zu den Grünalgen II). – Arch. Hydrobiol. [Suppl.] **60**, Algol. Stud. **41**: 345–374.

FELSENSTEIN, J., 1978: Cases in which parsimony and compatibility methods will be positively misleading. – Syst. Zool. **27**: 401–410.

– 1981: Evolutionary trees from DNA sequences: a maximum likelihood approach. – J. Mol. Evol. **17**: 368–376.

– 1985: Confidence limits on phylogenies: an approach using the bootstrap. – Evolution **39**: 66–70.

– 1993: PHYLIP (Phylogeny Inferences Package), Version 3.5c. Distributed by the author, Department of Genetics, University of Washington, Seattle.

FLOYD, G. L., O'KELLY, C. J., 1990: Phylum *Chlorophyta*. Class *Ulvophyceae*. – In MARGULISS, L., CORLISS, J. O., MELKONIAN, M., CHAPMAN, D. J., (Eds): Handbook of *Protoctista*, pp. 617–648. – Boston: John & Bartlett.

FOTT, B., 1971: Algenkunde. 2nd edn. – Jena: G. Fischer.

FRIEDL, T., 1995: Inferring taxonomic positions and testing genus level asignments in coccoid green lichen algae: a phylogenetic analysis of 18S ribosomal RNA sequences from *Dictyochloropsis reticulata* and from members of the genus *Myrmecia* (*Chlorophyta*, *Trebouxiophyceae* cl. nov.). – J. Phycol. **31**: 632–639.

– 1996: Evolution of the polyphyletic genus *Pleurastrum* (*Chlorophyta*): inferences from nuclear-encoded ribosomal DNA sequences and motile cell ultrastructure. – Phycologia **35**: 456–469.

– ZELTNER, C., 1994: Assessing the relationships of some coccoid green lichen algae and the *Microthamniales* (*Chlorophyta*) with 18S ribosomal RNA gene sequence comparisons. – J. Phycol. **30**: 500–506.

FRITSCH, F. E., 1935: The structure and reproduction of the algae, 1. – London: Cambridge University Press.

GRAHAM, L. E., 1984: An ultrastructural re-examination of putative multilayered structures in *Trentepohlia aurea*. – Protoplasma **123**: 1–7.

– 1993: Origin of land plants. – New York, Chichester, Brisbane, Toronto, Singapore: J. Wiley & Sons, Inc.

– McBride, G. E., 1975: The ultrastructure of multilayered structures associated with flagellar bases in motile cells of *Trentepohlia aurea.* – J. Phycol. **11**: 86–96.

Hillis, D. M., Huelsenbeck, J. P., Swofford, D., 1994: Hobglobin of phylogenetics. – Nature **369**: 363–364.

Huss, V. A. R., Sogin, M. L., 1990: Phylogenetic position of some *Chlorella* species within the *Chlorococcales* based upon complete small-subunit ribosomal RNA sequences. – J. Molec. Evol. **31**: 432–442.

Kies, L., 1979: Zur systematischen Einordnung von *Cyanophyora paradoxa*, *Gloeochaete wittrockiana* und *Glaucocystis nostochinearum*. – Ber. Deutsch. Bot. Ges. **92**: 445–454.

Kimura, M., 1980: A simple method for estimating evolutionary rates of base substitution through comparative studies of sequence evolution. – J. Mol. Evol. **16**: 111–120.

Kranz, H. D., Miks, D., Siegler, M.-L., Capesius, I., Sensen, C. W., Huss, V. A. R., 1995: The origin of land plants: phylogenetic relationships among charophytes, bryophytes, and vascular plants inferred from complete small-subunit ribosomal RNA gene sequences. – J. Molec. Evol. **41**: 74–84.

Lewis, L. A., Wilcox, L. W., Fuerst, P. A., Floyd, G. L., 1992: Concordance of molecular and ultrastructural data in the study of zoosporic chlorococcalean green algae. – J. Phycol. **28**: 375–380.

Lokhorst, G. M., Rongen, G. P. J., 1994: Comparative ultrastructural studies of division processes in the terrestrial green alga *Leptosira erumpens* (Deason et Bold) Lukeshová confirm the ordinal status of the *Pleurastrales*. – Crypt. Bot. **4**: 394–409.

– Stam, W., 1985: Ultrastructure of mitosis and cytokinesis in *Klebsormidium mucosum* nov. comb., formerly *Ulothrix verrucosa* (*Chlorophyta*). – J. Phycol. **21**: 466–476.

– Sluiman, H. J., Stam, W., 1988: The ultrastructure of mitosis and cytokinesis in the sarcinoid *Chlorokybus atmophyticus* (*Chlorophyta*, *Charophyceae*) revealed by rapid freeze fixation and freeze substitution. – J. Phycol. **24**: 237–248.

Mattox, K. R., Stewart, K. D., 1984: Classification of the green algae: a concept based on comparative cytology. – In Irvine, D. E. G., John, D. M., (Eds): Systematics of the Green Algae, pp. 29–72. – London: Academic Press.

McCourt, R. M., 1995: Green algal phylogeny. – TREE **10**: 159–163.

Melkonian, M., 1982: Two different types of motile cells within the *Chlorococcales* and *Chlorosarcinales*: taxonomic implications. – Br. Phycol. J. **17**: 236.

– 1989: Flagellar apparatus ultrastructure in *Mesostigma viride* (*Prasinophyceae*). – Pl. Syst. Evol. **164**: 93–122.

– 1990: Chlorophyte orders of uncertain affinities: Order *Microthamniales.* – In Marguliss, L., Corliss, J. O., Melkonian, M., Chapman, D. J., (Eds): Handbook of *Protoctista*, pp. 652–654. – Boston: John & Bartlett.

– Berns, B., 1983: Zoospore ultrastructure in the green alga *Friedmannia israelensis*: an absolute configuration analysis. – Protoplasma **114**: 67–84.

– Peveling, E., 1988: Zoospore ultrastructure in species of *Trebouxia* and *Pseudotrebouxia* (*Chlorophyta*). – Pl. Syst. Evol. **158**: 183–210.

– Surek, B., 1995: Phylogeny of the *Chlorophyta:* congruence between ultrastructural and molecular evidence. – Bull. Soc. Zool. France. **120**: 191–208.

– Marin, B., Surek, B., 1995: Phylogeny and evolution of the algae. – In Arai, R., Kato, M., Doi, Y., (Eds): Biodiversity and evolution, pp. 153–176. – Tokyo: The National Science Museum Foundation.

Moestrup, Ø., 1978: On the phylogenetic validity of the flagellar apparatus in green algae and other chlorophyll a and b containing plants. – BioSystems **10**: 117–144.

– 1991: Further studies of presumedly primitive green algae, including the description of *Pedinophyceae* class nov. and *Resultor* gen. nov. – J. Phycol. **27**: 119–133.

– THRONDSEN, J., 1988: Light and electron microscopical studies on *Pseudoscourfieldia marina*, a primitive scaly green flagellate (*Prasinophyceae*) with posterior flagella. – Can. J. Bot. **66**: 1415–1434.

MOLNAR, K. E., STEWART, K. D., MATTOX, K. R., 1975: Cell division in the filamentous *Pleurastrum* and its comparison with the unicellular *Platymonas* (*Chlorophyceae*). – J. Phycol. **11**: 287–296.

NAKANO, T., ISAGI, Y., 1987: *Dictyochloropsis irregularis* sp. nov. (*Chlorococcales, Chlorophyceae*) isolated from the surface of bark. – Phycologia **26**: 222–7.

NAKAYAMA, T., WATANABE, S., MITSUI, K., UCHIDA, H., INOUYE, I., 1996: The phylogenetic relationship between the *Chlamydomonadales* and *Chlorococcales* inferred from 18S rDNA sequence data. – Phycol. Res. **44**: 47–55.

OLSEN, G. J., MATSUDA, H., HAGSTROM, R., OVERBEEK, R., 1994: fastDNAml: a tool for construction of phylogenetic trees of DNA sequences using maximum likelihood. – CABIOS **10**: 41–48.

OLSEN, J. L., STAM, W. T., BERGER, S., MENZEL, D., 1994: 18S rDNA and evolution in the *Dasycladales* (*Chlorophyta*): modern living fossils. – J. Phycol. **30**: 729–744.

PASCHER, A., 1918: Von einer allen Algenreihen gemeinsamen Entwicklungsregel. – Ber. Deutsch. Bot. Ges. **36**: 390–409.

– 1924: Über die morphologische Entwicklung der Flagellaten zu den Algen. – Ber. Deutsch. Bot. Ges. **42**: 148–155.

– 1931: Systematische Übersicht über die mit Flagellaten in Zusammenhang stehenden Algenreihen und Versuch einer Einreihung dieser Algenstämme in die Stämme des Pflanzenreiches. – Beih. Bot. Centralbl., Abt. 2, **48**: 317–332.

PICKETT-HEAPS, J. D., 1975: Green algae: Structure, reproduction and evolution in selected genera. – Massachusetts: Sinauer.

ROBERTS, K. R., 1984: The flagellar apparatus in *Batophora* and *Trentepohlia* and its phylogenetic significance. – In IRVINE, D. E. G., JOHN, D. M., (Eds): Systematics of the Green Algae, pp. 331–341. – London: Academic Press.

ROGERS, C. E., DOMOZYCH, D. S., STEWART, K. D., MATTOX, K. R., 1981: The flagellar apparatus of *Mesostigma viride* (*Prasinophyceae*): multilayered structures in a scaly green flagellate. – Pl. Syst. Evol. **138**: 247–258.

SAITOU, N., NEI, M., 1987: The neighbor-joining method: a new method for reconstructing phylogenetic trees. – Molec. Biol. Evol. **4**: 406–425.

SLUIMAN, H. J., 1985: A cladistic evaluation of the lower and higher green plants (*Viridiplantae*). – Pl. Syst. Evol. **149**: 217–232.

– 1989: The green algal class *Ulvophyceae* – an ultrastructural survey and classification. – Crypt. Bot. **1**: 83–94.

– 1991: Cell division in *Gloeotilopsis planctonica*, a newly identified ulvophycean alga (*Chlorophyta*) studied by freeze fixation and freeze substitution. – J. Phycol. **27**: 291–298.

– GÄRTNER, G., 1990: Taxonomic studies on the genus *Pleurastrum* (*Pleurastrales, Chlorophyta*). I. The type species, *P. insigne*, rediscovered and isolated from soil. – Phycologia **29**: 133–138.

– ROBERTS, K. R., STEWART, K. R., MATTOX, K. R., 1983: Comparative cytology and taxonomy of the *Ulvophyceae*. IV. Mitosis and cytokinesis in *Ulothrix* (*Chlorophyta*). – Acta Bot. Neerl. **32**: 257–269.

STEINKÖTTER, J., BHATTACHARYA, D., SEMMELROTH, I., BIBEAU, C., MELKONIAN, M., 1994: Prasinophytes form independent lineages within the *Chlorophyta*: evidence from ribosomal RNA sequence comparisons. – J. Phycol. **30**: 340–345.

Stewart, K. D., Mattox, K. R., 1975: Comparative cytology, evolution and classification of the green algae with some considerations of the origin of other organisms with chlorophylls a and b. – Bot. Rev. **41**: 104–135.

– – 1978: Structural evolution in the flagellated cells of green algae and land plants. – BioSystems **10**: 145–152.

Surek, B., Beemelmanns, U., Melkonian, M., Bhattacharya, D., 1994: Ribosomal RNA sequence comparisons demonstrate an evolutionary relationship between *Zygnematales* and charophytes. – Pl. Syst. Evol. **191**: 171–181.

Swofford, D. L., 1993: Paup: Phylogenetic Analysis Using Parsimony, version 3.1.1. Computer program distributed by the Illinois Natural History Survey, Champaign.

– Olsen, G. J., Waddel, P. J., Hillis, D. M., 1996: Phylogenetic inference. – In Hillis, D. M., Moritz, C., Mable, B. K., (Eds): Molecular systematics. – Sunderland, MA: Sinauer.

Tschermak–Woess, E., 1984: Über die weite Verbreitung lichenisierter Sippen von *Dictyochloropsis* und die systematische Stellung von *Myrmecia reticulata* (*Chlorophyta*). – Pl. Syst. Evol. **147**: 299–322.

– Hoek, C., van den, Stam, W. T., Olsen, J. L., 1988: The emergence of a new chlorophytan system, and Dr. Kornmann's contribution thereto. – Helgol. Meeresunters. **42**: 339–383.

– – 1992: The *Chlorophyta*: systematics and phylogeny. – In Stabenau, H., (Ed.): Phylogenetic changes in peroxisomes of algae, phylogeny of plant peroxisomes, pp. 330–368. – Oldenburg: University of Oldenburg.

Wilcox, L. W., Lewis, L. A., Fuerst, P. A., Floyd, G. L., 1992: Assessing the relationships of autosporic and zoosporic chlorococcalean green algae with 18S rDNA sequence data. – J. Phycol. **28**: 381–386.

– Fuerst, P. A., Floyd, G. L., 1993: Phylogenetic relationships of four charophycean green algae inferred from complete nuclear-encoded small subunit rRNA gene sequences. – Amer. J. Bot. **80**: 1028–1033.

Zechman, F. W., Theriot, E. C., Zimmer, E. A., Chapman, R. L., 1990: Phylogeny of the *Ulvophyceae* (*Chlorophyta*): cladistic analysis of nuclear-encoded rRNA sequence data. – J. Phycol. **26**: 700–710.

Charophyte evolution and the origin of land plants

VOLKER A. R. HUSS and HARALD D. KRANZ

Key words: Charophytes, bryophytes, pteridophytes. – Land plants, systematics, evolution, phylogeny, 18S ribosomal RNA.

Abstract: Morphological, biochemical, and molecular data strongly suggest a common ancestry of land plants and *Charophyceae* sensu MATTOX & STEWART. Although it is now widely accepted that the *Charophyceae* are a sister group to the land plants, there is considerable disagreement over the systematics of different charophycean taxa and whether a *Chara*- or *Coleochaete*-like alga was ancestral to the land plants. Comparative analyses of complete 18S rRNA gene sequences of charophycean algae, together with sequences from bryophytes, pteridophytes and higher plants now confirm a common ancestry of charophytes and land plants. These data suggest that the *Charales* diverged early into a separate lineage, whereas other orders within the *Charophyceae* show close affinities to the bryophytes.

In 1469, PLINIUS, in his book "Historiae Naturalis", assigned the green alga *Chara* to the land plants as an aquatic horse-tail. It was thus more than 500 years ago, that man, for the first time, reflected upon the relationship between green algae and land plants. Since then, the systematic position of *Chara* and related algae within the plants has been controversial. Early this century, SCHENCK & al. (1917) placed the stoneworts (*Characeae*) in a common evolutionary lineage with the brown algae (*Phaeophyceae*). This lineage was independent of the green algae (*Chlorophyceae*) which were grouped with the red algae (*Rhodophyceae*) and fungi (*Eumycetes, Phycomycetes*). About 20 years later, KARSTEN & al. (1936) recognized a common ancestry of the 'green plants' (i.e. *Chlorophyceae, Bryophyta* and *Pteridophyta*), but the *Characeae* and *Conjugatae* were still excluded from the green algae. Instead, the *Rhodophyceae* were regarded as their closest relatives.

During the last 25 years, the earlier studies which only contained morphological data, have been supplemented by ultrastructural and biochemical characters, and more recently, by molecular data. The most important observation was made by PICKETT-HEAPS (1967, 1969, 1972a,b 1975) who showed that two radically different types of microtubule organization are involved in cytokinesis in the green algae. Species of the *Volvocales*, *Tetrasporales*, *Chlorococcales*, *Oedogoniales* and some *Ulotrichales* are characterized by the collapse of the interzonal spindle apparatus after mitosis and by developing a 'phycoplast' with the microtubules all oriented in the plane of cell division. In contrast, *Klebsormidiales*, *Coleochaetales*, *Charales* and *Conjugales* along with the land plants have a persistent spindle and develop a cleavage furrow (*Klebsormidium*) or a 'phragmoplast' in which the microtubules are oriented perpendicular to the eventual plane of cytoplasmic division. Based on the assumption that these features must be evolutionarily conservative, Pickett-Heaps & Marchant (1972) proposed a new phylogenetic scheme of green algal evolution. According to this scheme, the phycoplast- and the phragmoplast-containing green algae constitute two different evolutionary lineages. The latter lineage, subsequently designated as *Charophyceae* by Stewart & Mattox (1975), were suggested to be the progenitors of the bryophytes and higher land plants. Such a relationship is further supported by synapomorphic characters, such as the usage of glycolate oxidase rather than glycolate dehydrogenase in photorespiration (Frederick & al. 1973, Syrett & Al-Houty 1984), the possession of copper-zinc superoxide dismutase (De Jesus & al. 1989) and two class I aldolases (Jacobshagen & Schnarrenberger 1990), and flagellate spores and/or gametes with a multilayered structure (MLS). Whereas it is unlikely that all MLS's found so far are homologous structures (Sluiman 1983, Roberts 1984), the combined features of a persistent and open mitotic spindle, MLS, and a unique type of peroxisome structurally and biochemically similar to those of land plants (Stewart & al. 1972) are considered reliable indicators of a common ancestry (Graham 1993). To emphasize the proposed relationship of the *Charophyceae* to the land plants, the terms '*Streptophyta*' and '*Anthocerotophyta*' were created to combine the charophytes and the embryophytes into a single division (cf. Bremer 1985, Sluiman 1985).

The classification of the green algae into two classes, *Chlorophyceae* and *Charophyceae*, by Stewart & Mattox (1975) was later extended to a total of five classes as more taxa were studied by comparative cytology (Mattox & Stewart 1984). The concept of the *Charophyceae* remained unchanged with the exception that the *Chlorokybales*, with the single genus *Chlorokybus*, were added as a fifth order with the *Zygnematales*, *Klebsormidiales*, *Coleochaetales*, and *Charales* (Table 1). This classification has gained wide acceptance since then, although different views of the systematic position of various charophycean orders have been suggested, e.g., in the textbook "Introduction to the algae" by Bold & Wynne (1985) and in the "Handbook of Protoctista" (cf. Hoshaw & al. 1990, Graham 1990, Grant 1990). Bold & Wynne (1985) followed the view of Grambլast (1974) who, after extensive studies on the fossil records of charophytes (used here in the sense of stoneworts = *Charales* sensu Mattox & Stewart 1984), stated that these algae occupy an isolated position between green algae and bryophytes. Taking these data into consideration, together with the unique complex and reproductive structures, sperm morphology, and possession of a protonematal stage, they were classified as a separate division, *Charophyta*, whereas the remaining charophycean algae were assigned to different chlorophycean orders (Table 1). Still another view is proposed by Hoshaw & al. (1990) As the features characteristic for the classification of the *Charophyceae* (phragmoplast in cell division, glycolate and urea metabolism) have not yet been extensively studied in members of the *Zygnematales*, these algae were provisionally separated from the *Charophyceae* as a phylum *Conjugaphyta* (Table 1).

Not only the systematic position of the different charophycean taxa is controversial, but also which taxa eventually gave rise to the land plants, or more

Table 1. Different proposals for the classification of charophycean algae. *The *Chlorokybales* are not mentioned by BOLD & WYNNE (1985)

MATTOX & STEWART (1984)	BOLD & WYNNE (1985)	Handbook of *Protoctista* (1990) (see text for references)
Phylum: *Chlorophyta* Class: *Charophyceae* Order: *Zygnematales*	Division: *Chlorophyta* Class: *Chlorophyceae* Order: *Zygnematales*	Phylum: *Conjugaphyta* Class: *Conjugatophyceae* Order: *Zygnematales*
Order: *Chlorokybales*	Order:*	Phylum: *Chlorophyta* Class: *Chlorophyceae* Order: *Chlorokybales*
Order: *Klebsormidiales*	Order: *Ulotrichales* Genus: *Klebsormidium*	Order: *Klebsormidiales*
Order: *Coleochaetales*	Order: *Chaetophorales* Genus: *Coleochaete*	Order: *Coleochaetales*
Order: *Charales*	Division: *Charophyta* Order: *Charales*	Order: *Charales*

precisely, which of the extant taxa is a sister group to the land plants. Based on the results of cladistic analyses of morphological and biochemical data (MISHLER & CHURCHILL 1985, SLUIMAN 1985, BREMER & al. 1987, GRAHAM & al. 1991) and on molecular evidence (DEVEREUX & al. 1990; MANHART & PALMER 1990; CHAPMAN & BUCHHEIM 1991, 1992; WILCOX & al. 1993; BHATTACHARYA & al. 1994; MANHART 1994; RAGAN & al. 1994; SUREK & al. 1994; MCCOURT & al. 1996), only *Chara* or *Coleochaete* are seriously considered as sister taxa to the land plants. Interestingly, the morphological and biochemical data usually indicated a closer relationship of *Coleochaete* with the land plants. In contrast, cladograms based on distribution of $tRNA^{Ile}$ and $tRNA^{Ala}$ introns as well as sequence comparisons of *rbc*L and ribosomal RNA genes consistently placed the *Charales* (*Chara, Nitella*) or *Coleochaete* as sister taxon to land plants, although with rather low reliability as measured by bootstrap analyses. Doubtless, a drawback of some of the latter studies has been the use of sequences of too small size and thereby too low information content, such as 5S rRNA (DEVEREUX & al. 1990) or partial small subunit (SSU) or large subunit (LSU) rRNA sequences (CHAPMAN & BUCHHEIM 1991, 1992). When complete SSU rRNAs were used to infer phylogenetic relationships, the *Charophyceae* were only compared with higher plants (*Gymnospermae* and *Angiospermae*; WILCOX & al. 1993, BHATTACHARYA & al. 1994, RAGAN & al. 1994, SUREK & al. 1994) because data were not available from lower land plants, such as pteridophytes and especially bryophytes that are proposed to have a more recent common ancestry with some charophycean taxa (GRAHAM & al. 1991).

Recently, complete SSU rRNA gene sequences from several charophytes, bryophytes and pteridophytes have become available (Table 2; KRANZ & al. 1995, CAPESIUS 1995, BOPP & CAPESIUS 1995, KRANZ & HUSS 1996), affording, for the first

Table 2. List of organisms studied, and GENBANK accession numbers for complete 18S rRNA gene sequences. [a]sensu Friedl (1995); [b]sensu Melkonian (1990); [c]sensu Mattox & Stewart (1984)

Taxonomic position	Species	GENBANK Acc. No.
Chlorophyta		
Chlorophyceae[a]		
Volvocales	*Chlamydomonas reinhardtii* Dangeard	M32703
	Dunaliella salina (Dunal) Teodoresco	M84320
Trebouxiophyceae[a]		
Chlorococcales	*Chlorella vulgaris* Beijerinck	X13688
	Prototheca wickerhamii Tubaki & Soneda	X56099
Ulvophyceae		
Ulotrichales	*Gloeotilopsis planctonica* Iyengar & Philipose	Z28970
	Ulothrix zonata (Weber & Mohr) Kützing	Z47999
Prasinophyceae[b]		
Mamiellales	*Mantoniella squamata* (Manton & Parke) Desikachary	X73999
Pseudoscourfieldiales	*Nephroselmis olivacea* Stein	X74754
Charophyceae[c]		
Charales	*Chara foetida* Braun	X70704
	Nitella sp.	M95615
Zygnematales		
Mesotaeniaceae	*Mesotaenium caldariorum* (Lagerheim) Hansgirg	X75763
Zygnemataceae	*Mougeotia scalaris* Hassall	X70705
	Zygnemopsis circumcarinata (Czurda) H. Krieger	X79495
Desmidiaceae	*Sphaerozosma granulatum* Roy & Bisset	X79496
	Cosmarium botrytis Meneghini	X79498
	Genicularia spirotaenia (de Bary) de Bary	X74753
	Gonatozygon aculeatum Hastings	X91346
Klebsormidiales	*Klebsormidium flaccidum* (Kützing) Silva, Mattox & Blackwell	X75520
Coleochaetales	*Coleochaete scutata* de Brébisson	X68825
	Coleochaete orbicularis Pringsheim	M95611
Chlorokybales	*Chlorokybus atmophyticus* Geitler	M95612
Bryophyta		
Anthocerotopsida		
Anthocerotales	*Anthoceros agrestis* Paton	X80984
Marchantiopsida		
Marchantiidae		
Marchantiales	*Marchantia polymorpha* L.	X75521
Jungermanniopsida		
Metzgeriales	*Fossombronia pusilla* (L.) Nees	X78341
Bryopsida		
Bryidae		
Funariales	*Funaria hygrometrica* Hedwig	X74114

Table 2 (continued)

Taxonomic position	Species	GENBANK Acc. No.
Pteridophyta		
Sphenopsida		
Equisetales	*Equisetum hyemale* L.	X78890
Filicopsida		
Filicales	*Adiantum raddianum* PRESL	X78889
Spermatophyta		
Gymnospermopsida		
Cycadatae	*Zamia pumila* A.DC	M20017
Pinatae	*Pinus wallichiana* A. B. JACKS.	X75080
Magnoliopsida		
Brassicaceae	*Arabidopsis thaliana* (L.) HEYNH.	X16077
Solanaceae	*Lycopersicon esculentum* MILL.	X51576
Liliopsida		
Poaceae	*Oryza sativa* L.	X00755
	Zea mays L.	X00794

time, a more detailed molecular analysis of charophycean evolution and of the origin of land plants.

The phylogenetic position of the charophytes within the green plants (*Viridiplantae*)

The term '*Viridiplantae*' was formally proposed by CAVALIER-SMITH (1981) to comprise the 'true' green plants (= green algae+higher plants) as an 'arguably monophyletic' group. They have a 'stellate' (or 'star' and 'H-piece') structure in the flagellar transition region in common (MATTOX & STEWART 1984) as well as various plastid characters, such as the possession of chlorophylls a + b and an envelope consisting of two membranes, arrangement of thylakoids in pairs or stacks, and production of true starch as a reserve polysaccharide stored inside the chloroplast. Within this well characterized group, the prasinophytes comprise a diverse assemblage of motile chlorophytes whose cell body and flagella are covered by nonmineralized organic scales. They are considered to be descendants of the earliest group of chlorophytes which presumably gave rise to all other groups of green algae (MELKONIAN 1990). STEINKÖTTER & al. (1994) have demonstrated by 18S rRNA sequence analyses a paraphyletic origin of some representatives of the prasinophytes with a position basal to the *Ulvophyceae* and *Chlorophyceae* but not to the *Charophyceae*. An ancient lineage ancestral to all green algae might be represented by *Mantoniella squamata* as suggested in Fig. 1. However, the exact position of *Mantoniella* cannot be conclusively deduced from this and from a previous study (KRANZ & al. 1995) which yield inconsistent results with respect to the phylogenetic position of this taxon. In contrast, the *Charophyceae* are consistently placed in a common clade together with the land plants, in agreement

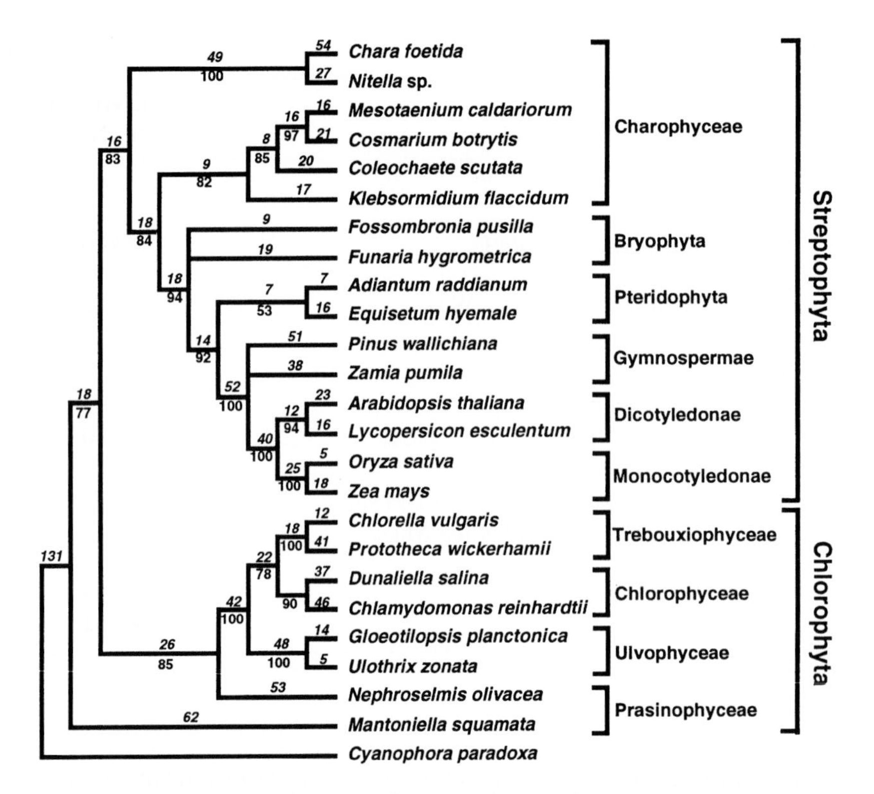

Fig. 1. 18S rRNA phylogeny of green plants. The tree shown is a maximum parsimony bootstrap analysis based on comparisons of 1676 nucleotide positions. The number of steps separating two nodes is taken from one of four most parsimonious trees that resulted from an heuristic search with random addition of sequences, and is shown in italics above the branches. Overall tree length was 1236 steps with a consistency index (CI) of 0.569, a homoplasy index (HI) of 0.431 and a retention index (RI) of 0.647. Bootstrap values based on 500 resamplings with 10 heuristic searches each are given below the branches. The tree includes all major lineages of green plants and demonstrates that chlorophytes are sister group to streptophytes (=land plants+charophytes). The glaucocystophyte *Cyanophora paradoxa* (EMBL Acc. No.: X68483) was used as the outgroup

with the ultrastructural and biochemical evidence outlined above. As shown for the *Prasinophyceae*, the *Charophyceae* likewise appear as a paraphyletic group. The *Charales*, often regarded as the more advanced *Charophyceae*, are shown to constitute a distinct and ancient lineage within this class (Figs. 1–3). The remaining charophycean taxa form a well supported monophyletic group which is divided into two lineages, albeit with only moderate bootstrap support (Fig. 2). One lineage contains the sarcinoid, unbranched and branched filamentous algae: the *Chlorokybales*, *Klebsormidiales*, and *Coleochaetales*, respectively. The other lineage consists of the *Zygnematales* which are characterized by their unique type of sexual reproduction (conjugation) and by the absence of flagellated cells throughout their life cycle.

Based on its sarcinoid thallus and pyrenoid structure, *Chlorokybus* is regarded as a relatively primitive member of the charophytes (Rogers & al. 1980). Our

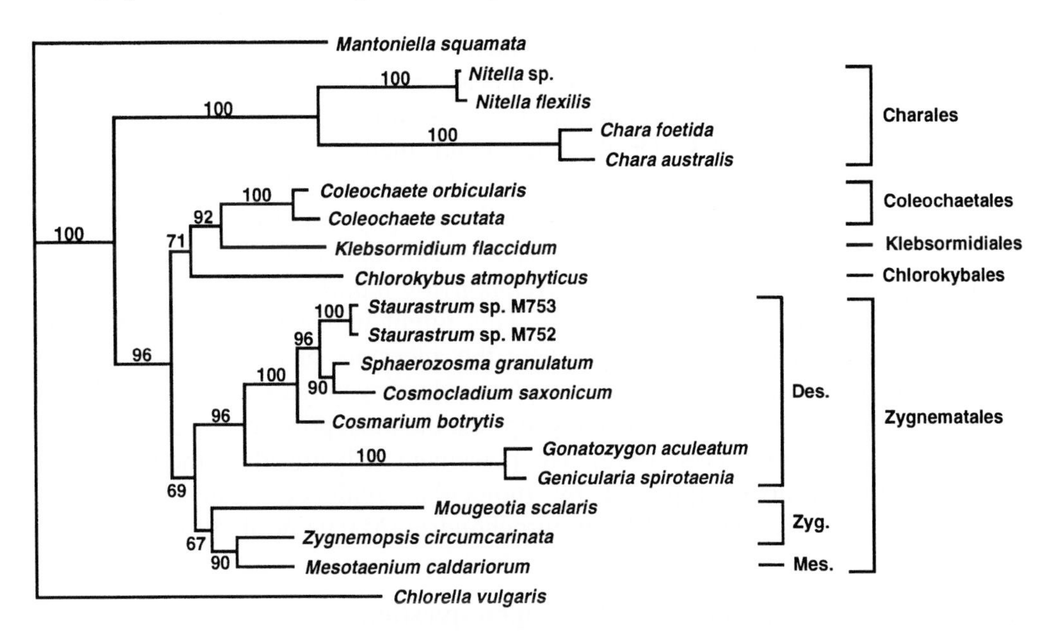

Fig. 2. Different evolutionary lineages within the charophytes. The neighbor joining bootstrap analysis based on 1000 resamplings shows at least three major evolutionary lineages within the *Charophyceae*, one comprising the *Charales*, one the *Coleochaetales*, *Klebsormidiales* and *Chlorokybales*, and one the *Zygnematales*. The *Zygnematales* themselves are separated into the *Desmidiaceae* (Des.) and *Zygnemataceae* (Zyg.) + *Mesotaeniaceae* (Mes.), which cannot be separated clearly. The tree is constructed with the Fitch & Margoliash (1967) algorithm based on structural distances (Kimura 1980) between 1680 nucleotides of the 18S ribosomal RNA gene. Bootstrap values are given above the branches. The sequence of the chlorophyte *Chlorella vulgaris* was used to root the tree

analyses (Figs. 2, 3) and those of others (Ragan & al. 1994, Surek & al. 1994, Wilcox & al. 1993) do not contradict this view, but none of these analyses are able to resolve the exact phylogenetic position of *Chlorokybus* with respect to the *Klebsormidiales*, *Coleochaetales*, and *Zygnematales*.

The *Zygnematales* usually are divided into three families based mainly on characteristics of their cell wall. Whereas the placoderm desmids, the *Desmidiaceae*, are shown to be monophyletic (Fig. 2), the *Zygnemataceae* are not. *Zygnemopsis circumcarinata* is more closely related to the saccoderm desmid *Mesotaenium caldariorum* (*Mesotaeniaceae*) rather than to *Mougeotia scalaris*. A similar result, indicating a polyphyletic origin of the *Zygnemataceae* and the *Mesotaeniaceae*, was obtained by *rbc*L sequence analyses (McCourt & al. 1995). The tree topology for the *Zygnematales* in Fig. 2 is essentially the same as that obtained by Bhattacharya & al. (1994) for a similar dataset, and is discussed in more detail in their report.

The *Charophyceae* are most closely related to the bryophytes which are the first to have emerged in the evolution of the land plants according to the 18S rRNA tree (Fig. 1). Such an ancient position of the bryophytes is not undisputed, as their fossil record dates back only to the Upper Devonian [*Pallavicinites devonians*, 374 million years (my) ago; SCHUSTER 1966]. Thus, this plant lived several million years later than the first tracheophytes that appear in the fossil record [*Cooksonia caledonica*, 420 my ago, and *Rhynia major* (=*Aglaophyton major*), 395 my ago; STEWART & ROTHWELL 1993]. On the other hand, there is evidence that primitive vascular plants (rhyniophytes) in the Lower Devonian had life cycles with an isomorphic alternation of gametophytic and sporophytic phases. Some gametophytic characteristics of the rhyniophyte species *Lyonophyton* and *Sciadophyton* as well as sporophytic characteristics of *Aglaophyton major* and *Horneophyton* are also found in bryophytes (REMY 1982). This might indicate that primitive land plants with an isomorphic life cycle were the ancestors of both bryophytes and tracheophytes, an idea which could explain why bryophytes cannot be clearly separated in 18S rRNA analyses from lycopsids, the most primitive members of the pteridophytes (KRANZ & HUSS 1996). The placement of lycopsids at the root of the tracheophytes is also supported by molecular evidence from chloroplasts (RAUBESON & JANSEN 1992) and from mitochondria (MALEK & al. 1996).

The ancestor of land plants – still an open question

After all, if it is not *Chara*, which green alga can then be regarded as most closely related to the land plants? Several authors suggest that *Coleochaete* rather than the stoneworts is the sister group of the land plants (PICKETT-HEAPS 1979; TAYLOR 1982; GRAHAM 1982, 1983; GRAHAM & WILCOX 1983; MISHLER & CHURCHILL 1985; GRAHAM & al. 1991). Other characters that have mainly served as arguments for a possible close relationship of *Coleochaete* and the embryophytes, include the multicellular development of the male gametangia (antheridia) and retention of the zygote on the gametophyte, as well as the presence of localized wall ingrowths in the vegetative cells adjacent to the zygote that resemble the gametophytic placental transfer cells of the land plants. Unfortunately, only some *Coleochaete* species (e.g., *C. orbicularis*) possess the characters listed above (BREMER 1985). Investigations using the *tuf*A gene indicate that the *Zygnematales* might be the closest sister group to land plants (BALDAUF & al. 1990). The *tuf*A gene, encoding protein synthesis elongation factor Tu, was transferred from the chloroplast to the nucleus within the charophyte lineage. This gene is present in chloroplasts of all *Chlorophyceae* and *Ulvophyceae*, but could be detected only in some members of the *Charophyceae*, and is absent in all land plant chloroplasts examined. Within the *Charophyceae*, only the *Charales* showed strong hybridization signals with a *tuf*A-specific probe (consistent with our view that the *Charales* are the most basal charophyte lineage). A weak signal resulting from a very divergent and probably nonfunctional *tuf*A gene was obtained with *Coleochaete*, whereas no signal was detected with chloroplasts from the *Zygnematales* (*Spirogyra* and *Sirogonium*).

Unfortunately, the rRNA data are not able to resolve the question of which extant green alga is most closely related to the land plants. Extensive bootstrap analyses with the three most widely used methods for phylogenetic reconstruction, neighbor joining, maximum parsimony and maximum likelihood, (trees A, B, and C in Fig. 3, respectively) all give the same result. They indicate that the different charophyte lineages (except the *Charales*) diverged within a short period of time and should together be regarded as a sister group to the bryophytes. There is no

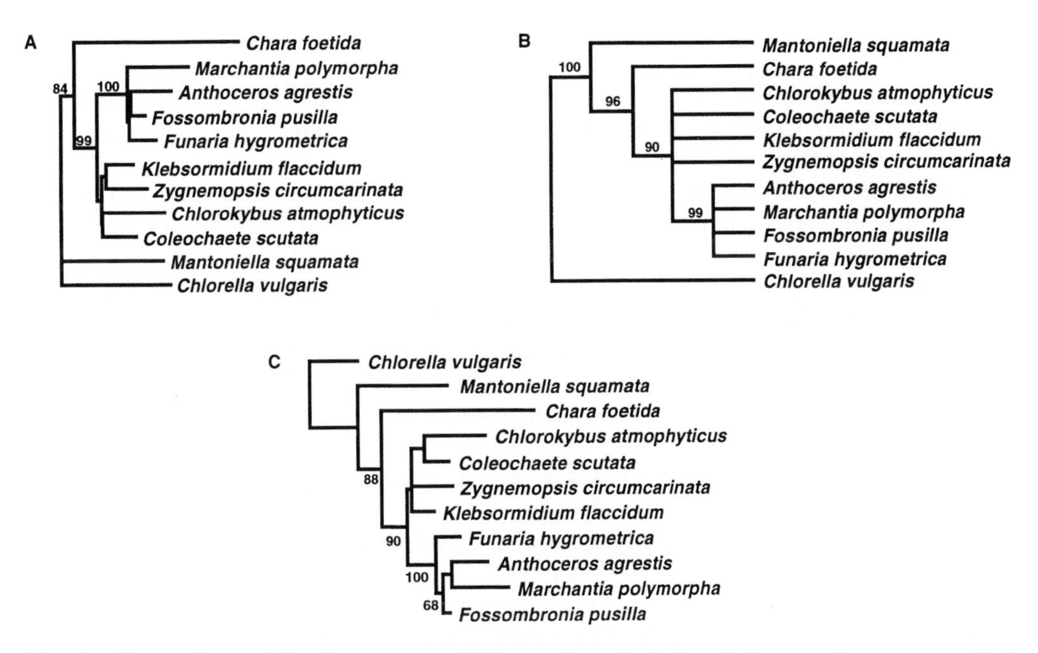

Fig. 3. Comparison of the three most widely used methods for phylogenetic reconstruction to show the relationships between charophytes and their closest relatives, the bryophytes. All trees (bootstrap analyses) are based on comparisons of 1575 nucleotide positions. Only bootstrap values >50% are shown. *Chlorella vulgaris* was used as the outgroup. *A* Neighbor joining analysis (1000 resamplings). *B* Maximum parsimony analysis (1000 resamplings). *C* Maximum likelihood analysis (250 resamplings)

indication that any of the charophytes investigated (including *Coleochaete orbicularis*; refer to Fig. 2) is significantly more closely related to the bryophytes than the others. There is also no indication, that some bryophytes might have independently evolved from different green algae, such as the hornworts from a *Coleochaete*-like alga as suggested by SLUIMAN (1983). Thus, features of *C. orbicularis*, such as zygote-associated cells that resemble the placental transfer cells of embryophytes, might well be the product of parallel evolution. Lack of resolution within the charophytes and also within the different lineages of bryophytes is indicated by bootstrap values $< 50\%$ in the distance and the maximum likelihood analysis, and by the multifurcation in the maximum parsimony analysis.

Why do our results contradict previous molecular studies even with complete 18S rRNA sequences, which favoured the *Charales* as closest relatives to the land plants? We think that the large 'evolutionary' gap between the charophytes and the higher plants coupled with the fact that *Charales* and higher plants have an accelerated mutation rate compared with the remaining charophytes, bryophytes, and pteridophytes (KRANZ & al. 1995, KRANZ & HUSS 1996) might have been responsible for artificially clustering the *Charales* with the higher plants, a phenomenon known as 'long branch attraction' (cf. LI & al. 1987, FELSENSTEIN 1988). Including sequences of organisms much more closely related to the charophytes, such as bryophytes and pteridophytes, may provide the means to avoid such artefacts.

References

Baldauf, S. L., Manhart, J. R., Palmer, J. D., 1990: Different fates of the chloroplast *tuf*A gene following its transfer to the nucleus in green algae. – Proc. Natl. Acad. Sci. USA **87**: 5317–5321.

Bhattacharya, D., Surek, B., Rüsing, M., Damberger, S., 1994: Group I introns are inherited through common ancestry in the nuclear-encoded rRNA of *Zygnematales* (*Charophyceae*). – Proc. Natl. Acad. Sci. USA **91**: 9916–9920.

Bold, H. C., Wynne, M. J., 1985: Introduction to the algae. Structure and reproduction. 2nd edn. – Englewood Cliffs, New Jersey: Prentice-Hall.

Bopp, M., Capesius, I., 1995: New aspects of the systematics of bryophytes. – Naturwissenschaften **82**: 193–194.

Bremer, K., 1985: Summary of green plant phylogeny and classification. – Cladistics **1**: 369–385.

– Humphries, C. J., Mishler, B. D., Churchill, S. P., 1987: On cladistic relationships in green plants. – Taxon **36**: 339–349.

Capesius, I., 1995: A molecular phylogency of bryophytes based on the nuclear encoded 18S rRNA genes. – J. Pl. Physiol. **146**: 59–63.

Cavalier-Smith, T., 1981: Eukaryote kingdoms: seven or nine? – BioSystems **14**: 461–481.

Chapman, R. L., Buchheim, M. A., 1991: Ribosomal RNA gene sequences: analysis and significance in the phylogeny and taxonomy of green algae. – Crit. Rev. Pl. Sci. **10**: 343–368.

– 1992: Green algae and the evolution of land plants: inferences from nuclear-encoded rRNA gene sequences. – BioSystems **28**: 127–137

De Jesus, M. D., Tabatabai, F., Chapman, D. J., 1989: Taxonomic distribution of copper-zinc superoxide dismutase in green algae and its phylogenetic importance. – J. Phycol. **25**: 767–772.

Devereux, R., Loeblich III, A. R., Fox, G. E., 1990: Higher plant origins and the phylogeny of green algae. – J. Molec. Evol. **31**: 18–24.

Felsenstein, J., 1988: Phylogenies from molecular sequences: inference and reliability. – Annu. Rev. Genet. **22**: 521–565.

Fitch, W. M., Margoliash, E., 1967: Construction of phylogenetic trees. – Science **155**: 279–284.

Frederick, S. E., Gruber, P. J., Tolbert, N. E., 1973: The occurrence of glycolate dehydrogenase and glycolate oxidase in green plants. – Pl. Physiol. **52**: 318–323.

Friedl, T., 1995: Inferring taxonomic positions and testing genus level assignments in coccoid green algae: a phylogenetic analysis of 18S ribosomal RNA sequences from *Dictyochloropsis reticulata* and from members of the genus *Myrmecia* (*Chlorophyta, Trebouxiophyceae* cl. nov.). – J. Phycol. **31**: 632–639.

Graham, L. E., 1982: The occurrence, evolution and phylogenetic significance of parenchyma in *Coleochaete* Breb. – Amer. J. Bot. **69**: 447–454.

– 1983: *Coleochaete*: advanced green alga or primitive embryophyte? – Amer. J. Bot. **70**: 5 (Abstract).

– 1990: Phylum *Chlorophyta*; class *Charophyceae*; orders *Chlorokybales., Klebsormidiales, Coleochaetales*. – In Margulis, L., Corliss, J. O., Melkonian, M., Chapman, D. J., (Eds): Handbook of *Protoctista*, pp. 636–640. – Boston: Jones & Bartlett.

– 1993: Origin of land plants. – New York: Wiley.

– Wilcox, L. W., 1983: The occurrence and phylogenetic significance of putative placental transfer cells in the green alga *Coleochaete*. – Amer. J. Bot. **70**: 113–120.

– Delwiche, C. F., Mishler, B., 1991: Phylogenetic connections between the 'green algae' and the 'bryophytes'. – Adv. Bryol. **4**: 213–244.

GRAMBLAST, L. J., 1974: Phylogeny of the *Charophyta*. – Taxon **23**: 463–481.

GRANT, M. C., 1990: Phylum *Chlorophyta*; class *Charophyceae*; order *Charales*. – In MARGULIS, L., CORLISS, J. O., MELKONIAN, M., CHAPMAN, D. J., (Eds): Handbook of *Protoctista*, pp. 641–648. – Boston: Jones & Bartlett.

HOSHAW, R. W., MCCOURT, R. M., WANG, J.-C., 1990: Phylum *Conjugaphyta*. – In MARGULIS, L., CORLISS, J. O., MELKONIAN, M., CHAPMAN, D. J., (Eds): Handbook of *Proctoctista*, pp. 119–131. – Boston: Jones & Bartlett.

JACOBSHAGEN, S., SCHNARRENBERGER, C., 1990: Two class I aldolases in *Klebsormidium flaccidum (Charophyceae)*: An evolutionary link from chlorophytes to higher plants. – J. Phycol. **26**: 312–317.

KARSTEN, G., FITTING, H., SIERP, H., HARDER, R., 1936: Strasburger – Lehrbuch der Botanik für Hochschulen. 19th edn. – Stuttgart: G. Fischer.

KIMURA, M., 1980: A simple method for estimating evolutionary rates of base substitution through comparative studies of nucleotide sequences. – J. Molec. Evol. **16**: 111–120.

KRANZ, H. D., HUSS, V. A. R., 1996: Molecular evolution of pteridophytes and their relationship to seed plants: evidence from complete 18S rRNA gene sequences. – Pl. Syst. Evol. **202**: 1–11.

– MIKŠ, D., SIEGLER, M.-L., CAPESIUS, I., SENSEN, C. W., HUSS, V. A. R., 1995: The origin of land plants: phylogenetic relationships among charophytes, bryophytes, and vascular plants inferred from complete small-subunit ribosomal RNA gene sequences. – J. Molec. Evol. **41**: 74–84.

LI, W.-H., WOLFE, K. H., SOURDIS, J., SHARP, P. M., 1987: Reconstruction of phylogenetic trees and estimation of divergence times under nonconstant rates of evolution. – Cold Spring Harb. Symp. Quant. Biol. **52**: 847–856.

MALEK, O., LÄTTIG, K., HIESEL, R., BRENNICKE, A., KNOOP, V., 1996: RNA editing in bryophytes and a molecular phylogeny of land plants. – EMBO J. **15**: 1403–1411.

MANHART, J. R., 1994: Phylogenetic analysis of green plant *rbc*L sequences. – Molec. Phylogenet. Evol. **3**: 114–127.

– PALMER, J. D., 1990.: The gain of two chloroplast tRNA introns marks the green algal ancestors of land plants. – Nature **345**: 268–270.

MATTOX, K. R., STEWART, K. D., 1984: Classification of the green algae: a concept based on comparative cytology. – In IRVINE, D. E. G., JOHN, D. M., (Eds): Systematics of the green algae, pp. 29–72. – London, Orlando: Academic Press.

MCCOURT, R. M., KAROL, K. G., KAPLAN, S., HOSHAW, R. W., 1995: Using *rbc*L sequences to test hypotheses of chloroplast and thallus evolution in conjugating green algae (*Zygnematales*, *Charophyceae*). – J. Phycol. **31**: 989–995.

– KAROL, K. G., GUERLESQUIN, M., FEIST, M., 1996: Phylogeny of extant genera in the family *Characeae* (*Charales, Charophyceae*) based on *rbc*L sequences and morphology. – Amer. J. Bot. **83**: 125–131.

MELKONIAN, M., 1990: Phylum *Chlorophyta*; class *Prasinophyceae*. – In MARGULIS, L., CORLISS, J. O., MELKONIAN, M., CHAPMAN, D. J., (Eds): Handbook of *Protoctista*, pp. 641–648. – Boston: Jones & Bartlett.

MISHLER, B. D., CHURCHILL, S. P., 1985: Transition to a land flora: phylogenetic relationships of the green algae and bryophytes. – Cladistics **1**: 305–328.

PICKETT-HEAPS, J. D., 1967: Ultrastructure and differentiation in *Chara* sp. II. Mitosis. – Austral. J. Biol. Sci. **20**: 883–894.

– 1969: The evolution of the mitotic apparatus: an attempt at comparative ultrastructural cytology in dividing plant cells. – Cytobios **1**: 257–280.

– 1972a: Variation in mitosis and cytokinesis in plant cells: its significance in the phylogeny and evolution of ultrastructural systems. – Cytobios **5**: 59–77.

– 1972b: Cell division in *Klebsormidium subtilissimum* (formerly *Ulothrix subtilissima*) and its possible phylogenetic significance. – Cytobios **6**: 167–184.
– 1975: Green algae. Structure, reproduction and evolution in selected genera. – Sunderland, Mass.: Sinauer.
– 1979: Electron microscopy and the phylogeny of green algae and land plants. – Amer. Zool. **19**: 545–554.
– Marchant, H. J., 1972: The phylogeny of the green algae: a new proposal. – Cytobios **6**: 255–264.

Plinius, S. C., 1469: Historiae Naturalis libri XXXVII. – Venice.

Ragan, M. A., Parsons, T. J., Sawa, T., Straus, N. A., 1994: 18S ribosomal DNA sequences indicate a monophyletic origin of *Charophyceae*. – J. Phycol. **30**: 490–500.

Raubeson, L. A., Jansen, R. K., 1992: Chloroplast DNA evidence on the ancient evolutionary split in vascular land plants. – Science **255**: 1697–1699.

Remy, W., 1982: Lower devonian gametophytes: relation to the phylogeny of land plants. – Science **215**: 1625–1627.

Roberts, K. R., 1984: The flagellar apparatus in *Batophora* and *Trentepohlia* and its phylogenetic significance. – In Irvine, D. E. G., John, D. M., (Eds): Systematics of the green algae, pp. 331–341. – London, Orlando: Academic Press.

Rogers, C. E., Mattox, K. R., Stewart, K. D., 1980: The zoospore of *Chlorokybus atmophyticus*, a charophyte with sarcinoid growth habit. – Amer. J. Bot. **67**: 774–783.

Schenck, H., Karsten, G., Jost, L., Fitting, H., 1917: Strasburger – Lehrbuch der Botanik für Hochschulen. 13th edn. – Stuttgart: G. Fischer.

Schuster, R., 1996: The *Hepaticae* and *Anthocerotae* of North America east of the hundredth meridian. – New York: Columbia University Press.

Sluiman, H. J., 1983: The flagellar apparatus of the zoospore of the filamentous green alga *Coleochaete pulvinata*: absolute configuration and phylogenetic significance. – Protoplasma **115**: 160–175.
– 1985: A cladistic evaluation of the lower and higher green plants (*Viridiplantae*). – Pl. Syst. Evol. **149**: 217–232.

Steinkötter, J., Bhattacharya, D., Semmelroth, I., Bibeau, C., Melkonian, M., 1994: Prasinophytes form independent lineages within the *Chlorophyta*: evidence from ribosomal RNA sequence comparisons. – J. Phycol. **30**: 340–345.

Stewart, K. D., Mattox, K. R., 1975: Comparative cytology, evolution and classification of the green algae with some consideration of the origin of other organisms with chlorophylls *a* and *b*. – Bot. Rev. **41**: 104–135.
– Floyd, G. L., Mattox, K. R., Davis, M. E., 1972: Cytochemical demonstration of a single peroxisome in a filamentous green alga. – J. Cell Biol. **54**: 431–434.

Stewart, W. N., Rothwell, G. W., 1993: Paleobotany and the evolution of plants. 2nd edn. – New York: Cambridge University Press.

Surek, B., Beemelmanns, U., Melkonian, M., Bhattacharya, D., 1994: Ribosomal RNA sequence comparisons demonstrate an evolutionary relationship between *Zygnematales* and charophytes. – Pl. Syst. Evol. **191**: 171–181.

Syrett, P. J., Al-Houty, F. A. A., 1984: The phylogenetic significance of the occurrence of urease/urea amidolyase and glycollate oxidase/glycollate dehydrogenase in green algae. – Brit. Phycol. J. **19**: 11–21.

Taylor, T. N., 1982: The origin of land plants: a paleobotanical perspective. – Taxon **31**: 155–177.

Wilcox, L. W., Fuerst, P. A., Floyd, G. L., 1993: Phylogenetic relationships of four charophycean green algae inferred from complete nuclear-encoded small subunit rRNA gene sequences. – Amer. J. Bot. **80**: 1028–1033.

A molecular perspective on red algal evolution: focus on the *Florideophycidae*

G. W. Saunders and G. T. Kraft

Key words: *Florideophycidae, Rhodophyta.* – Molecular phylogeny, rbcL, red algae, small-subunit rRNA, systematics, taxonomy.

Abstract: This chapter assesses advances in red algal systematics that are attributable to molecular studies. The first attempts to examine phylogenetic relationships among red algae were by analyses of 5.8S rDNA, *rbc*L and SSU rDNA sequences. These initial studies established the foundation on which subsequent, increasingly sophisticated investigations have developed. We explore recent phylogenetic hypotheses published for the red algae. Emphasis is placed on reconciling conflicts between inconsistent phylogenies inferred from different gene sequences, and between the molecular and non-molecular analyses. This chapter closes with an overview of the most recent SSU rDNA data and a discussion of the relationships between the major red algal lineages.

Red algal species have traditionally been assigned to one of the two subclasses *Bangiophycidae* and *Florideophycidae*, based on morphological, anatomical and life-history differences (Kraft 1981: 7, Bold & Wynne 1985: 30). Although the *Bangiophycidae* are generally regarded as a paraphyletic taxon (Garbary & Gabrielson 1990: 487), the *Florideophycidae* are widely agreed to be monophyletic (Gabrielson & Garbary 1987, Ragan & al. 1994). The composition and systematic relationships of orders within the *Florideophycidae*, however, are still very uncertain and have been the objects of intense speculation for over 100 years.

The apotheosis of classical-morphological system-building was Kylin's (1956) taxonomic treatment of florideophycidean genera, families and orders, and it is on his framework that later modifications have largely been grafted. A new age of ultrastructural ordinal criteria, based on features of pit-plugs, was ushered in by Pueschel & Cole (1982) with the segregation of two new orders and the confirmation of four previously contentious orders. Most recently, anatomical observations have been combined with molecular data in the proposal of the orders *Plocamiales* (Saunders & Kraft 1994) and *Halymeniales* (Saunders & Kraft 1996). The review that we present here considers relationships between the orders of the *Florideophycidae* in light of continuing molecular systematic investigations.

We limit ourselves to the *Florideophycidae* for two reasons: firstly, we are actively investigating relationships among the orders of this group and can therefore extend our reportage beyond published accounts; and secondly, recent reviews of molecular systematics within the *Bangiophycidae*, and for the

Rhodophyta itself as a lineage among other eukaryotic lines, by and large remain current (OLIVEIRA & al. 1995, RAGAN & al. 1994, RAGAN & GUTELL 1995). These studies serve to highlight the infancy of our knowledge of higher-level rhodophytan systematics. In addition, RAGAN & GUTTELL (1995) have summarized the molecular evidence for monophyly of the *Rhodophyta* as a whole (also see BHATTACHARYA & al. 1995) and canvassed work done on gene-sequences other than those that we will be discussing. These additional genes may provide valuable systems for supporting and extending the implications of data arising from SSU and *rbc*L sequence comparisons that have constituted the bulk of research to date.

Traditional approaches

Modern red-algal systematics traces its roots to SCHMITZ's (1892) and OLTMANNS' (1904–1905) radical emphases on female reproductive anatomy and postfertilization development – in other words, on the events leading to the formation of the zygote and those of its subsequent embryogenesis. KYLIN (1923, 1928, 1931, 1932) extended SCHMITZ's and OLTMANNS' theories and laid the observational and philosophical bases for red algal classification that have held mostly intact through to the early 1980s. KYLIN's impact was so profound that some 25 years later KRAFT (1981) was able to write that KYLIN's posthumous book (1956) was still "the single most authoritative work on the group... the point of departure for any revisionist attempts to alter the classification of red algae at and above the genus level."

KYLIN (1932, 1956) distributed the genera of *Florideophycidae* among the orders *Nemaliales* (as *Nemalionales*), *Gelidiales*, *Cryptonemiales*, *Gigartinales*, *Rhodymeniales* and *Ceramiales*. He conceived of these orders in a roughly ascending scale of advancement, with the *Nemaliales* and the *Gelidiales* at the base because of their lack of generative auxiliary cells (cells that initiate gonimoblasts following diploidization by discrete structures – cells, tubes, filaments – that convey zygote nuclei from the original sites of syngamy). The *Nemaliales* differed from the *Gelidiales* in supposedly having haplobiontic, rather than diplobiontic, life histories. The *Cryptonemiales, Gigartinales* and *Rhodymeniales* occupied a rather unspecified median position in KYLIN's red-algal tree distinguished by the fact that in all three groups the generative auxiliary cells are produced regardless of whether fertilization has taken place. In both the *Cryptonemiales* and *Rhodymeniales* auxiliary cells are borne as or in some sort of supernumerary structure, whereas in the *Gigartinales* such cells are intercalary in "normal" vegetative filaments. The *Ceramiales* was considered the most advanced group owing to the seemingly clever husbanding of energy by its members that permitted auxiliary-cell formation at a site immediately adjacent to a carpogonium only when that carpogonium had been fertilized. For a number of years it seemed reasonable to defend the prima-facie case for an evolutionarily "advanced" *Ceramiales* by analogy with the fertilization processes of higher plants. Water- and wind-pollinated seed plants seemed to stand in contrast to the highly derived animal-vectored pollination mechanisms of many "specialized" flowering plants, the former requiring a profligate expenditure of energy, the latter a far more efficient and intricately engineered one.

KYLIN's orders were taken as virtually given by most workers, although DIXON (1961, 1973) argued that the *Gelidiales* should be subsumed in the *Nemaliales* in light of investigations by FELDMANN & FELDMANN (1939a, b, 1942) and MAGNE (1960, 1961, 1967) on certain members of the *Nemaliales* that demonstrated diplobiontic, rather than the previously assumed haplobiontic, life histories, thus compromising the major criterion by which the *Nemaliales* and *Gelidiales* had been separated. DIXON's proposal did nothing to alter the fact that the *Nemaliales* of KYLIN's (1956) system already contained what appeared on anatomical grounds to be highly disparate families, and in fact added to the heterogeneity. Nonetheless, attempts to achieve monophyly for the *Nemaliales* by the removal of discordant elements to their own orders – the *Bonnemaisoniales* for the *Bonnemaisoniaceae* and *Naccariaceae* by FELDMANN & FELDMANN (1942); the *Acrochaetiales* for the *Acrochaetiaceae* by CHEMIN (1937) and FELDMANN (1953); and the *Chaetangiales* for the *Chaetangiaceae* by DESIKACHARY (1958, 1963) – generally received little support from phycologists. Dissatisfaction also grew with KYLIN's distinction between the *Cryptonemiales* and *Gigartinales* (SEARLES 1968, KRAFT & ROBINS 1985) and led to proposals as diverse as the merging of the two orders (KRAFT & ROBINS 1985) to the establishment of segregate orders – the *Hypneales* by STOLOFF (1962); the *Corallinales* by JOHANSEN (1973) – but these too were controversial and often ignored.

The first attempt at promoting greater monophyly in KYLIN's florideophycidean system to receive widespread acceptance was that of GUIRY (1978), who established the *Palmariales* for several taxa previously included in the *Rhodymeniales*. At first criticized, presumably for being based on a completely non-Kylinian character (the ontogeny of tetrasporangia), GUIRY's proposal was proven correct by subsequent life-history (VAN DER MEER & TODD 1980) and pit-plug ultrastructure (PUESCHEL & COLE 1982) investigations. Nevertheless, at this time the Kylinian system remained dominant despite its highly speculative underpinnings and untested domain assumptions as to the primary importance of carposporophyte architecture and of what constitute relatively primitive and advanced morphological characters.

At least three explanations for the strength of the allegiance to KYLIN's system can be offered: (1) the system was fundamentally true and thus tended to be supported by repeated further investigations; (2) the system was widely seen to be flawed but no better alternative could be seen on the horizon; or (3) the plausibility of the system as an abstract construct overwhelmed dissenting opinion or discordant data in the minds of most researchers. Probably all three would have their defenders or adherents. BOLD & WYNNE (1985) considered the latter possibility when they quoted ARNOLD (1948) as saying "Once a system of classification becomes widely adopted, it takes on many of the attributes of a creed. Not only does it constitute the framework about which the botanist does his thinking but it rapidly becomes a substitute for it... To function properly, all systems must be kept in a fluid and flexible state." It was only with the beginning of the 1980's that the first changes in the Kylinian ordinal classification began, changes that rapidly led to the present fluid state of red algal taxonomy.

Ultrastructural studies

KRAFT (1981) iterated some of the seeming shortcomings of the Kylinian system that had been noted by others and foreshadowed that major changes to red-algal classification were likely in the offing, although as yet the existing edifice was "relatively untouched by data from recent life history, biochemical or ultra-structural studies." In this he far underestimated the extent and pace of the changes that were so shortly to take place, as he admitted a decade later (KRAFT 1992). The announcement that pivotal change was now upon us came with the publication of PUESCHEL & COLE's (1982) paper in which ultrastructural evidence was convincingly marshaled to support the previously proposed *Bonnemaisoniales, Corallinales, Gelidiales* and *Palmariales*, as well as providing justification for the establishment of two new orders, the *Batrachospermales* and *Hildenbrandiales*. Important in itself for introducing an independent data set against which phylogenetic speculation based on anatomy could be tested, PUESCHEL & COLE's work was instrumental in bringing into question the Kylinian ordinal criteria.

The cladistic data

Heightened interest in new avenues for approaching higher-level taxonomy in red algae did not immediately result in major restructuring. The new data continued to be subjected to "armchair" speculation as to their implications, as had the morphological observations of the preceding decades. By the mid-1980's, however, a new approach to integrating the data and generating testable hypotheses was adopted by GARBARY & GABRIELSON based on the phylogenetic-systematic concepts and methods of cladistic analysis (GARBARY 1978; GABRIELSON & al. 1985; GABRIELSON & GARBARY 1986, 1987; GARBARY & GABRIELSON 1987). Another method was thus introduced to allow a total re-evaluation of red-algal systematics.

Like ourselves in this review, GABRIELSON & GARBARY (1987) confined their ultimate synthesis of data to the *Florideophycidae*. Using their data alone, we have constructed a strict-consensus tree (Fig. 1a). A major conclusion of their analysis was that the *Acrochaetiales* are a distinct assemblage that is ancestral to, or at least the sister of the remaining *Florideophycidae*. This conclusion supported the widely-held view (CHEMIN 1937, FELDMANN 1952, FRITSCH 1945, GARBARY 1978) that species of the *Acrochaetiales* represent the prototypical florideophyte assemblage due to their anatomical simplicity, and as such the group clearly merited ordinal status.

Succeeding the *Acrochaetiales* clade on GABRIELSON & GARBARY's analysis was a branch to a polychotomous node including the *Batrachospermales, Corallinales, Gelidiales, Hildenbrandiales, Nemaliales* and *Palmariales* (Fig. 1a). Within this group, the *Palmariales, Hildenbrandiales* and *Gelidiales* form a clade by virtue of their lack of filamentous germling states – a character common to many other orders – with the *Hildenbrandiales* and *Gelidiales* allied by the common presence of pit-plugs with a single 'inner' cap layer. Evidence for homology of such pit-plug caps does not exist (PUESCHEL & COLE 1982, PUESCHEL 1989, TRICK & PUESCHEL 1991) therefore undermining the utility of this character as indicative of a close relationship between these two orders (GABRIELSON & GARBARY 1987). Further-more, the 'inner' cap is distributed among a wide diversity of florideophycidean orders, as well as being present in the bangiophycidean order *Bangiales*, so that even if homology were demonstrated, it would constitute the sharing of an ancestral feature and thus provide no evidence for a recent common ancestry for

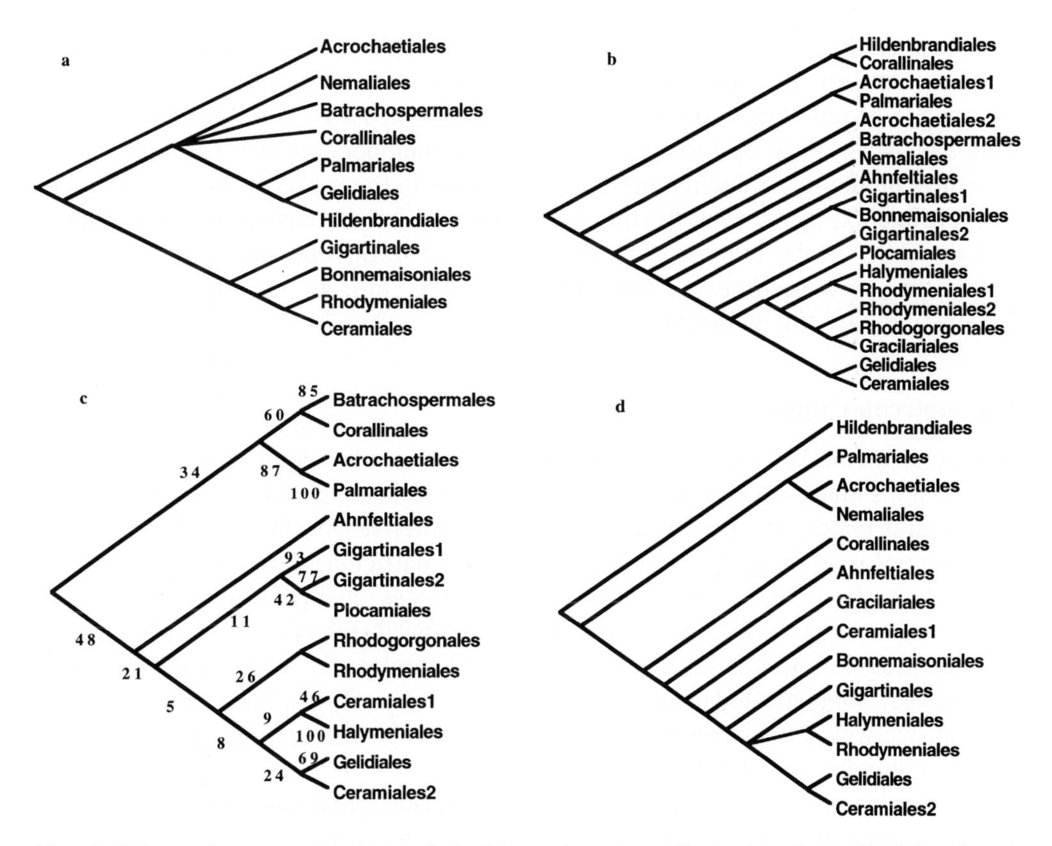

Fig. 1. Schematic representations of phylogenetic trees collapsed to the ordinal level. *a* A tree derived from cladistic analysis of non-molecular data (Gabrielson & Garbary 1987). *b* Simplified version of a published tree from *rbc*L data (Freshwater & al. 1994). *c* Cursory re-analysis (see text) of the *rbc*L data by distance methods. Values along branches represent bootstrap support (terminal branches lacking bootstrap values indicate that only a single representative was included for that order). *d* Earliest SSU phylogeny in summary (Ragan & al. 1994)

the *Hildenbrandiales* and *Gelidiales*. Even at that time (Gabrielson & Garbary 1987), given the multitude of anatomical and reproductive differences between members of these two orders, it was considered unlikely that they constituted true sister taxa.

Another outcome of cladistic handling of traditional anatomical data was the emergence of a group of "higher florideophytes" consisting of the *Bonnemaisoniales, Ceramiales, Gigartinales* (including Kylin's *Cryptonemiales*) and *Rhodymeniales* which together "crowned" the evolutionary tree (Fig. 1a). These orders share a complete absence of pit-plug caps and, with the exception of the *Bonnemaisoniales*, possess generative auxiliary cells. The characters that resolved relationships within this large assemblage, however, are variably expressed among members of the given orders, so that Gabrielson & Garbary (1987) recognized the tree topology as a preliminary result.

Perhaps the most unfortunate result of cladistic analyses of strictly anatomical (both light- and electron-microscope level) characters, apart from embellishing the view that the *Acrochaetiales* constitute the most primitive florideophycidean order, is the support given to the virtually axiomatic Kylinian view of the *Ceramiales* as the apogee of red-algal evolution (KYLIN 1932, 1956; FRITSCH 1945). This conclusion has never been self-evident to us. That members of the *Ceramiales* are generally united in producing auxiliary cells only in the presence of, and succeeding, fertilization is indeed a striking synapomorphy, but to construe this as evidence that the order is the highest rung on a ladder of florideophycidean evolution of which the *Acrochaetiales* is the bottom rung is not, in our opinion, clearly demonstrated by the evidence.

The molecular data

Molecular methods have enjoyed enormous success when applied to virtually all the divisions and classes of life, but it was particularly inevitable that they would impact in a major way on the algae, if for no other reason than the fact that the fossil record is so meager for most phyla. This is particularly true of the red algae, where the earliest fossils represent both the putatively most primitive (CAMPBELL & al. 1979, CAMPBELL 1980) and comparatively advanced (EDWARDS & al. 1993: p. 35) architectures known in the group. Powerful molecular tools, coupled with increasingly sophisticated methods of phylogenetic analyses, have revolutionized the way we weigh evidence for common ancestry and changed, probably forever, the ways in which we frame taxonomic hypotheses.

Molecular systematic investigations of the Florideophycidae have thus-far been based on three different genes which we will discuss in turn.

5S ribosomal RNA sequences. The first application of molecular-systematic techniques to florideophycidean species used the 5S rRNA gene (LIM & al. 1986, HORI & OSAWA 1987). These initial studies suffered from several drawbacks, most notably the small sample size (a total of eight species) and the fact (which soon became clear) that the 5S gene is too short to have phylogenetic significance (cf. STEELE & al. 1991). Nevertheless, three broad aspects of the 5S trees constructed by LIM & al. (1986) and HORI & OSAWA (1987) are consistent with the more robust phylogenies based on other gene systems that are presented later in this review: (1) there is an indication of an early divergence of species currently assigned to the *Batrachospermales* and *Palmariales* from those presently included in the *Gelidiales, Gigartinales, Gracilariales* and *Halymeniales*; (2) representatives of the *Palmariales (Palmaria)* and *Batrachospermales (Batrachospermum)* are closely related; and, (3) the single representative of the newly erected order *Halymeniales (Carpopeltis)* was only distantly related to the one member of the *Gigartinales (Gloiopeltis)* analyzed (cf. HORI & OSAWA 1987 with SAUNDERS & KRAFT 1996). Beyond these broad, sweeping indications, however, 5S sequences have not advanced red-algal phylogenetics at the family and ordinal levels. Although in retrospect the 5S tree can be considered as resolving sensible relationships among the included taxa, at the time this tree was first published (HORI & OSAWA 1987) the results would have been so unorthodox as to cause serious doubt about the utility of molecular data in algal systematic investigations.

***rbc*L phylogenies.** More recently, a phylogeny of the florideophyte orders based on *rbc*L sequence data (the large subunit of ribulose-1-5-bisphosphate carboxylase/oxygenase) was published by FRESHWATER & al. (1994). Consistent with the 5S analyses, the *rbc*L tree (summarized in Fig. 1b) resolved a monophyletic *Florideophycidae* but contains a number of internal branchings that are difficult to reconcile with the evolutionary implications of virtually all traditional, ultrastructural and other molecular data. A clade incorporating the *Acrochaetiales, Batrachospermales, Nemaliales* and *Palmariales* was not resolved, as would be predicted from ultrastructural (PUESCHEL & COLE 1982, TRICK & PUESCHEL 1991, PUESCHEL 1994) and other molecular analyses (HORI & OSAWA 1987, RAGAN & al. 1994, SAUNDERS & al. 1995). The recently described *Rhodogorgonales*, which is affiliated with the *Batrachospermales* and/or *Corallinales* on anatomical (NORRIS & BUCHER 1989, FREDERICQ & NORRIS 1995), ultrastructural (PUESCHEL & al. 1992) and SSU molecular data (SAUNDERS & BAILEY, 1997), was incongruously allied to the *Gracilariales* (Fig. 1b). The family *Endocladiaceae* (Gigartinales 1) was grouped with the *Bonnemaisoniales* rather than with the *Gigartinales* as strongly supported by the SSU data (SAUNDERS & KRAFT 1996).

Although the *rbc*L phylogeny of FRESHWATER & al. (1994) does support the recently described *Plocamiales* (SAUNDERS & KRAFT 1994) and *Halymeniales* (SAUNDERS & KRAFT 1996), the seeming anomalies in the *rbc*L tree (Fig. 1b) caution against considering these data as confirmation of the ordinal proposals of SAUNDERS & KRAFT. Questioning the validity of molecular data for phylogenetic inferences is obviously warranted in light of the significant discrepancies between *rbc*L and SSU trees (cf. Fig. 1b with Figs. 1d and 2), but a strong case can be made that the SSU trees are in better agreement with consistent external data sets provided by anatomy and ultrastructure. Re-analysis of the *rbc*L data could offer the means to resolve the inconsistencies.

When published (FRESHWATER & al. 1994), the *rbc*L data were analyzed exclusively by parsimony methods and estimates of robustness were not determined to assess how well the data supported the inferred topology. We obtained 41 *rbc*L sequences placed in GenBank by FRESHWATER & al. (1994) and concluded that, perhaps being more conservative in our approach, we would have regarded a number as not being suited for the phylogenetic analyses. Many of the genes were only partially sequenced and, in some instances, as little as 40% of the coding region had been determined (e.g., GenBank U04034, U04043). A multiple alignment was constructed and we noted several regions where data were missing for as many as 15 of the included sequences. We then excluded species from our alignment in which sequence was determined for less than ca. 80% of the gene, while at the same time attempting to maintain ordinal representation. This reduced our alignment from 41 to 25 species. Regions where many of the included sequences had missing data were then removed, reducing the original 1417 bp multiple alignment to 950 bp. This conservative alignment was next subjected to neighbor-joining distance analysis (Fig. 1c) with subsequent bootstrap sampling (100 replicates). Although a cursory analysis of this data, it is obvious from Fig. 1c that few relationships are supported by this *rbc*L analysis (low bootstrap support), but that for the few associations that are supported, congruence with the 5S and SSU (Fig. 2) data has been achieved.

Small-subunit ribosomal RNA sequence. The first SSU rDNA-based phylogeny to compare florideophycidean orders was that of RAGAN & al. (1994), an extremely valuable contribution that laid the foundation for the increasing number of subsequent investigations that have employed this gene system. The relationships resolved by Ragan & al.'s (1994) SSU analysis (Fig. 1d) are intuitively satisfying in that they tend to support the general thrusts of anatomical and ultrastructural observations. Being a pioneer study, however, it was unavoidably beset by low taxon representation. Thus members of the orders *Plocamiales, Rhodogorgonales* and *Batrachospermales* were missing, the first two because they had not yet been proposed, the third because of its restriction to freshwater habitats outside the scope of the study. Of the included taxa sharing the ultrastructural feature for two pit-plug cap layers, the *Acrochaetiales, Palmariales* and *Nemaliales* formed a monophyletic lineage, with the *Corallinales* on a separate line. Apart from this anomaly, their tree is largely consistent with later SSU phylogenies (e.g., Saunders & Bailey, 1997; Fig. 2 this chapter) with the exception of the positions of members of the family *Ceramiaceae* (discussed below).

Subsequent to Ragan & al. (1994), a number of more restricted phylogenies have addressed particular familial and ordinal relationships within the *Florideophycidae* (Saunders & Kraft 1994, Chiovitti & al. 1995, Lluisma & Ragan 1995, Saunders & al. 1995, Goff & al. 1996, Millar & al. 1996, Saunders & Kraft 1996, Saunders & al. 1996). Saunders & Bailey (1997) have expanded sampling for lineages in which species have double-layered pit-plug caps (Fig. 2, Lineage 2). Data were added to those existing for the *Nemaliales* (Ragan & al. 1994) and *Batrachospermales* (Saunders & al. 1995), as well as providing the first SSU data for the *Rhodogorgonales*. Although the resulting tree topology (Fig. 2) is not expected to be definitive, it does point in an exciting way to the congruence of two totally different data sets, in this case the molecular and the ultrastructural data (Pueschel & Cole 1982, Pueschel 1989, Trick & Pueschel 1991, Pueschel 1994).

As a result of the SSU indications to date, we recognize four distinct lineages within the *Florideophycidae* (Fig. 2), although relationships between these lineages are not yet fully resolved (Saunders & Bailey 1997). This ambiguity will either be overcome with the addition of more taxa to the analyses, or the actual sequence of branching may prove indeterminable by molecular methods. Because we are not yet at a position to evaluate the outcomes, we will not speculate on the relative primitiveness of these lineages. Representatives of lineages 2, 3 and 4 all have character states capable of being interpreted as either derived or ancestral relative to one another and to the various orders of the *Bangiophycideae*.

At the most fundamental level, we now know that the time-honored view of red algal evolution as a series of steps leading from the "primitive" *Acrochaetiales* to the "advanced" *Ceramiales* (Kylin 1932, 1956; Fritsch 1945; Gabrielson & Garbary 1987) is not supported. This linear evolutionary sequence is actually counterintuitive from a more contemporary perspective in any case, as in order for the *Acrochaetiales* to be ancestral to the remaining *Florideophycidae* the other orders would have to branch from within, or be sister to, the *Acrochaetiales* leading in a linear progression toward the *Ceramiales*. The bushy tree generated by the SSU phylogeny (Fig. 2) runs counter to such a perspective, nor does it support repeated claims (Chemin 1937, Feldmann 1952, Fritsch 1945, Garbary 1978, cf. Garbary & Garbrielson 1987) that the ancestral florideophycidean alga was likely to have had an acrochaetiaceous morphology. The SSU data indicate that the

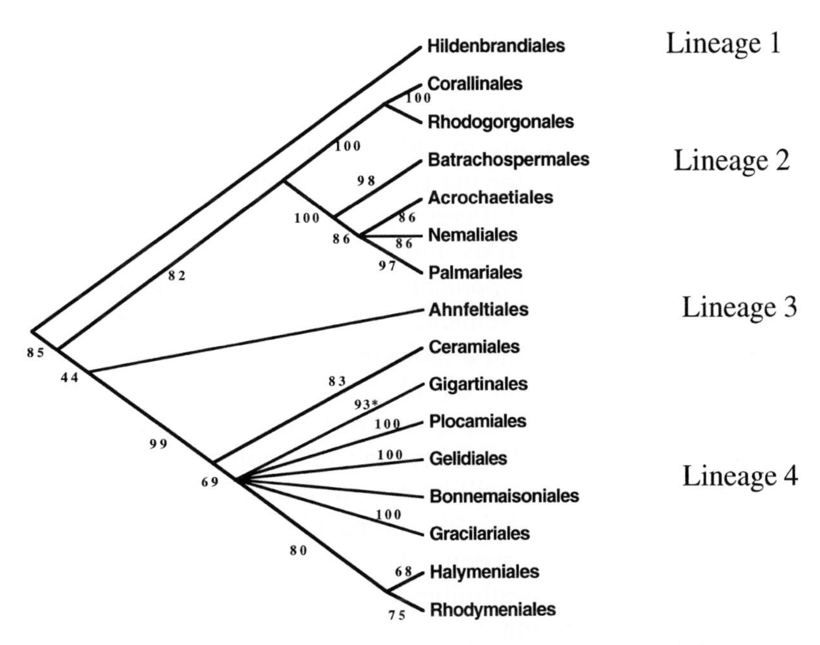

Fig. 2. Simplification of a recent parsimony tree derived from SSU data (SAUNDERS & BAILEY, 1997). Values along branches represent bootstrap support (terminal branches lacking bootstrap values indicate that only a single representative was included for that order),* *Schizymeniaceae* is variable in positioning and was not included in this tree (see text)

Acrochaetiales are a highly derived order within one of the florideophycidean lineages, as was once speculated by KRAFT (1981). The notion that the *Ceramiales* are the pinnacle of florideophycidean evolution has been virtually taken for granted since the time of OLTMANNS (1922) and is still echoed today. This view is not supported in the SSU trees (Fig. 2). SAUNDERS & KRAFT (1994: 261) discuss the important effect that the assumption of a highly-evolved *Ceramiales* has had on the polarization of morphological characters generally within the *Florideophycidae* (i.e., uni- vs. multiaxiality; procarpy vs. non-procarpy).

The major taxonomic lineages identified by molecular sequence comparisons

Lineage 1 – *Hildenbrandiales.* A single order (Fig. 2) and family (*Hildenbrandiaceae*) constitute this lineage which has always been problematically related to other florideophycidean orders in classical treatments. This is because an emphasis on zygote formation and embryogenesis is inapplicable to the *Hildenbrandiales*, its dozen or so members reproduce exclusively by either asexual propagules or presumably apomeiotic tetrasporangia. KYLIN (1956) allied the *Hildenbrandiaceae* with the order *Cryptonemiales*, presumably because its vegetative construction seemed analogous to that found in the *Peyssonneliaceae* (as the *Squamariaceae*) for which sexual reproduction is well documented.

PUESCHEL & COLE (1982) investigated cell ultrastructure within the genus *Hildenbrandia* and found that its pit-plugs differed from other members of the *Cryptonemiales* in having single caps and cap membranes. This and the unique way that tetrasporangial production leads to enlargement of the conceptacles (fertile pits) were regarded as evidence for the independent ordinal status of the *Hildenbrandiales* (PUESCHEL & COLE 1982, PUESCHEL 1982, 1988a, 1988b).

Preliminary indications of SSU analysis further justify recognition of the *Hildenbrandiales* as distinct from the *Cryptonemiales/Gigartinales* complex and support its isolated position relative to the remainder of the *Florideophycideae*. As only a single species has been studied from a molecular standpoint, however, the data are still limited. Besides the other 10–11 boreal/antiboreal species of *Hildenbrandia* still to be considered, there is a bizarre freshwater representative that will be of particular interest to analyze. Forming thin crusts in habitats as diverse as shallow, fast-flowing streams to still lakes at well over 20 m depths, this species apparently reproduces only by disk-like propagules termed "gemmae" (cf. NICHOLS 1965). Its status as a single species throughout an extreme range of habitats worldwide, as well as its actual affinities to the marine members of the genus, are completely unstudied at present.

Apophlaea, with two endemic New Zealand species, is generally regarded as the second genus of the *Hildenbrandiales* and displays a number of peculiar features including a consistent mutualism with an ascomycete fungus (KOHLMEYER & HAWKES 1983). It too should be added to the data set to resolve further internal and external relationships of the *Hildenbrandiales*.

Lineage 2 – *Rhodogorgonales, Corallinales, Batrachospermales, Nemaliales, Acrochaetiales* and *Palmariales*. PUESCHEL (1994) proposed that these six orders (Fig. 2), with their shared pit-plug features, form a monophyletic lineage. Earlier cytochemical investigations which established the homology of the outer pit-plug caps in species of these orders fueled this hypothesis (TRICK & PUESCHEL 1991). Phylogenetic support was not forthcoming from cladistic analyses (Fig. 1a) of non-molecular data (GABRIELSON & GARBARY 1987) or the initial molecular (Fig. 1b, d) investigations (FRESHWATER & al. 1994, RAGAN & al. 1994). Recent molecular investigations by SAUNDERS & BAILEY (1997), however, do provide at least weak support for this lineage and the evolutionary hypothesis of PUESCHEL (1994).

From a classical standpoint, few regroupings of Kylinian orders could be more counter-intuitive than Lineage 2. Although KYLIN (1956) did subsume the *Batrachospermaceae, Nemaliaceae* (as the *Helminthocladiaceae*) and *Acrochaetiaceae* (as the *Chantransiaceae*) within the one order *Nemaliales* (as the *Nemalionales*), he considered the *Corallinaceae* to be a family of the *Cryptonemiales* and those present-day members of the palmarialean *Palmariaceae, Rhodothamniellaceae* and *Rhodophysemataceae* of which he was aware to be members of the *Rhodymeniales, Acrochaetiales* and *Cryptonemiales* and/or *Acrochaetiales* respectively (GUIRY 1974, 1978, SAUNDERS & MCLACHLAN 1989, 1991, SAUNDERS & al. 1995). These orders represented widely divergent lines to KYLIN, possessing very different vegetative structures, life history pattterns, reproductive (particularly auxiliary-cell) features, and carposporophyte developments. The *Rhodogorgonaceae*, of course, were yet to be even discovered in

KYLIN's time. As a result of new approaches that lay less emphasis on strictly morphological characters, however, a 'feel' for the possible unity of this lineage begins to emerge.

Rhodogorgonales. This order was erected by FREDERICQ & NORRIS (1995) for two bizarre, recently discovered Caribbean genera that combine the lubricous, worm-like habit of some *Nemaliales*, several seeming gonimoblast characters of the *Batrachospermales*, and the deposition of calcite (NORRIS & BUCHER 1989). Although species of the *Rhodogorgonales* are the only marine algae outside the family *Corallinaceae* to form calcite, affiliating these with the *Corallinaceae* would have struck most phycologists as absurd given the total disparity of their textures and anatomies. Clearly too distinct to be included in any existing order, it seemed a radical enough notion that *Rhodogorgon* species could possibly be direct descendants of the marine ancestors of the entirely freshwater *Batrachospermales* (NORRIS & BUCHER 1989), a view that nevertheless was substantially strengthened by PUESCHEL & al.'s (1992) discovery that, like *Batrachospermum, Rhodogorgon* has pit-plugs with two cap layers, the outer one distinctively inflated. Only in the *Corallinales* is this particular feature also displayed, although PUESCHEL & al. (1992) hesitated to conclude firm links between the *Corallinales* and *Rhodogorgon*. The SSU-gene results of SAUNDERS & BAILEY (1997) strongly support the anatomical and ultrastructural grounds on which FREDERICQ & NORRIS (1995) erected the *Rhodogorgonales* (see Fig. 2), but they also provide evidence that the *Corallinales* and *Rhodogorgonales* may be affiliated.

Corallinales. Although independent ordinal status for the family *Corallinaceae* was being promoted from an early date (JOHANSEN 1973, GUIRY 1978), the grounds became compelling following the ultrastructural observations of PUESCHEL & COLE (1982; see also PUESCHEL 1987). SILVA & JOHANSEN (1986) finally recognized the order *Corallinales* which had been informally adopted by many phycologists in the preceding years (e.g., KRAFT & ROBINS 1985). The *Corallinales* is an excellent example of how a phylogenetic impasse can be broken by the emergence of entirely new approaches and sets of data. Under the Kylinian system, the underlying similarities of vegetative architecture clearly seemed to link the *Corallinaceae* with the *Peyssonneliaceae* and *Hildenbrandiaceae*. What then became an issue was finding and interpreting the position of the generative auxiliary cell, which devolved into the difficult cytological task of following the progress of zygote and zygote-derived nuclei from syngamy to gonimoblast initiation. On this process presumably hinged the critical question of whether the auxiliary cell was accessory or nonaccessory, and consequently whether affinities of the *Corallinaceae* were with the *Cryptonemiales* or the *Gigartinales* (JOHANSEN 1973, LEBEDNIK 1977).

PUESCHEL & COLE (1982) speculated that the *Corallinaceae* evolved from a common ancestor to the *Nemaliales* and *Palmariales* rather than from anything specifically ancestral to the *Cryptonemiales/Gigartinales* complex. Molecular data weakly support this conclusion (SAUNDERS & BAILEY 1997) but much more strongly indicate that the closest relative to the *Corallinales* is the *Rhodogorgonales* (Fig. 2). Wherever the *Corallinales* ultimately positions itself in the phylogeny of florideophycidean orders, the overwhelming indications are that it will be far removed from the associations posited by classical taxonomy.

Batrachospermales. This order was proposed by PUESCHEL & COLE (1982) for three exclusively freshwater families, the *Batrachospermaceae, Lemaneaceae* and *Thoreaceae*, that KYLIN (1956) had included in the *Nemaliales* because of their supposedly haplobiontic life histories and lack of generative auxiliary cells. From the work of many phycologists it became known that the life-histories of the nemalialean algae sensu lato are actually diplobiontic, with the macroscopic gametophytes usually alternating with diminutive filamentous sporophytes (cf. WEST & HOMMERSAND 1981). For species of the *Batrachospermales*, however, gametophytes arise directly from cryptic diploid phases by somatic meiosis (MAGNE 1972, VON STOSCH & THEIL 1979), a condition greatly at odds with the isomorphic or heteromorphic alternation of free-living gametophytes and tetrasporophytes characteristic of the other families placed by KYLIN (1956) in the *Nemaliales*. PUESCHEL & COLE (1982) found, in addition, that the pit plugs of the two uniaxial freshwater families *Batrachospermaceae* and *Lemaneaceae* differ from other *Nemaliales* in having domed outer caps, although this feature has been disputed in the multiaxial family *Thoreaceae* (SCHNEPF 1992).

The *Batrachospermales* are generally considered to be monophyletic (GARBARY & GABRIELSON 1990, ENTWISLE & NECCHI 1992), despite the anatomical and possible ultrastructural distinctiveness of the *Thoreaceae*. PUESCHEL (1989, 1994) and TRICK & PUESCHEL (1991) have speculated that domed outer caps and an absence of cap membranes are ancestral characters leading into Lineage 2; thus those character states would not in and of themselves constitute grounds for uniting the *Batrachospermaceae, Lemaneaceae* and *Thoreaceae* into a single order even if these feature proved to be uniformly expressed by species of all these families (cf. SAUNDERS & BAILEY 1997). Molecular data are in the process of being generated to test monophyly for the *Batrachospermales*, but are not as yet sufficiently complete to allow a prediction of the outcome (VIZ, SAUNDERS, ENTWISLE & SHEATH, unpubl.).

Nemaliales. Of all the orders recognized by KYLIN (1956), the *Nemaliales* (as the *Nemalionales*) has had the most event-filled subsequent history. Of the eight families credited to it by KYLIN, only the *Liagoraceae* (as the *Helminthocladiaceae*) and *Galaxauraceae* (as the *Chaetangiaceae*) now remain. From a morphological standpoint, there are few synapomorphies in this vestige. Members display a range of life-histories from alternating macroscopic isomorphic or heteromorphic gametophytes and tetrasporophytes to alternating macroscopic gametophytes and virtually microscopic tetrasporophytes. All members are multiaxial, produce gonimoblasts directly from fertilized carpogonia, and have cruciate tetrasporangia.

SSU analyses to date have been conducted on only a single species of each family: *Nemalion helminthoides* (VELLEY in WITHERING) BATTERS (*Liagoraceae*) by RAGAN & al. (1994) and *Galaxaura marginata* (ELLIS & SOLANDER) LAMOUROUX (*Galaxauraceae*) by SAUNDERS & BAILEY (1997). Preliminary results indicate that the group is a monophyletic sister to the *Acrochaetiales/Palmariales* complex (SAUNDERS & al. 1995), although the relationships among these three orders are equivocal (Fig. 2).

The *Liagoraceae* itself has been subdivided by some authors into as many as three component families, a situation that is not supported by a cladistic analysis of morphological characters (KRAFT 1989). The *Galaxauraceae* have been elevated to ordinal status by DESIKACHARY (1958, 1963), a move that is not supported by either morphological (HUISMAN 1985) or preliminary molecular (Fig. 2) data. SSU analyses of more members of both families will further test the internal coherence of the *Nemaliales* and clarify its relationships to other orders.

Acrochaetiales. A common ground for the recognition of the *Acrochaetiales* as a distinct order is the claim that it represents the most primitive forideophycidean levels of vegetative and reproductive organization (CHEMIN 1937, FRITSCH 1945, GARBARY 1978, GARBARY & GABRIELSON 1987). Cladistic analyses of mostly morphological data seem to support this assumption by resolving the *Acrochaetiales* as the earliest florideophycidean lineage (GARBARY 1978; GABRIELSON & al. 1985; GABRIELSON & GARBARY 1986, 1987; GARBARY & GABRIELSON 1987). From the classical-taxonomic standpoint, this is a satisfactory result.

With the coming of molecular analyses, however, a very different and seemingly counter-intuitive perspective on the *Acrochaetiales* begins to emerge. SSU data support an alliance between the *Acrochaetiales* and the *Palmariales* (SAUNDERS & al. 1995) and indicate that the *Acrochaetiales* (Fig. 2) is a comparatively recent derivative of a lineage that includes the corticated uniaxial and multiaxial algae of the *Batrachospermales, Corallinales, Nemaliales* and *Palmariales*. If this is an actual reflection of phylogenetic relationship, the most parsimonious conclusion would be that the uniaxial, uncorticated, diminutive composition of the acrochaetialian gametophyte is a derived state. On the other hand, the acrochaetialian-type sporophyte occurs throughout Lineage 2, as well as in the *Bangiophycidae*, and would be probably ancestral.

Some acrochaetiacean genera have recently been transferred to the *Palmariales* based on a suite of morphological, life-history and SSU characters (SAUNDERS & MACLACHLAN 1989, 1991; SAUNDERS & al. 1995). Implicit in such moves is a recognition that the *Acrochaetiales* may be paraphyletic by the exclusion of the *Palmariales*. Additional complication is introduced by the documentation of both domed and plate-like outer pit-plug caps (PUESCHEL 1989) and the presence of both B and R phycoerythrins (GLAZER & al. 1982) in various species, features that are found unvaryingly in one state or the other in the other red algal orders. To date the SSU data vary in their degree of support for monophyly of the *Acrochaetiales* depending on the taxa sampled and the method of phylogenetic inference employed. An extensive survey of the *Acrochaetiales/Palmariales* complex is underway (HARPER & SAUNDERS, unpubl.) in an attempt to clarify the relationships of this important group.

Palmariales. Recognition and confirmation of the order *Palmariales* was the first major break with Kylinian ordinal criteria and, in retrospect, now appears the forerunner of a necessary paradigm shift in our approach to red-algal higher classification. GUIRY (1974, 1978) established the order for former members of the *Rhodymeniales* in which tetrasporangia grew in successive crops from a basal cell included within the sporangial wall. At first considered premature for its total lack of reference to gonimoblast processes, the *Palmariales* was soon to be vindicated by the discovery of a unique female gametophyte from which the tetrasprophyte developed directly without the intermediate carposporophyte (VAN DER MEER & TODD 1980).

Recognition that species previously placed in at least three florideophycidean orders – ranging from the putatively most primitive (*Acrochaetiales*) to some of the supposedly most advanced (*Gigartinales, Rhodymeniales*) – all share two-layered pit-plug caps (PUESCHEL & COLE 1982), have similar life-histories (cf. SAUNDERS & MCLACHLAN 1991), and are grouped together in SSU-derived phylogenies (SAUNDERS & al. 1995) has enlarged and confirmed a definition of the *Palmariales* as a distinct rhodophyte order. Nevertheless, if molecular data indicate that the *Acrochaetiales* are paraphyletic in excluding the *Palmariales*, there will be systematists who will argue that the palmarialean species must of necessity be included in the *Acrochaetiales*. Although obviously an option, we would counter that paraphyly is not always a taxonomic transgression, particularly in a case like this one where members of the *Palmariales* have in common many features removed from those of typical *Acrochaetiales*. In any event, an extensive investigation of the *Acrochaetiales* is required before any serious proposal to subsume the *Palmariales* should be considered.

Lineage 3 – *Ahnfeltiales*. The single genus *Ahnfeltia*, with three cool- to cold-temperate species, comprises the recently established family *Ahnfeltiaceae* and order *Ahnfeltiales* (MAGGS & PUESCHEL 1989). Long recognized as reproductively unusual, *Ahnfeltia* was nevertheless placed by KYLIN (1932, 1956) in the family *Phyllophoraceae* (*Gigartinales*) based primarily on incomplete studies by ROSENVINGE (1931) that emphasized its vegetative similarities to members of that group. MAGGS & PUESCHEL (1989) showed that true *Ahnfeltia* species display a number of unique features: completely external carposporophytes; heteromorphic life histories involving crustose *Porphyrodiscus*-like tetrasporophytes with terminal zonate tetrasporangia; naked pit-plugs (i.e., lacking both caps and membranes); and production of agar rather than carrageenan. Species of similar habit but with internal cystocarps, heteromorphic life histories involving crustose *Erythrodermis*-like tetrasporophytes with chains of cruciate tetrasporangia, pit-plugs with cap membranes, and carrageenans as their main non-fibrillar polysaccharides are still retained in the *Phyllophoraceae* under the genus *Ahnfeltiopsis* (cf. MAGGS & al. 1989).

In support of its distinctive ultrastructure and reproductive anatomy, *Ahnfeltia* is shown by molecular analysis to be only distantly associated (Fig. 2) with other florideophycidean lineages (FRESHWATER & al. 1994; RAGAN & al. 1994; SAUNDERS & BAILEY 1997). With sequencing of the remaining *Ahnfeltia* species, and the addition of more representatives of the *Hildenbrandiales*, the true affinities of these two small and seemingly isolated lineages will hopefully become clearer.

Lineage 4 – *Gelidiales, Bonnemaisoniales, Gracilariales, Gigartinales/Rhodymeniales/Halymeniales/Plocamiales* and *Ceramiales*. With the exception of the *Gelidiales* and, for KYLIN, the *Bonnemaisoniales*, this lineage contains those florideophycidean orders that KYLIN (1956) and others (GARBARY & GARRIELSON 1990) have regarded as "higher" rhodophytes. In light of recent molecular investigations, however, we prefer to treat this group as Lineage 4, no higher or lower on the evolutionary "ladder" than the other three lineages, although certainly characterized by suites of derived features on which most taxonomic emphasis has been traditionally placed.

Gelidiales. KYLIN (1923) proposed this order for a small family, the *Gelidiaceae*, that was formerly included in the *Nemaliales*. Then and later (KYLIN 1928, 1932) he emphasized the diplobiontic, isomorphic life history of the gelidioid algae and the carposporophytes that emerged directly from fertilized carpogonia as a series of branching filaments that connected with isolated clusters of distinctive "nurse" cells. This contrasted to the supposedly haplobiontic life histories of the remainder of the *Nemaliales* and their different carposporophyte ontogenies and anatomies.

DIXON (1961, 1973: 239) vigorously argued for the return of the *Gelidiaceae* to the *Nemaliales* on the grounds that algae of both groups are diplobiontic (FELDMANN & MAZOYER 1937; FELDMAN & FELDMANN 1939a, 1939b, 1942; MAGNE 1960, 1961, 1967; cf. WEST & HOMMERSAND 1981). With the notable exception of PAPENFUSS (1966), workers largely accepted DIXON's recommendation until the ultrastructural investigations of PUESCHEL & COLE (1982) provided compelling new evidence for recognizing an independent *Gelidiales*. Meticulous anatomical studies by HOMMERSAND & FREDERICQ (1988) added further support and definition to the order.

Acceptance of the *Gelidiales* is strongly promoted by every molecular study that has included the group to date (Fig. 2, RAGAN & al. 1994, FRESHWATER & al. 1994, SAUNDERS & al. 1996, SAUNDERS & BAILEY 1997). Cladistic analysis of non-molecular features indicates an affinity with the *Hildenbrandiales* (GABRIELSON & GARBARY 1987), but there are no grounds for this association in the molecular data at present. Suggestions that the order may be related to the *Gracilariales* (HOMMERSAND & FREDERICQ 1988, FREDERICQ & HOMMERSAND 1989) are also not supported by molecular data. Relationships between orders of Lineage 4 are still poorly resolved, however, so this hypothesis cannot be wholly rejected at present. The suggestion that the three orders in which agars are the dominant non-fibrillar polysaccharide – the *Gelidiales, Gracilariales* and *Ahnfeltiales* – constitute a monophyletic assemblage (FREDERICQ & HOMMERSAND 1989) is also not supported by current molecular results (Fig. 2; RAGAN & al. 1994; FRESHWATER & al. 1994; SAUNDERS & al. 1996; SAUNDERS & BAILEY 1997).

Bonnemaisoniales. Culture methods were used by J. FELDMANN and G. FELDMANN (née MAZOYER) to unravel life-history patterns for the genera *Bonnemaisonia* and *Asparagopsis* (FELDMANN & MAZOYER 1937; FELDMANN & FELDMANN 1939a, b, 1942). Species proved to be diplobiontic with a heteromorphic alternation of generations, the larger gametophytes being succeeded by much smaller and differently constructed tetrasporophytes that had previously been named as three separated genera (*Falkenbergia, Hymenoclonium, Trailiella)* in the *Ceramiales*. As a result, they proposed that the *Bonnemaisoniaceae*, along with the *Naccariaceae*, should be placed in the new order *Bonnemaisoniales* as distinct from the remaining, presumably haplobiontic, *Nemaliales*.

Differences in anatomy, cytology and biochemistry (reviewed in CHIHARA & YOSHIZAKI 1972) have continued to set the *Bonnemaisoniales* apart from the *Nemaliales* despite suggestions that the latter is also largely composed of species with diplobiontic life histories (MAGNE 1961, 1967; cf. WEST & HOMMERSAND 1981). Cladistic analysis (GABRIELSON & GARBARY 1987) of non-molecular features has suggested that, as proposed by some workers (e.g., CHIHARA & YOSHIZAKI

1978), the *Bonnemaisoniales* are closely related to the *Ceramiales*, although, as pointed out by GARBARY & GABRIELSON (1990: 492), the only synapomorphy between the two is possibly bipolar spore germination. Investigation of pit-plug anatomy by PUESCHEL & COLE (1982) showing a complete lack of cap layers is compatible with such a hypothesis, although all groups of Lineage 4, with the exception of the *Gelidiales*, share this feature.

Available molecular data for members of the *Bonnemaisoniales* do not support a close association with the *Ceramiales* (Fig. 2), or with any other Lineage 4 order for that matter (FRESHWATER & al. 1994; RAGAN & al. 1994; SAUNDERS & BAILEY 1997). More taxa, particularly from Australasia where the greatest number of genera and species are concentrated (WOMERSLEY 1996), must be included in molecular investigations in order to establish the affinities of the *Bonnemaisoniales*. Of particular interest is the small family *Naccariaceae* which FELDMANN & FELDMANN (1942) included in the order based on similarities in nutritive cells (see also HOMMERSAND & FREDERICQ 1990) and carpospore germination. Differences in vegetative structure and carposporophyte anatomy from species of the *Bonnemaisoniaceae*, however, have suggested to GABRIELSON & GARBARY (1987) and ABBOTT (1997) that the *Naccariaceae* are better placed near the *Gigartinales*. Rare though they generally appear to be, representatives to the *Naccariaceae* remain a top priority for inclusion in molecular analysis.

Gracilariales. The family *Gracilariaceae* never fit comfortably into KYLIN's (1932, 1956) concept of the *Gigartinales* as it required an extreme stretching of his definition of the generative auxiliary cell in order to allow inclusion. Nevertheless, its positioning within the *Gigartinales* was unquestioned until FREDERICQ & HOMMERSAND (1989) proposed the order *Gracilariales* following detailed studies of cytology and anatomy in the type genus and species. Molecular data support the ordinal status of the *Gracilariales* (RAGAN & al. 1994; FRESHWATER & al. 1994; SAUNDERS & al. 1996; SAUNDERS & BAILEY 1997), although not in a monophyletic association with the two other agarophyte orders *Ahnfeltiales* and *Gelidiales* as proposed by FREDERICQ & HOMMERSAND (1989). Considering the broad distribution of agarocolloids (possibly the ancestral state) among both subclasses of the *Rhodophyceae* (CRAIGIE 1990), it is perhaps not surprising that this feature is diplayed independently in several lineages.

Gigartinales/Rhodymeniales/Halymeniales/Plocamiales. KYLIN (1956) assigned the multitude of red algae with auxiliary cells that form before fertilization to one of the three orders *Cryptonemiales, Gigartinales* or *Rhodymeniales*. Species with auxiliary cells terminal or intercalary on special lateral filaments (accessory branches) were placed in the *Cryptonemiales*, those with auxiliary cells intercalary in normal cortical filaments were placed in the *Gigartinales*, and those with auxiliary cells on a one- or two-celled lateral branchlet borne on the supporting or supra-supporting cell were assigned to the *Rhodymeniales*.

KRAFT & ROBINS (1985) examined the various interpretations that phycologists had read into KYLIN's ordinal criteria and concluded from linguistic as well as phenomenological standpoints that there were no bases for the separation of the *Cryptonemiales* and *Gigartinales* as then constituted. And although, unlike the *Cryptonemiales* and *Gigartinales*, the *Rhodymeniales* certainly appeared to be a monophyletic group, KRAFT & ROBINS (1985) foreshadowed that there might be little justification for that order as well (also see GARBARY & GABRIELSON 1990). KRAFT & ROBINS (1985) advocated subsuming the *Cryptonemiales* in the *Gigartinales* as a first step to a re-evaluation of the entire complex of families.

The *Gigartinales* as a mega-order of some 40 families was soon to lose some of its unwieldliness with the removal of several discordant groups to other orders. The *Corallinaceae* was even then regarded as unrelated and soon was formally elevated to the *Corallinales* (SILVA & JOHANSEN 1986), the *Gracilariales* were then established for the *Gracilariaceae* (FREDERICQ & HOMMERSAND 1989), and the *Ahnfeltiales* was erected for the genus *Ahnfeltia* (MAGGS & PUESCHEL 1989).

This flurry of activity was based on detailed anatomical and ultrastructural evidence and was followed by a brief hiatus in which several authors accepted the remainder of the KRAFT & ROBINS (1985) merger as a provisional advance over KYLIN's system (e.g., WOMERSLEY 1994). Then followed the first conclusions of SSU analysis, which supported establishment of the order *Plocamiales* for the family *Plocamiaceae* (SAUNDERS & KRAFT 1994). SAUNDERS & KRAFT (1996) then subjected a large suite of families from the former *Cryptonemiales/Gigartinales* complex to molecular analyses and concluded that there was strong support for a single order *Gigartinales* with the exception of the old cryptonemialean type family *Halymeniaceae* and gigartinalean *Sebdeniaceae*, which form a monophyletic group distinct from the remainder of the *Gigartinales* s.l. They thus proposed the new order *Halymeniales* for these two families.

The genus *Sarcodia* (*Sarcodiaceae*), also does not ally with the *Gigartinales*, and although its precise ordinal affinities are still to be determined, it currently shows weak affinities to the *Plocamiales* in SSU phylogenies (SAUNDERS & KRAFT, unpubl.). Preliminary indications (SAUNDERS, pers. obs.) are that the genera *Corynomorpha* (*Corynomorphaceae*) and *Tsengia* (*Nemastomataceae*) will also prove to be most closely affiliated to the *Halymeniales*. Of the remaining gigartinalean families for which the SSU data have been determined (ca. 27 of the remaining 37 or so), all appear to belong to a monophyletic assemblage with the possible exception of the *Schizymeniaceae*, which is inconsistently affiliated with Lineage 4 taxa on current analyses (SAUNDERS & KRAFT 1994, 1996, unpubl. data; SAUNDERS & BAILEY 1997).

The analyses of SAUNDERS & KRAFT (1996) provide support for continued recognition of the *Rhodymeniales* as distinct from the *Gigartinales*. In conflict with traditional taxonomic schemes is the apparent position of the *Rhodymeniales* relative to other orders, for its affinities lie with the *Halymeniales* in molecular phylogenies (FRESHWATHER & al. 1994, RAGAN & al. 1994, SAUNDERS & KRAFT 1996). Although this result may receive support from the non-fibrillar polysaccharides characteristic of the *Halymeniales* and *Rhodymeniales* (being neither "simple" agars nor carrageenans but a diverse collection of structures possibly including hybrids of the two informally termed "carragars", CHOPIN & al. 1994, MILLER & al. 1995, TAKANO & al. 1994, USOV & KLOCHKOVA (1992)), there is little apparent similarity in vegetative anatomy and, especially, carposporophyte ontogeny, to suggest a strong taxonomic link on classical grounds. The region of the phylogenetic tree (Fig. 2) in which these groups are rooted is one that requires more detailed investigation.

Recently, a proposal has been made to divide the core *Gigartinales* (i.e., those families that form a monophyletic assemblage in our SSU analyses) into three smaller orders (FREDERICQ & al. 1996). This foreshadowed move is based on *rbc*L data, but it runs counter to all present indications based on SSU results (SAUNDERS & KRAFT 1996, unpubl.). SSU phylogenies present a monophyletic *Gigartinales* that includes the proposed segregates (the *Dumontiales* and *Sphaerococcales*) rendering them superfluous at best or not supported at worst. In view of such conflicting evidence, it is our opinion that FREDERICQ & al.'s (1996) proposals should not be formalized until more data are forthcoming. We do not consider one gene inherently superior to another as a basis for phylogenetic determinations, but the data of 'equally informative' genes should be mutually supportive. We recommend that both data sets be carefully re-analyzed, the taxon sampling increased, and additional information obtained that will hopefully lead to a consensus before new orders are advocated.

Ceramiales. This order is widely regarded as the "most advanced", clearly circumscribed and firmly established of all the red algal orders (KRAFT 1981, BOLD & WYNNE 1985, GABRIELSON & GARBARY 1987, GARBARY & GABRIELSON 1990). Following KYLIN's (1928) removal of the *Bonnemaisoniaceae*, the four component families (the *Ceramiaceae, Delesseriaceae, Dasyaceae* and *Rhodomelaceae*) have maintained a taxonomic unity that has been repeatedly confirmed (GARBARY & GARBRIELSON 1990). The earliest molecular studies (RAGAN & al. 1994) questioned the monophyly of the *Ceramiales*, but these data have been reassessed and no longer support this proposal (SAUNDERS & al. 1996).

The *Ceramiales* of KYLIN (1956) are characterized by a suite of features that includes uniaxial structure, mostly tetrahedrally (rarely cruciately) divided tetrasporangia and, with few exceptions (GORDON-MILLS & KRAFT 1981), the production of an auxiliary cell on the supporting cell of the carpogonial branch after (and only after) fertilization of the carpogonium has taken place. This latter feature, in particular, is generally regarded as a synapomorphy (GARBARY & GABRIELSON 1987), but as such all that can be concluded is that it unites the included species in a monophyletic clade. It should not, a priori, be construed as evidence that the *Ceramiales* are the most derived florideophycidean order. The molecular data indicate that the *Ceramiales* diverge deep within Lineage 4 (Fig. 2), and there is no compelling evidence that this lineage is the culmination of florideophyte evolution. As pointed out by PUESCHEL (1994), the taxa of Lineage 2 are derived with regard to pit-plug ultrastructure, yet this fact alone does not render them the most advanced red algae. Until evidence contradicting the current indications of the molecular data are convincingly presented, we recommend that we cease referring to the *Ceramiales* and other members of Lineage 4 as "higher" florideophytes.

Nevertheless, given that the *Ceramiales* contains almost half the genera and over a third of all red-algal species (KRAFT & WOELKERLING 1990), including the largest single genus (*Polysiphonia*) and many other genera that occur nearly ubiquitously (e.g., *Ceramium, Antithamnion, Callithamnion, Herposiphonia, Laurencia*), the current lack of studies of the phylogenetic relationships between the component families, their tribes and their species, is a pointer to what will surely be an increasing preoccupation of molecular systematists. Our preliminary indications are that the greatest heterogeneity resides in the paraphyletic *Ceramiaceae* (Saunders & al. 1996).

Conclusions

In the 15 years since the introduction of ultrastructural markers of ordinal-level definitions by Pueschel & Cole (1982) and the 12 years since Kraft & Robins (1985) brought into question one of the classical Kylinian ordinal foundations, profound changes have taken place in studies of red-algal taxonomy and phylogeny. Applications of ultrastructural techniques, cladistic methods of character analysis and, most recently, molecular systematics have all contributed greatly to this transition. These tools are powerful in the information each can separately generate in the course of evolutionary investigations, and all are even more powerful when combined. The revolution and renaissance in red-algal systematics is already well underway, and the promise for productive future directions is manifest.

We thank Drs P. Gabrielson and M. Wynne for thoughtful reviews of this manuscript and A. Chiovitti for helpful discussions concerning red algal polysaccharides. During preparation of this review GWS was supported by funds from the Natural Sciences and Engineering Research Council of Canada.

References

Abbott, I. A., 1997: The red algal flora of Hawaii. – Honolulu: Bishop Museum and University of Hawaii Press (in press).

Arnold, C. A., 1948: Classification of gymnosperms from the viewpoint of paleobotany. – Bot. Gaz. **110**: 2–12.

Bhattacharya, D., Helmchen, T., Bibeau, C., Melkonian, M., 1995: Comparisons of nuclear-encoded small-subunit ribosomal RNAs reveal the evolutionary position of *Glaucocystophyta*. – Molec. Biol. Evol. **12**: 415–420.

Bold, H. C., Wynne, M. J., 1985: Introduction to the algae, 2nd edn. – Englewood Cliffs: Prentice-Hall.

Campbell, S. E., 1980: *Palaeoconchocelis starmachii*, a carbonate boring microfossil from the Upper Silurian of Poland (425 million years old): implications for the evolution of the *Bangiaceae* (*Rhodophyta*). – Phycologia **19**: 25–36.

– Kazmierczak, J., Golubic, S., 1979: *Paleoconchocelis starmachii* n. gen., n. sp., a Silurian endolithic rhodophyte (*Bangiaceae*). – Acta Palaeontol. Polonica **24**: 405–408.

Chemin, M. E., 1937: Le développement des spores chez les Rhodophyées. – Revue Gén. Bot. **49**: 300–327.

Chihara, M., Yoshizaki, M., 1972: *Bonnemaisoniaceae*: their gonimoblast development, life history and systematics. – In Abbott, I. A., Kurogi, M., (Eds): Contributions to the systematics of benthic marine algae of the North Pacific, pp. 243–251. – Kobe: Japanese Society of Phycology.

Chiovitti, A., Kraft, G. T., Saunders, G. W., Liao, M.-L., Bacic, A., 1995: A revision of the systematics of the *Nizymeniaceae* (*Gigartinales, Rhodophyta*) using classical and molecular markers. – J. Phycol. **31**: 153–166.

CHOPIN, T., HANISAK, M. D., CRAIGIE, J. S., 1994: Carrageenans from *Kallymenia westii* (*Rhodophyceae*) with a review of the phycocolloids produced by the *Cryptonemiales*. – Bot. Mar. **37**: 433–444.

CRAIGIE, J. S., 1990: Cell walls. – In COLE, K. M., SHEATH, R. G., (Eds): Biology of the red algae, pp. 221–257. – New York: Cambridge University Press.

DESIKACHARY, T. V., 1958: Taxonomy of algae. – Mem. Indian Bot. Soc. **1**: 52–62.

– 1963: Status of the order *Chaetangiales* (*Rhodophyta*). – J. Indian Bot. Soc. **42A**: 16–26.

DIXON, P. S., 1961: On the classification of the *Florideae* with particular reference to the position of the *Gelidiaceae*. – Bot. Mar. **3**: 1–16.

– 1973: Biology of the *Rhodophyta*. – Edinburgh: Oliver & Boyd.

EDWARDS, D., BALDAUF, J. G., BROWN, P. R., DORNING, K. J., FEIST, M., GALLAGHER, L. T., GRAMBAST-FESSARD, N., HART, M. B., POWELL, A. J., RIDING, R., 1993: Algae. – In BENTON, M. J., (Ed.): The fossil record **2**. – London, New York: Chapman & Hall.

ENTWISLE, T. J., NECCHI, O. Jr., 1992: Phylogenetic systematics of the freshwater red algal order *Batrachospermales*. – Japan. J. Phycol. **40**: 1–12.

FELDMANN, J., 1952: Les cycles de reproduction des algues et leurs rapports avec la phylogénie. – Rev. Cytol. Cytolphysiol. Vég. **13**: 1–49.

– 1953: L'évolution des organes femelles chez Floridées. – Proc. Int. Seaweed Symp. **1**: 11–12.

– FELDMANN, G., 1939a: Sur le développement des carpospores et l'alternance de générations de l'*Asparagopsis armata* HARVEY. – Compt. Rend. Acad. Sci. Paris, sér. D., **208**: 1240–1242.

– 1939b: Sur l'alternance de générations chez les Bonnemaisiacées. – Compt. Rend. Acad. Sci. Paris, sér. D, **208**: 1425–1427.

– 1942: Recherches sur les Bonnemaisoniacées et leur alternance de générations. – Ann. Sci. Nat. Bot. Ser., **11**: 75–175.

– MAZOYER, G., 1937: Sur l'identité de l'*Hymenoclonium serpens* (CROUAN) BATTERS et du protonéma du *Bonnemaisonia asparagiodes* (WOODW.) C. AG. – Compt. Rend. Acad. Sci. Paris, sér. D, **205**: 1084–1085.

FREDERICQ, S., HOMMERSAND, M. H., 1989: Proposal of the *Gracilariales* ord. nov. (*Rhodophyta*) based on an analysis of the reproductive development of *Gracilaria verrucosa*. – J. Phycol. **25**: 213–227.

– NORRIS, J. N., 1995: A new order (*Rhodogorgonales*) and family (*Rhodogorgonaceae*) of red algae composed of two tropical calciferous genera, *Renouxia* gen. nov. and *Rhodogorgon*. – Crypt. Bot. **5**: 316–331.

– ZIMMER, E. A., FRESHWATER, D. W., HOMMERSAND, M. H., 1996: Proposal of the *Dumontiales* ord. nov. and reinstatement of the *Sphaerococcales* SJOSTEDT emend. based on family complexes previously placed in the marine red algal order *Gigartinales*. – J. Phycol. **32** [Suppl.]: 16.

FRESHWATER, D. W., FREDERICQ, S., BUTLER, B. S., HOMMERSAND, M. H., CHASE, M. W., 1994: A gene phylogeny of the red algae (*Rhodophyta*) based on plastid *rbc*L. – Proc. Natl. Acad. Sci. USA **91**: 7281–7285.

FRITSCH, F. E., 1945: The structure and reproduction of the algae. II. Foreword, *Phaeophyceae, Rhodophyceae, Myxophyceae*. – Cambridge: Cambridge University Press.

GABRIELSON, P. W., GARBARY, D., 1986: Systematics of red algae (*Rhodophyta*). – CRC Crit. Rev. Pl. Sci. **3**: 325–366.

– 1987: A cladistic analysis of *Rhodophyta*: Florideophycidean orders. – Brit. Phycol. J. **22**: 125–138.

– GARBARY, D. J., SCAGEL, R. F., 1985: The nature of the ancestral red alga: inferences from a cladistic analysis. – BioSystems **18**: 335–346.

GARBARY, D. J., 1978: On the phylogenetic relationships of the *Acrochaetiaceae* (*Rhodophyta*). – Brit. Phycol. J. **13**: 247–254.

– GABRIELSON, P. W., 1987: *Acrochaetiales* (*Rhodophyta*): taxonomy and evolution. – Crypt. Algol. **8**: 241–252.

– 1990: Taxonomy and evolution. – In COLE, K. M., SHEATH, R. G., (Eds): Biology of the red algae, pp. 477–498. – New York: Cambridge University Press.

GLAZER, A. N., WEST, J. A., Chan, C., 1982; Phycoerythrins as chemotaxonomic markers in red algae: a survey. – Biochem. Syst. Ecol. **10**: 203–215.

GOFF, L. J., MOON, D. A., NYVALL, P., STACHE, B., MANGIN, K., ZUCCARELLO, G., 1996: The evolution of parasitism in the red algae: molecular comparisons of adelphoparasites and their hosts. – J. Phycol. **32**: 297–312.

GORDON-MILLS, E. M., KRAFT, G. T., 1981: The morphology of *Radiathamnion speleotis* gen. et sp. nov., representing a new tribe in the *Ceramiaceae* (*Rhodophyta*) from southern Australia. – Phycologia **20**: 122–130.

GUIRY, M. D., 1974: A preliminary consideration of the taxonomic position of *Palmaria palmata* (LINNAEUS) STACKHOUSE=*Rhodymenia palmata* (LINNAEUS) GREVILLE. – J. Mar. Biol. Assoc. U.K. **54**: 509–528.

– 1978: The importance of sporangia in the classification of the *Florideophycidae*. – In IRVINE, D. E. G., PRICE, J. H., (Eds): Modern approaches to the taxonomy of red and brown algae, pp. 111–144. – London: Academic Press.

HOMMERSAND, M. H., FREDERICQ, S., 1988: An investigation of cystocarp development in *Gelidium pteridifolium* with a revised description of the *Gelidiales* (*Rhodophyta*). – Phycologia **27**: 254–272.

– FREDERICQ, S., 1990: Sexual reproduction and cystocarp development. – In COLE, K. M., SHEATH, R. G., (Eds): Biology of the red algae, pp. 305–345. – New York: Cambridge University Press.

HORI, H., OSAWA, S., 1987: Origin and evolution of organisms as deduced from 5S ribosomal RNA sequences. – Molec. Biol. Evol. **4**: 445–472.

HUISMAN, J. M., 1985: The *Scinaia* assemblage (*Galaxauraceae, Rhodophyta*): a reappraisal. – Phycologia **24**: 403–418.

JOHANSEN, H. W., 1973: Ontogeny of sexual conceptacles in a species of *Bosiella* (*Corallinaceae*). – J. Phycol. **9**: 141–148.

KOHLMEYER, J., HAWKES, M. W., 1983: A suspected case of mycophycobiosis between *Mycosphaerella apophlaeae* (*Ascomycetes*) and *Apophlaea* spp. (*Rhodophyta*). – J. Phycol. **19**: 257–260.

KRAFT, G. T., 1981: *Rhodophyta*: morphology and classification. – In LOBBAN, C. S., WYNNE, M. J., (Eds): The biology of seaweeds, pp. 6–51. – Oxford: Blackwell.

– 1989: *Cylindraxis rotundatus* gen. et sp. nov. and its generic relationships within the *Liagoraceae* (*Nemaliales, Rhodophyta*). – Phycologia **28**: 275–304.

– 1992: Book review: Biology of the red algae, edited by COLE, K. M., SHEATH, R. G. – Phycologia **31**: 368–371.

– ROBINS, P. A., 1985: Is the order *Cryptonemiales* (*Rhodophyta*) defensible? – Phycologia **24**: 67–77.

– WOELKERLING, W. J., 1990: *Rhodophyta*. – In CLAYTON, M. N., KING, R. J., (Eds): Biology of marine plants, pp. 41–85. – Melbourne: Longman Cheshire.

KYLIN, H., 1923: Studien über die Entwicklungsgeschichte der Florideen. – Kongl. Sven. Sk. Vetenskapsakad. Handl. **63**: 1–139.

– 1928: Entwicklungsgeschichtliche Florideenstudien. – **24**(4): 1–127.

– 1931: Die Florideenordnung *Rhodymeniales.* – Acta Univ. Å Lund. **27**: 1–48.
– 1932: Die Florideenordnung *Gigartinales.* – Acta Univ. Å Lund. **28**(8): 1–88, Plates 1–28.
– 1956: Die Gattungen der Rhodophyceen. – Lund: Gleerups.
LEBEDNIK, P. A., 1977: Postfertilization development in *Clathromorphum, Melobesia* and *Mesophyllum* with comments on the evolution of the *Corallinaceae* and the *Cryptonemiales* (*Rhodophyta*). – Phycologia **16**: 379–406.
LIM, B. L., KAWAI, H., HORI, H., OSAWA, S., 1986: Molecular evolution of 5S ribosomal RNA from red and brown algae. – Japan. J. Genet. **61**: 169–176.
LLUISMA, A. O., RAGAN, M. A., 1995: Relationships among *Eucheuma denticulatum, Eucheuma isiforme* and *Kappaphycus alvarezii* (*Gigartinales, Rhodophyta*) based on nuclear SSU–rRNA gene sequences. – J. Appl. Phycol. **7**: 471–477.
MAGGS, C. A., PUESCHEL, C. M., 1989: Morphology and development of *Ahnfeltia plicata* (*Rhodophyta*): proposal of *Ahnfeltiales* ord. nov. – J. Phycol. **25**: 333–351.
– MCLACHLAN, J. L., SAUNDERS, G. W., 1989: Infrageneric taxonomy of *Ahnfeltia* (*Ahnfeltiales, Rhodophyta*). – J. Phycol. **25**: 351–368.
MAGNE, F., 1960: Sur le lieu de la méiose chez le *Bonnemaisonia asparagoides* (WOODW.) C. AG. – Compt. Rend. Acad. Sci. Paris, sér. D, **250**: 2742–2744.
– 1961: Sur le cycle cytologique du *Nemalion helminthoides* (VELLEY) BATTERS. – Compt. Rend. Acad. Sci. Paris, sér. D, **252**: 157–159.
– 1967: Sur l'existence, chez les *Lemanea* (Rhodophycées, Némalionales), d'un type de cycle de développement encore inconnu chez les Alges rouges. – Compt. Rend. Acad. Sci. (Paris), sér. D, **264**: 2632–2633.
– 1972: Le cycle de développement des Rhodophycées et son évolution. – Soc. Bot. France Mém. **1972**: 247–268.
MILLAR, A. J. K., SAUNDERS, G. W., STRACHAN, I. M., KRAFT, G. T., 1996: The morphology, reproduction and small-subunit rRNA gene sequence of *Cephalocystis* (*Rhodymeniaceae, Rhodophyta*), a new genus based on *Cordylecladia furcellata* J. AGARDH from Australia. – Phycologia **35**: 48–60.
MILLER, I. J., FALSHAW, R., FURNEAUX, R. H., 1995: Structural analysis of the polysaccharide from *Pachymenia lusoria* (*Cryptonemiaceae, Rhodophyta*): – Carbohydrate Res. **268**: 219–232.
NICHOLS, H. W., 1965: Culture and development of *Hildenbrandia rivularis* from Denmark and North America. – Amer. J. Bot. **52**: 9–15.
NORRIS, J. N., BUCHER, K. E., 1989: *Rhodogorgon*, an anomolous new red algal genus from the Caribbean Sea. – Proc. Biol. Soc. Washington **102**: 1050–1066.
OLIVEIRA, M. C., KURNIAWAN, J., BIRD, C. J., RICE, E. L., MURPHY, C. A., SINGH, R. K., GUTELL, R. R., RAGAN, M. A., 1995: A preliminary investigation of the order *Bangiales* (*Bangiophycidae, Rhodophyta*) based on sequences of nuclear small-subunit ribosomal RNA genes. – Phycol. Res. **43**: 71–79.
OLTMANNS, F., 1904–1905: Morphologie und Biologie der Algen, 1, 2. – Jena: G. Fischer.
– 1922: Morphologie und Biologie der Algen, 2nd edn. – Jena: G. Fischer.
PAPENFUSS, G. F., 1966: A review of the present system of classification of the *Florideophycidae*. – Phycologia **5**: 247–255.
PUESCHEL, C. M., 1982: Ultrastructural observations of tetrasporangia and conceptacles in *Hildenbrandia* (*Rhodophyta: Hildenbrandiales*). – Brit. Phycol. J. **17**: 333–341.
– 1987: Absence of cap membranes as a characteristic of pit plugs of some red algal orders. – J. Phycol. **23**: 150–156.
– 1988a: Cell sloughing and chloroplast inclusions in *Hildenbrandia rubra* (*Rhodophyta, Hildenbrandiales*). – Brit. Phycol. J. **23**: 17–23.

– 1988b: Secondary pit-connections in *Hildenbrandia* (*Rhodophyta, Hildenbrandiales*). – Brit. Phycol. J. **23**: 25–32.
– 1989: An expanded survey of the ultrastructure of red algal pit plugs. – J. Phycol. **25**: 625–636.
– 1994: Systematic significance of the absence of pit plug cap membranes in the *Batrachospermales* (*Rhodophyta*). – J. Phycol. **30**: 310–315.
– Cole, K. M., 1982: Rhodophycean pit plugs: an ultrastructural survey with taxonomic implications. – Amer. J. Bot. **69**: 703–720.
– Trick, H. N., Norris, J. N., 1992: Fine structure of the phylogenetically important marine alga *Rhodogorgon carriebowensis* (*Rhodophyta, Batrachospermales?*). – Protoplasma **166**: 78–88.

Ragan, M. A., Gutell, R. R., 1995: Are red algae plants? – Bot. J. Linn. Soc. **118**: 81–105.
– Bird, C. J., Rice, E. L., Gutell, R. R., Murphy, C. A., Singh, R. K., 1994: A molecular phylogeny of the marine red algae (*Rhodophyta*) based on the nuclear small-subunit rRNA gene. – Proc. Natl. Acad. Sci. USA **91**: 7276–7280.

Rosenvinge, L. K., 1931: The reproduction of *Ahnfeltia plicata*. – Biol. Meddel. Kongel. Danske Vidensk. Selsk. **10**: 1–29.

Saunders, G. W., Bailey, C. J., 1997: Phylogenesis of pit-plug associated features in the Rhodophyta: inferences from molecular systematic data. – Canad. J. Bot. **75**: (in press).
– Kraft, G. T., 1994: Small-subunit rRNA gene sequences from representatives of selected families of the *Gigartinales* and *Rhodymeniales* (*Rhodophyta*). 1. Evidence for the *Plocamiales* ord. nov. – Canad. J. Bot. **72**: 1250–1263.
– 1996: Small-subunit rRNA gene sequences from representatives of selected families of the *Gigartinales* and *Rhodymeniales* (*Rhodophyta*). 2. Recognition of the *Halymeniales* ord. nov. – Canad. J. Bot. **74**: 694–707.
– McLachlan, J. L., 1989: Taxonomic considerations of the genus *Rhodophysema* and the *Rhodophysemataceae* fam. nov. (*Rhodophyta, Florideophycidae*). – Proc. Nova Scotia Inst. Sci. **39**: 19–26.
– 1991: Morphology and reproduction of *Meiodiscus spetsbergensis* (Kjellman) gen. et comb. nov., a new genus of *Rhodophysemataceae* (*Rhodophyta*). – Phycologia **30**: 272–286.
– Bird, C. J., Ragan, M. A., Rice, E. L., 1995: Phylogenetic relationships of species of uncertain taxonomic position within the *Acrochaetiales/Palmariales* complex (*Rhodophyta*): inferences from phenotypic and 18S rDNA sequence data. – J. Phycol. **31**: 601–611.
– Strachan, I. M., West, J. A., Kraft, G. T., 1996: Nuclear small-subunit ribosomal RNA gene sequences from representative *Ceramiaceae* (*Ceramiales, Rhodophyta*). – Eur. J. Phycol. **31**: 23–29.

Schmitz, F., 1892: [6. Klasse *Rhodophyceae*] 2. Unterklasse *Florideae*. – In Engler, A., (Ed.): Syllabus der Vorlesungen über specielle und medicinish-pharmaceutische Botanik...Grosse Ausgabe, pp. 16–23. – Berlin: Borntraeger.

Schnepf, E., 1992: Electron microscopical studies of *Thorea ramosissima* (*Thoreaceae, Rhodophyta*): taxonomic implications of *Thorea* pit plut ultrastructure. – Pl. Syst. Evol. **181**: 233–244.

Searles, R. B., 1968: Morphological studies of red algae of the order *Gigartinales*. – Univ. Calif. Publ. Bot. **43**: 1–86.

Silva, P. C., Johansen, H. W., 1986: A reappraisal of the order *Corallinales* (*Rhodophyceae*). – Brit. Phycol. J. **21**: 245–254.

STEELE, K. P., HOLSINGER, K. E., JANSEN, R. K., TAYLOR, D. W., 1991: Assessing the reliability of 5S rRNA sequence data for phylogenetic analysis in green plants. – Molec. Biol. Evol. **8**: 240–248.

STOLOFF, L., 1962: Algal classification – an aid to improved industrial untilization. – Econ. Bot. **16**: 86–94.

TAKANO, R., NOSE, Y., HAYASHI, K., HARA, S., HIRASE, S., 1994: Agarose-carrageenan hybrid polysaccharides from *Lomentaria catenata*. – Phytochemistry **37**: 1615–1619.

TRICK, H. N., PUESCHEL, C. M., 1991: Cytochemical evidence for homology of the outer cap layer of red algal pit plugs. – Phycologia **30**: 196–204.

USOV, A. I., KLOCHKOVA, N. G., 1992: Polysaccharides of algae. 45. Polysaccharide composition of red seaweeds from Kamchatka coastal waters (Northwestern Pacific) studied by reductive hydrolysis of biomass. – Bot. Mar. **35**: 371–378.

VAN DER MEER, J. P., TODD, E. R., 1980: The life history of *Palmaria palmata* in culture. A new type for the *Rhodophyta*. – Canad. J. Bot. **58**: 1250–1256.

VON STOSCH, H. A., THEIL, G., 1979: New mode of life history in the freshwater red algal genus *Batrachospermum*. – Amer. J. Bot. **66**: 105–107.

WEST, J. A., HOMMERSAND, M. H., 1981: *Rhodophyta*: life histories. – In LOBBAN, C. S., WYNNE, M. J., (Eds): The biology of seaweeds, pp. 133–193. – Oxford: Blackwell.

WOMERSLEY, H. B. S., 1994: The marine benthic flora of Southern Australia. *Rhodophyta*. IIIA. *Bangiophyceae* and *Florideophyceae* (*Acrochetiales, Nemalieales, Gelidiales, Hildenbrandiales* and *Gigartinales* sensu lato). – Canberra: Australian Biological Resources Study.

– 1996: The marine benthic flora of Southern Australia. *Rhodophyta*. IIIB. *Gracilariales, Rhodymeniales, Corallinales* and *Bonnemaisoniales*. – Canberra: Australian Biological Resources Study.

Division *Glaucocystophyta*

Debashish Bhattacharya and Heiko A. Schmidt

Key words: *Glaucocystophyta*. – Actin, cyanelle, endosymbiosis, evolution, phylogeny, plastids, small subunit ribosomal DNA.

Abstract: The algal lineage *Glaucocystophyta* is distinguished by its cyanelles (=plastids). Cyanelles were once thought to represent recent endosymbioses involving coccoid cyanobacteria since these organelles lack chlorophyll-b or -c, contain phycobilins, phycobilisomes, carboxysomes and concentric (unstacked) thylakoids and retain a bounding peptidoglycan wall. Phylogenetic analyses of the nuclear- and plastid-encoded small subunit ribosomal DNA of four members of the *Glaucocystophyta* provide evidence for a monophyletic origin of the glaucocystophyte host cell within the eukaryotic "crown group" radiation. The cyanelles of the glaucocystophytes are also of a monophyletic origin within this lineage, are "advanced" plastids, and form the earliest divergence within phylogenetic analyses including sequences from all other major plastid groups (excluding the dinoflagellates) and the cyanobacteria.

The *Glaucocystophyta* (synonym *Glaucophyta,* Skuja 1954) are a small (ca. nine genera and 13 species) group of photosynthetic protists that includes flagellate, palmelloid and coccoid cells whose unique set of characters has led to their classification into a distinct division or phylum (Kies 1979, Kies & Kremer 1986, 1990). The evolutionary relationship of glaucocystophytes to other eukaryotes is unknown though the existence of a cruciate flagellar root system with multilayered structures (MLS) has suggested a relationship to prasinophyte green algae (Melkonian 1983) whereas plastid characters have been used to position glaucocystophytes as a sister group to red algae (*Biliphyta*, Cavalier-Smith 1982, 1987). *Cyanophora paradoxa*, the best-studied glaucocystophyte, has also been included in the *Cryptophyta* on the basis of cell morphology (Bourrelly 1970, Gillott 1990).

Glaucocystophyta and other cyanelle-containing taxa (e.g., *Paulinella chromatophora, Filosea*) have been closely associated with the theory of endosymbiosis since they are characterized by such 'primitive' plastid characters as the existence of only chlorophyll-a and phycobilins, phycobilisomes, carboxysomes and concentric (unstacked) thylakoids. *Paulinella chromatophora* was the first protist to be identified as containing a "cyanobacterial-like endosymbiont" as its photosynthetic organelle (Lauterborn 1895). Cyanelle characters within glaucocystophytes have been interpreted as supporting a relatively recent endosymbiosis between a modified coccoid cyanobacterium and a nonphotosynthetic eukaryote

(GEITLER 1959, HALL & CLAUS 1963). The existence of a peptidoglycan wall surrounding the cyanelles of all glaucocystophytes (and the photosynthetic organelle of *P. chromatophora*), except for *Glaucosphaera vacuolata* (KIES & KREMER 1990, KRAUS & al. 1990), further suggested that cyanelles must be of a recent origin. The cyanelles of some glaucocystophytes (i.e., *Cyanophora*, spp. *Glaucocystis nostochinearum*) were even raised to the rank of cyanobacterial taxa to reflect the distinctive ultrastructure and putative polyphyletic origin of these organelles (HALL & CLAUS 1963, 1967). This latter view regarding *Cyanophora* is no longer accepted since analyses of the cyanelle genome size in this taxon show it to be comparable to other plastids (HERDMAN & STANIER 1977; LÖFFELHARDT & al. 1997, this volume). Analyses of gene content, gene order and phylogeny [e.g., small subunit ribosomal DNA (SSU rDNA), rpoC1] of the *Cyanophora* cyanelle show this organelle to be more closely related to extant plastids than to cyanobacteria (BOHNERT & al. 1982, GIOVANNONI & al. 1988, DOUGLAS & TURNER 1991, PALENIK & HASELKORN 1992).

In order to delineate the phylogeny of the glaucocystophytes (sensu KIES & KREMER 1986) and to gain insights into the origin of the characters that define this algal division, the nuclear- and plastid-encoded SSU rDNA sequences have been determined from *Cyanophora paradoxa* KORSH. (KIES strain, Sammlung von Algenkulturen Göttingen, SAG B 45.84, SCHLÖSSER 1984), *Glaucocystis nostochinearum* ITZIGS. (SAG 45.88), *Glaucosphaera vacuolata* KORSH. (SAG B 13.82) and *Gloeochaete wittrockiana* LAGERHEIM (SAG B46.84, see also BHATTACHARYA & al. 1995, HELMCHEN & al. 1995). The nuclear- and plastid-encoded SSU rDNA nucleotide sequences, respectively, of the glaucocystophyte taxa have been released to the Genbank/EMBL/DDBJ databases with the following accession numbers; *Cyanophora* (X68483, X81840), *Glaucocystis* (X70803, X82496), *Glaucosphaera* (X81902, X81903), *Gloeochaete* (X81901, X82495). All other rDNA sequences analysed in this study are available from these sequence databases.

The single copy genomic actin gene of *Cyanophora* has also been characterized to position this taxon in the actin phylogeny (BHATTACHARYA & WEBER 1997). The *Cyanophora* actin sequence was found to contain five spliceosomal introns (four at novel positions in the actin intron catalogue, WEBER & KABSCH 1994). Actin is a multiple copy gene family in many eukaryotes (e.g., angiosperms, animals) but exists as a single copy in several protist lineages (e.g., green algae, red algae, ciliates) and in the fungi (BHATTACHARYA & al. 1991, BHATTACHARYA & EHLTING 1995). When in multiple copies, the actin gene is often a valuable tool for studying concerted gene evolution, gene function diversification and the evolutionary relationships of eukaryotic lineages (for review see SHETERLINE & SPARROW 1994). This last hypothesis is supported by the observation that all actin genes (whether in single or multiple copy) form lineage-specific monophyletic groups in phylogenies suggesting that actin gene duplications have occurred after the divergence of eukaryotic lineages (BHATTACHARYA & EHLTING 1995). That the archezoan, *Giardia lamblia*, contains a single-copy actin gene (DROUIN & al. 1995) is consistent with the hypothesis that all acting coding regions are descendants of a single sequence found in the ancestor of this primitive eukaryote. When in single copies actin may be used as a phylogenetic marker.

Actin coding regions used in the phylogenetic analyses, with the exception of the *Coleochaete scutata* and *Scherffelia dubia* sequences (Bhattacharya, unpubl.), are available from the Genbank/EMBL/DDBJ databases.

The glaucocystophyte nuclear lineage

SSU rDNA. The position of the glaucocystophyte host cell within the SSU rDNA phylogeny was determined using the maximum likelihood (pfastDNAml V. 2.0, Schmidt & al., unpubl.), maximum parsimony (PAUP V. 3.1.1, Swofford 1993) and LogDet (Lockhart & al. 1994) methods. Bootstrap resampling was done with each phylogenetic method (except the LogDet analysis) to test the stability of monophyletic groups within the trees. The phylogenies inferred from these methods are summarized in Fig. 1.

The SSU rDNA analyses provide evidence for an origin of the glaucocystophyte host cell within the eukaryote crown group radiation. The bootstrap analyses with the rDNA data set support the monophyly of the *Glaucocystophyta* as does the relatively long common branch length that unites them in the maximum likelihood (Fig. 1A) and LogDet analyses (Fig. 1B). All major algal groups (except euglenophytes) trace their origins to this radiation that has primarily been identified with SSU rDNA analyses (Sogin & al. 1986). The phylogenetic analyses are also consistent with a sister group relationship between glaucocystophytes and cryptophyte algae (see also Marin & al. 1996).

The monophyly of the glaucocystophyte/cryptophyte lineage was previously tested with user-defined tree analyses with the maximum likelihood method (Bhattacharya & al. 1995). Topologies were created with the RETREE program of PHYLIP (*Felsenstein* 1993) that address three alternate hypotheses regarding the evolutionary position of the glaucocystophytes, (1) as an independent lineage within the crown group radiation, (2) on the branch uniting them with the rhodophytes and, (3) on the branch uniting them with the chlorophytes. The log-likelihoods of these alternate trees and that of the 'best' tree (i.e., as a sister group to the cryptophytes, and see Bhattacharya & al. 1995: Fig. 1A) were compared with the likelihood ratio test (LRT, Kishino & Hasegawa 1989). The user-defined tree analyses were consistent with a monophyletic origin of the glaucocystophytes/ cryptophytes since the disruption of this clade resulted in a significantly 'worse' tree as did the positioning of the glaucocystophytes with either the red or green algae.

A sister group relationship between glaucocystophytes and cryptophytes is surprising since these protists do not share obvious morphological/ultrastructural characters (apart from the possession of flattened mitochondrial cristae, a character that is shared with red and green algae, animals/fungi). This relationship is taken as a preliminary result and must be further tested with protein sequence comparisons (e.g., actin, Bhattacharya, unpubl.). There are no host cell characters that unequivocally position the glaucocystophytes as a sister group to any other eukaryotic lineage (and, therefore, the confusing taxonomy). The crown group radiation is characterized by the divergence of many (presently) morphologically distinct groups; resolving the interrelationships of these lineages poses one of the most important challenges in molecular evolution. Our analyses also suggest that the

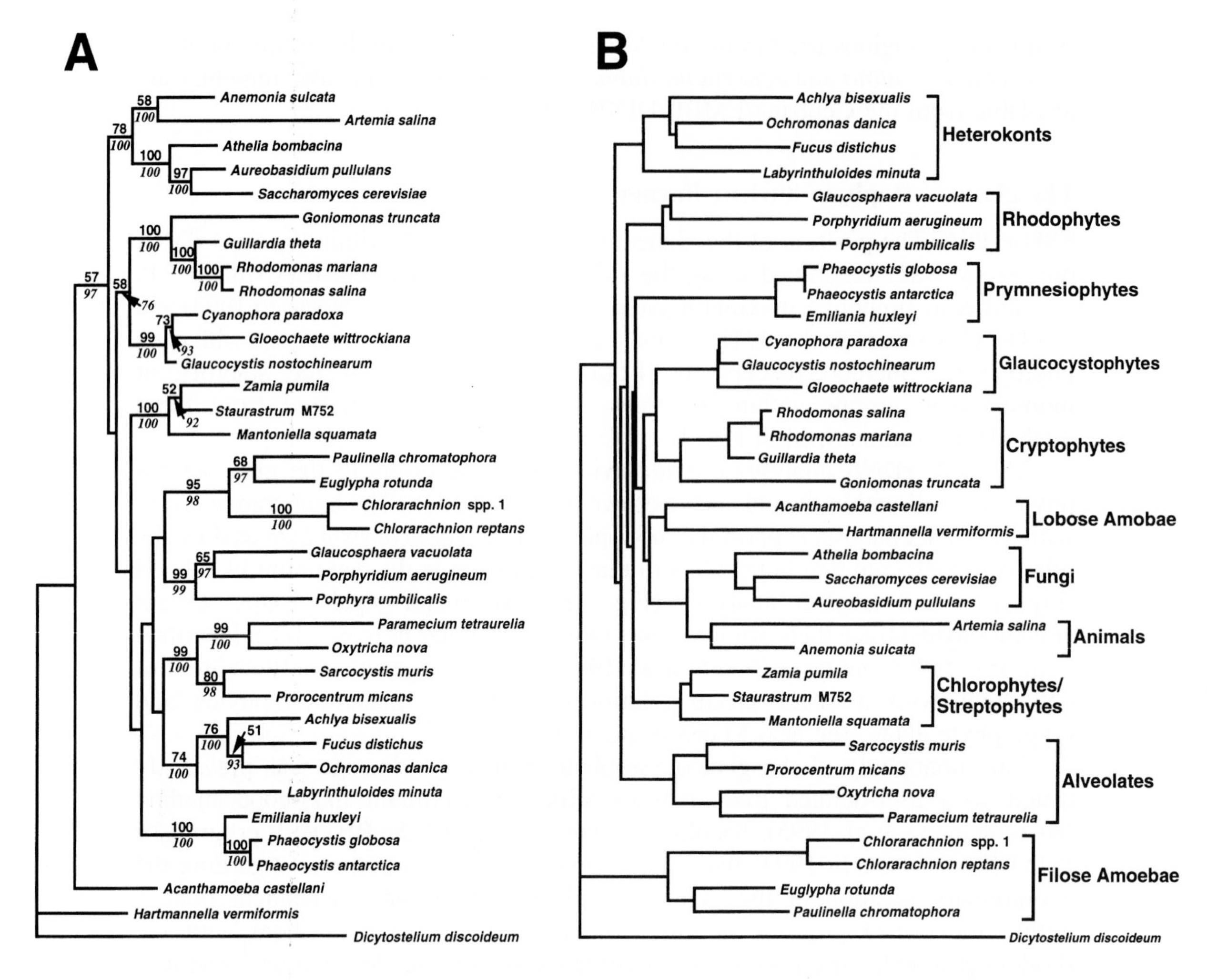

Fig. 1. Phylogeny of eukaryotes based on nuclear-encoded SSU rDNA sequence comparisons. *A* Tree inferred with the parallel maximum likelihood computer program (pfast DNAml V. 2.0, Schmidt & al., unpubl.) using 1565 aligned nucleotides. The bootstrap (Felsenstein 1985) values (100 replications) above the internal nodes are inferred from the maximum likelihood analysis whereas the bootstrap values (200 replications) shown below the internal nodes are inferred from a weighted (RC index over the interval 1–1000) maximum parsimony analysis (PAUP V. 3. 1. 1, Swofford 1993) using a heuristic search procedure with a branch-swapping algorithm (TBR, tree bisection-reconnection). The maximum parsimony tree has a consistency index = 0.648. Arrows are used to show bootstrap values where these do not fit on the branches. Only bootstrap values above 50% are shown. *B* Tree inferred with the neighbor-joining method (Saitou & Nei 1987) using a LogDet matrix (Lockhart & al. 1994) as input. Only parsimony sites (485 nt) were included in the LogDet analysis, gaps were excluded. These phylogenies are rooted within the branch leading to *Dictyostelium discoideum*

existence of MLS and cortical alveoli (a system of vesicles underlying the plasma membrane), found in all glaucocystophytes, do not alone distinguish this algal group. Both characters are presumably of a primitive origin and were found in the common ancestor of the crown group eukaryotes; MLS are also found in the zooflagellate *Jakoba* sp. and in dinoflagellates (O'KELLY 1992) and cortical alveoli are found in the dinoflagellates (included in the alveolates in Fig. 1A). There is no evidence for a specific relationship between glaucocystophytes and alveolates in the rDNA analyses.

The phylogenetic analyses do not position *Glaucosphaera* within the *Glaucocystophyta* but instead show that this taxon is a red alga. This is an important result since *Glaucosphaera* was hypothesized to be an evolutionary 'link' between the glaucocystophytes and the red algae on the basis of plastid characters (i.e., with loss of the peptidoglycan wall in the *Glaucosphaera* plastid, CAVALIER-SMITH 1982, 1987); rhodoplasts also have two envelope membranes and contain phycobilisomes and unstacked thylakoids. The positioning of *Glaucosphaera* in Fig. 1 as a member of the red algae is supported, however, by other morphological and biochemical data [e.g., presence of R-phycocyanin in the *Glaucosphaera* plastid (the glaucocystophyte cyanelles contain C-phycocyanin, RICHARDSON & BROWN 1970) and the complete absence of flagellae and basal bodies]. On the basis of electron microscopic studies, MCCRACKEN & al. (1980) described only one deeply lobed rhodoplast without a pyrenoid instead of several lens shaped cyanelles (KORSHIKOV 1930) in *Glaucosphaera*. These authors suggested placing *Glaucosphaera* in the red algal order *Porphyridiales*. Our phylogenetic analyses support the hypothesis of MCCRACKEN & al. (1980) and suggest that the following characters, together, may be used to characterize the *Glaucocystophyta*; presence of flattened microtubule-associated cortical vesicles (lacunae) under the plasma membrane, cruciate flagellar roots with associated MLS and a cyanelle bound by a peptidoglycan wall.

Actin. The actin phylogeny was inferred with the maximum likelihood and maximum parsimony methods (Fig. 2). The apicomplexan (*Plasmodium falciparum, Toxosplasma gondii*) actin sequences are used to root this phylogeny of crown group eukaryotes since this protist lineage represents an early divergence in actin phylogenies which use the actin-related proteins (products of actin gene duplications in the common ancestor of all eukaryotes) as the outgroup root (BHATTACHARYA, unpubl.). As shown above with the SSU rDNA sequence comparisons, phylogenetic analyses of actin coding regions are also consistent with the origin of many eukaryotes within a near simultaneous divergence (BHATTACHARYA & al. 1991, BHATTACHARYA & EHLTING 1995, DROUIN & al. 1995). Phylogenetic analyses of the *Cyanophora* single-copy actin coding region positions this sequence within this crown group radiation (Fig. 2). The sister-group relationship between the red and green algae in Fig. 2 has some bootstrap support in the maximum parsimony analysis (69%). This is an interesting result since the red and green algae possess so-called 'simple' plastids with 2-bounding membranes (see BHATTACHARYA & MEDLIN 1995). These plastids, along with the cyanelles, form a paraphyletic radiation of closely related lineages in SSU rDNA phylogenies (see below). That the host cells of the rhodophyte and chlorophyte plastids are also closely related in the actin trees is consistent with a monophyletic origin of the

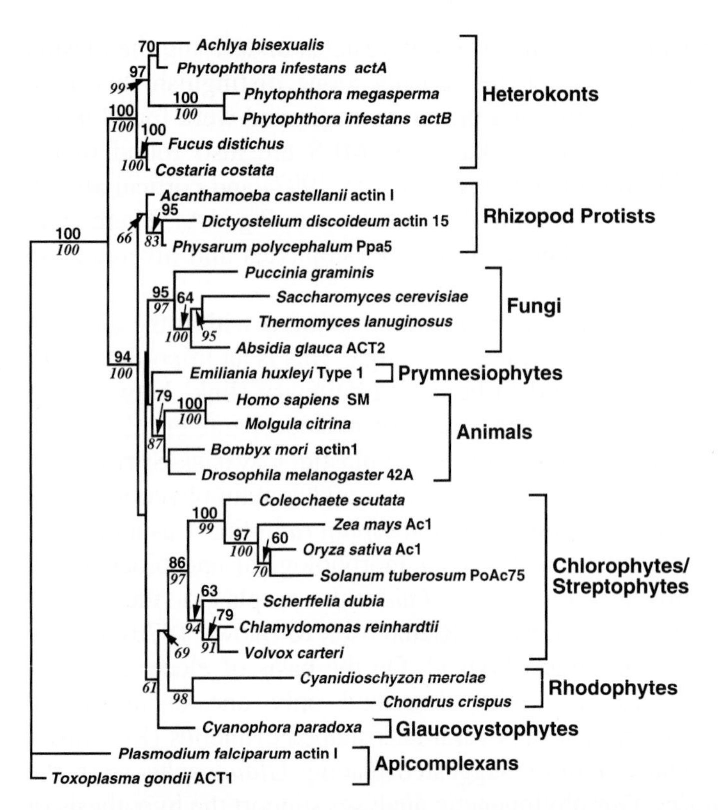

Fig. 2. Phylogeny of eukaryotes based on actin coding sequence comparisons. Tree inferred with the parallel maximum likelihood computer program (pfast DNAml bln. 2.0, Schmidt & al., unpubl.) using 744 aligned nucleotides. The bootstrap (Felsenstein 1985) values (100 replications) above the internal nodes are inferred from the maximum likelihood analysis whereas the bootstrap values (200 replications) shown below the internal nodes are inferred from a weighted (RC index over the interval 1–1000) maximum parsimony analysis (PAUP V. 3.1.1, Swofford 1993) using a heuristic search procedure with a branch-swapping algorithm (TBR, tree bisection-reconnection). Only bootstrap values above 60% are shown. The maximum parsimony bootstrap consensus phylogeny has a consistency index = 0.736. Arrows are used to show bootstrap values where these do not fit on the branch

plastids in these groups. Further actin sequences from glaucocystophytes, red and green algae and cryptophytes are required to test this interesting finding.

Phylogeny of the glaucocystophyte cyanelles

The phylogeny of the glaucocystophyte cyanelle SSU rDNA coding regions was determined using maximum likelihood, maximum parsimony and LogDet methods. These analyses are consistent with a monophyletic origin of the cyanelle in the common ancestor of the glaucocystophytes, *Cyanophora, Glaucocystis* and *Gloeochaete* (Fig. 3). Variation in cyanelle size and shape and ultrastructure (e.g., carboxysome shape, Hall & Claus 1967) within these taxa is not an indication of polyphyletic origin. The positioning of the *Glaucosphaera* plastid SSU rDNA as a

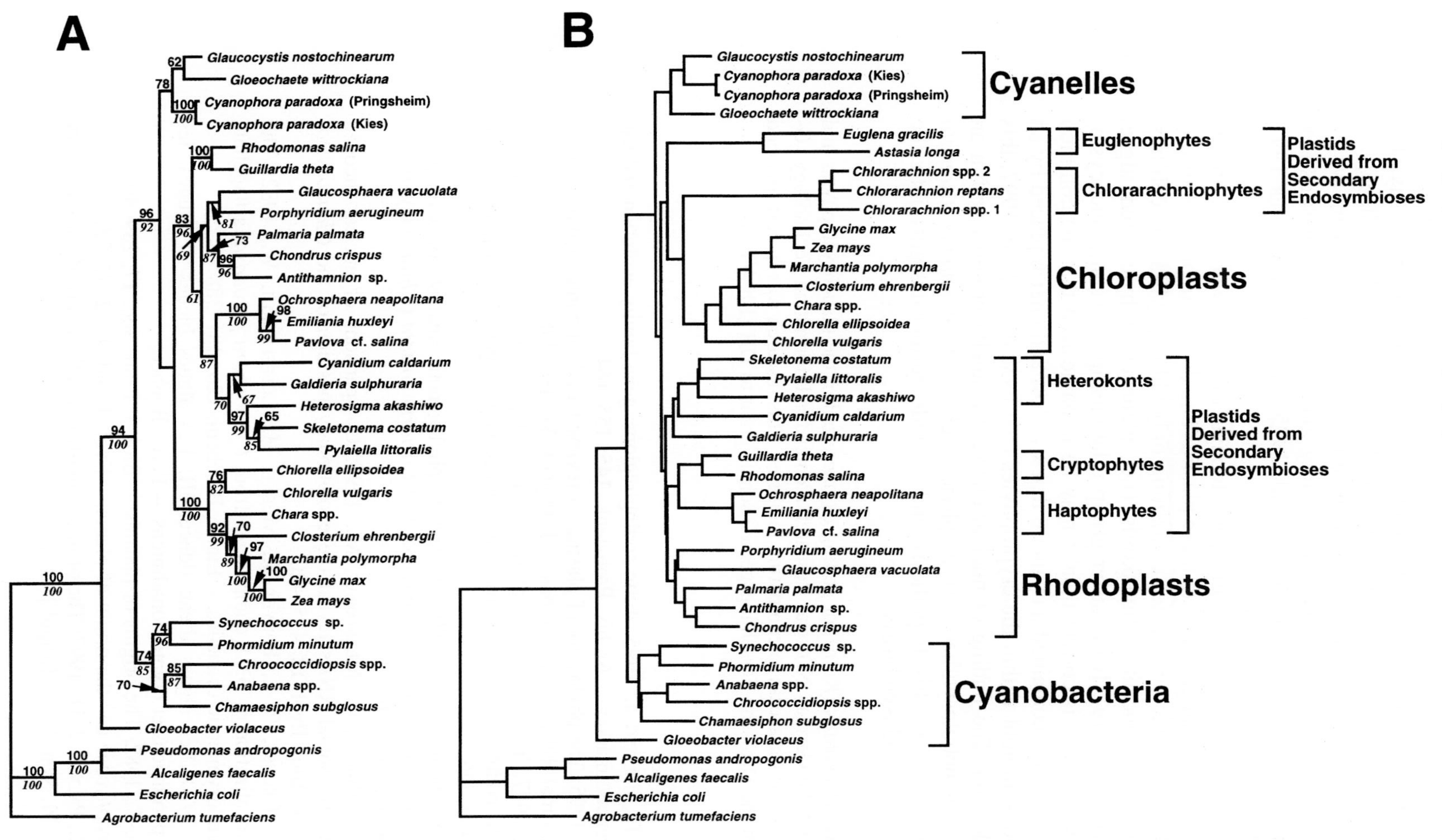

Fig. 3. Phylogeny of plastids based on SSU rDNA sequence comparisons. *A* Tree inferred with the parallel maximum likelihood computer program (pfast DNAml V. 2.0, SCHMIDT & al., unpubl.) using 1393 aligned nucleotides. The bootstrap (FELSENSTEIN 1985) values (100 replications) above the internal nodes are inferred from the maximum likelihood analysis whereas the bootstrap values (200 replications) shown below the internal nodes are inferred from a weighted (RC index over the interval 1–1000) maximum parsimony analysis (PAUP V. 3.1.1, SWOFFORD 1993) using a heuristic search procedure with a branch-swapping algorithm (TBR, tree bisection-reconnection). Only bootstrap values above 60% are shown. The maximum parsimony bootstrap consensus phylogeny has a consistency index = 0.632. Arrows are used to show bootstrap values where these do not fit on the branch. *B* Tree inferred with the neighbor-joining method (SAITOU & NEI 1987) using a LogDet matrix (LOCKHART & al. 1994) as input. Only parsimony sites (518 nt) were included in the LogDet analysis, gaps were excluded. These phylogenies are rooted within the branch leading to *Agrobacterium tumefaciens*

sister group to rhodoplasts is also consistent with the host cell phylogeny. The positioning of the glaucocystophyte cyanelles at the base of the plastid lineage in the maximum likelihood and LogDet phylogenetic analyses is consistent with a single loss of the peptidoglycan wall surrounding glaucocystophyte plastids (see BHATTACHARYA & MEDLIN 1995, HELMCHEN & al. 1995, DELWICHE & PALMER 1997, this volume, for further discussion).

The 'primitive' characters of the glaucocystophyte cyanelles were likely found in the cyanobacterium that gave rise to these plastids and do not support a common ancestry of photosynthetic eukaryotes containing only chlorophyll-a, phycobilins, phycobilisomes and concentric thylakoids (i.e. glaucocystophytes, red algae, *Paulinella chromatophora*). If the sister group relationship between glaucocystophytes and cryptophytes is correct then it may be hypothesized that the common ancestor of this lineage contained a cyanelle that was lost and later replaced by an eukaryotic endosymbiont within the cryptophytes (see FRAUNHOLZ & al. 1997, this volume, for details). The finding of nuclear-encoded cyanelle proteins within an early-diverging nonphotosynthetic cryptophyte such as *Goniomonas truncata* would lend credence to this hypothesis and provide further insights into the dynamics of plastid endosymbiosis.

We thank M. MELKONIAN (Köln) for getting us interested in the glaucocystophytes and in *Paulinella* and the German National Research Center for Information Technology (GMD, Sankt Augustin) for access to computing facilities. This research was in part financed by a grant from the Deutsche Forschungsgemeinschaft grant to DB (Bh 4/1-2).

References

BHATTACHARYA, D., EHLTING, J., 1995: Actin coding regions: gene family evolution and use as a phylogenetic marker. – Archiv. Protistenk. **145**: 155–164.

– MEDLIN, L., 1995: The phylogeny of plastids: a review based on comparisons of small subunit ribosomal RNA coding regions. – J. Phycol. **31**: 489–498.

– WEBER, K., 1977: The actin gene of the glaucocystophyte *Cyanophora paradoxa*: analysis of the coding region and introns, and an action phylogeny of eukaryotes. Curr. Genet. **31**: 439–446.

– STICKEL, S. K., SOGIN, M. L., 1991: Molecular phylogenetic analysis of actin genic regions from *Achlya bisexualis* (*Oomycota*) and *Costaria costata* (*Chromophyta*). – J. Molec. Evol. **33**: 535–536.

– HELMCHEN, T., BIBEAU, C., MELKONIAN, M., 1995: Comparisons of nuclear-encoded small-subunit ribosomal RNAs reveal the evolutionary position of the *Glaucocystophyta*. – Molec. Biol. Evol. **12**: 415–420.

BOHNERT, H. J., CROUSE, E. J., POUYET, J., MUCKE, H., LÖFFELHARDT, W., 1982: The subcellular localization of DNA components from *Cyanophora paradoxa*, a flagellate containing endosymbiotic cyanelles. – Eur. J. Biochem. **126**: 381–388.

BOURRELLY, P., 1970: Les algues d'eau douce. III: Les algues bleues et rouges, les eugleniens, peridiniens et cryptomonadiniens. – Paris: Boubée.

CAVALIER-SMITH, T., 1982: The origins of plastids. – Biol. J. Linn. Soc. **17**: 289–306.

– 1987: *Glaucophyceae* and the origin of plants. – Evol. Trends Pl. **2**: 75–78.

DELWICHE, C. R., PALMER, J. D., 1997: The origin of plastids and their spread via secondary symbiosis. – Pl. Syst. Evol., [Suppl.] **11**: 53–86.

DOUGLAS, S. E., TURNER, S., 1991: Molecular evidence for the origin of plastids from a cyanobacterium-like ancestor. – J. Molec. Evol. **33**: 267–273.

DROUIN, G., MONIZ DE SÁ, M., ZUKER, M., 1995: The *Giardia lamblia* actin gene and the phylogeny of the eukaryotes. – J. Molec. Evol. **41**: 841–849.

FELSENSTEIN, J., 1985: Confidence limits on phylogenies: an approach using the bootstrap. – Evolution **39**: 783–91.

– 1993: PHYLIP Manual, version 3.5c. – University of Washington: Department of Genetics.

GEITLER, L., 1959: Syncyanosen. – In RUHLAND, W., (Ed.): Handbuch der pflanzenphysiologie **11**, pp. 530–545. – Berlin Göttingen Heidelberg: Springer.

GILLOTT, M., 1990: Phylum *Cryptophyta* (Cryptomonads). – In Margulis, L., CORLISS, J. O., MELKONIAN, M., CHAPMAN, D. J., (Eds): Handbook of *Protoctista*, pp. 139–151. – Boston: Jones & Bartlett.

GIOVANNONI, S., TURNER, S., OLSEN, G., BARNS, S., LANE, D., PACE, N., 1988: Evolutionary relationships among cyanobacteria and green chloroplasts. – J. Bacteriol. **170**: 3584–3592.

HALL, W. T., CLAUS, G., 1963: Ultrastructural studies on the blue-green algal symbiont in *Cyanophora paradoxa* KORSHIKOFF. – J. Cell Biol. **19**: 551–563.

– – 1967: Ultrastructural studies on the cyanelles of *Glaucocystis nostochinearum* ITZIGSOHN. – J. Phycol. **19**: 37–51.

HELMCHEN, T., BHATTACHARYA, D., MELKONIAN, M., 1995: Analyses of ribosomal RNA sequences from glaucocystophyte cyanelles provide new insights into the evolutionary relationships of plastids. – J. Molec. Evol. **41**: 203–210.

HERDMAN, M., STANIER, R., 1977: The cyanelle: chloroplast or endosymbiotic procaryote?- FEMS Lett. **1**: 7–12.

FRAUNHOLZ, M. J., WASTL, J., ZAUNER, S., RENSING, S. A., SCHERZINGER, M. M., MAIER, U.-G., 1997: The evolution of cryptophytes. – Pl. Syst. Evol., [Suppl.] **11**: 163–174.

KIES, L., 1979: Zur systematischen Einordnung von *Cyanophora paradoxa*, *Gloeochaete wittrockiana* und *Glaucocystis nostochinearum*. – Ber. Deutsch. Bot. Ges. **92**: 445–454.

– KREMER, B. P., 1986: Typification of the *Glaucocystophyta*. – Taxon **35**: 128–133.

– – 1990: Phylum *Glaucocystophyta*. – In MARGULIS, L., CORLISS, J. O., MELKONIAN, M., CHAPMAN, D. J., (Eds): Handbook of *Protoctista*, pp. 152–166. – Boston: Jones & Bartlett.

KISHINO, H., HASEGAWA, M., 1989: Evaluation of the maximum likelihood estimate of the evolutionary tree topologies from DNA sequence data, and the branching order of the *Hominoidea*. – J. Molec. Evol. **29**: 170–179.

KORSHIKOV, A. A., 1930: *Glaucosphaera vacuolata*, a new member of the *Glaucophyceae*. – Archiv Protistenk. **70**: 217–222.

KRAUS, M., GÖTZ, M., LÖFFELHARDT, W., 1990: The cyanelle *str* operon from *Cyanophora paradoxa*: sequence analysis and phylogenetic implications. – Pl. Molec. Biol. **15**: 561–573.

LAUTERBORN, R., 1895: Protozoenstudien. II. *Paulinella chromatophora* nov. gen., nov. spec., ein beschalter Rhizopode des Süsswassers mit blaugrünen chromatophorenartigen Einschlüssen. – Z. Wiss. Zool. **59**: 537–544.

LOCKHART, P. J., STEEL, M. A., HENDY, M. D., PENNY, D., 1994: Recovering evolutionary trees under a more realistic model of sequence evolution. – Molec. Biol. Evol. **11**: 605–612.

LÖFFELHARDT, W., BOHNERT, H. J., BRYANT, D. A., 1997: The complete sequence of the *Cyanophora paradoxa* cyanelle genome (*Glaucocystophyaera*). – Pl. Syst. Evol., [Suppl.] **11**: 149–162.

MARIN, B., HELMCHEN, T., BHATTACHARYA, D., MELKONIAN, M., 1996: Analyses of ribosomal RNA sequences reveal the phylogeny of glaucocystophytes and the origin of plastids. – 1st Eur. Phycol. Cong. (Abstracts), Aug. 11–18, Cologne, p. 42.

MCCRACKEN, D. A., NADAKAVUKAREN, M. J., CAIN, J. R., 1980: A biochemical and ultrastructural evaluation of the taxonomic position of *Glaucosphaera vacuolata* KORSH. – New Phytol. **86**: 39–44.

MELKONIAN, M., 1983: Evolution of green algae in relation to endosymbiosis. In SCHENK, H. E. A., SCHWEMMLER, W., (Eds): Endocytobiology II, pp. 1003–1007. – Berlin: de Gruyter.

O'KELLY, C. J., 1992: Flagellar apparatus architecture and the phylogeny of "green" algae: chlorophytes, euglenoids, glaucophytes. – In MENZEL, D., (Ed.): The cytoskeleton of the algae, pp. 315–345. – Boca Raton: CRC Press.

PALENIK, B., HASELKORN, R., 1992: Multiple evolutionary origins of prochlorophytes, the chlorophyll b-containing prokaryotes. – Nature **355**: 265–267.

RICHARDSON, F. L., BROWN, T. E., 1970: *Glaucosphaera vacuolata*, its ultrastructure and physiology. – J. Phycol. **6**: 165–171.

SAITOU, N., NEI, M., 1987: The neighbor-joining method: a new method for reconstructing phylogenetic trees. – Molec. Biol. Evol. **4**: 406–425.

SCHLÖSSER, U. G., 1984: Sammlung von Algenkulturen: additions to the collection since 1982. – Ber. Deutsch. Bot. Ges. **97**: 465–475.

SHETERLINE, P., SPARROW, J. C., 1994: Actin. – Protein Profile **1**: 1–121.

SKUJA, H., 1954: *Glaucophyta*. – In MELCHER, H., WERDERMANN, E., (Eds): Syllabus der Pflanzenfamilien, p. 56. – Berlin: Borntraeger.

SOGIN, M. L., ELWOOD, H. J., GUNDERSON, J. H., 1986: Evolutionary diversity of eukaryotic small subunit rRNA genes. – Proc. Natl. Acad. Sci. USA **83**: 1383–1387.

SWOFFORD, D. L., 1993: PAUP: Phylogenetic analysis using parsimony, V3.1.1. – Washington, D. C.: Smithsonian Institution.

WEBER, K., KABSCH, W., 1994: Intron positions in actin genes seem unrelated to the secondary structure of the protein. – EMBO J. **13**: 1280–1286.

The complete sequence of the *Cyanophora paradoxa* cyanelle genome (*Glaucocystophyceae*)

W. Löffelhardt, H. J. Bohnert, and D. A. Bryant

Key words: *Glaucocystophyceae*, – *Cyanophora paradoxa*. Cyanelles, peptidoglycan, genome organization, endosymbiosis, protoplastid.

Abstract: The obligatorily autotrophic protist, *Cyanophora paradoxa*, harbors cyanelles, primitive plastids with the morphology and surrounding peptidoglycan sacculus of endosymbiotic cyanobacteria. The complete nucleotide sequence of the 135.6 kb cyanelle genome leaves no doubt about the plastid status of these unusual organelles. Peculiarities of genome organization that are found in cyanelles as well as in algal and higher plant chloroplasts support the hypothesis that all plastid types have a common origin (i.e., that only a singular primary endosymbiotic event occurred). The ancestral semiautonomous organelle, the protoplastid, would still have retained the prokaryotic wall in this scenario.

The two different isolates of *Cyanophora paradoxa:* strains or individual species?

Cyanophora paradoxa Korsh., first reported in 1924, has only been encountered again twice during the past 50 years (Kies 1992). The two independent isolates presently available, the Pringsheim- and the Kies-isolates, respectively, are indistinguishable in morphology and in vivo spectra. The structures of their unique peptidoglycan walls are also identical as evidenced by both the characteristic muropeptide patterns and the presence of an N-acetylputrescine substituent at the C-1 carboxyl group of the D- glutamyl residues (Pfanzagl & al. 1996a,b) as well as the size and number of their penicillin-binding proteins (Berenguer & al. 1987). However, they differ significantly in the size and restriction pattern of their plastid (cyanelle) DNAs, although the overall sequence identity is above 85% and the gene loci appear to be conserved (Löffelhardt & al. 1983, 1987; Breiteneder & al. 1988). In general, comparisons of plastid DNA restriction patterns in rhodophytes, chromophytes, and chlorophytes show little or no variation at the species level but considerable differences (much more pronounced than in higher plants) at the genus level (Coleman & Goff 1991). The 16S ribosomal RNA genes of the two *C. paradoxa* isolates deviate from each other in several positions (Helmchen & al. 1995) with an identity score of 97.4%. However, there is a gap of 13 bp in the alignment. Thus far, the genera grouped together as *Glaucocystophyceae* appear to be monophyletic (Kies 1992, Bhattacharya & al. 1995). Therefore it would be interesting to look for additional criteria, now at the level of the nuclear genome, that would allow a classification of the two isolates either as different strains of *C. paradoxa* or as different species of the genus *Cyanophora*.

The organization of the repetitive rDNA units has been used as a differentiating trait in the genus *Chlamydomonas* (MARCO & ROCHAIX 1981). In the case of *C. reinhardtii* and *C. callosa* the complete restriction pattern of the rDNA unit appeared to be identical supporting the proposed conspecificity of these two organisms. More distantly related species showed differences in size and restriction pattern of the internal transcribed spacer (ITS) as well as the intergenic spacer (IGS), whereas closely related species only differed with respect to the latter. With the two *Cyanophora* isolates, the restriction patterns in the coding regions and the ITS appeared to be identical, although they differed in the intergenic spacer that accounted for the size difference in the rDNA units: 4.3 kb for the Pringsheim-isolate and 7.4 kb for the Kies-isolate. The rDNA unit size was determined to be 10.4 kb and 13.4 kb for the Pringsheim- and Kies-isolates, respectively (D. ARYEE, unpubl.). Attempts to compare the chromosomal location of the rDNA in both strains using in situ hybridization were unsuccessful due to the small size of *Cyanophora* chromosomes (ARYEE 1989). Thus, we concentrated on the determination of the respective genome sizes via flow cytometry. The DNA content per nucleus was determined as 0.26 pg for the Pringsheim-strain and 0.28 pg for the Kies-strain, respectively (H. J. BOHNERT, unpubl.). This difference falls within the standard error. Thus, assuming 2C nuclei in *Cyanophora*, the genome sizes from both isolates correspond to approximately 140,000 kb. Upon screening of about 36,000 clones, 2–3% of positive signals have been obtained amounting to 3000–5000 kb. With a repeat size of 10 kb the copy number of the *Cyanophora* rRNA genes is estimated to be 300–500.

In summary, after application of two independent criteria to probe the relationship of the two isolates of *C. paradoxa*, we propose to ascribe the status of an individual species to each of them. In the following section the complete cyanelle sequence of the Pringsheim-isolate (555UTEX) is described; it will be interesting to see if the increment in cyanelle genome size of the Kies-isolate (1555) can account for about 10 additional genes that are nuclear-encoded in the Pringsheim-isolate.

The 135,599 bp cyanelle genome of *Cyanophora paradoxa* 555UTEX harbors 192 genes

The gene map of the completely sequenced cyanelle genome (STIREWALT & al. 1995) is shown in Fig. 1. Gene nomenclature follows the guidelines for chloroplast genes (HALLICK & BOTTOMLEY 1983, HALLICK 1989, HALLICK & BAIROCH 1994) or for (cyano)bacterial genes. The 192 identified genes and open reading frames (10 of which are duplicated) are approximately evenly distributed between the two strands. A representative list of genes, given in Table 1, contains more than 60 genes that are nuclear-encoded or missing in higher plants. This gene content is typical for "primitive" plastids, i.e. those from algae devoid of chlorophyll *b*. The cyanelle gene content falls between the 174 genes present on the 120 kb plastome from the diatom, *Odontella sinensis* (KOWALLIK & al. 1995), and the 251 genes found on the 191 kb plastome from the red alga *Porphyra purpurea* (REITH &

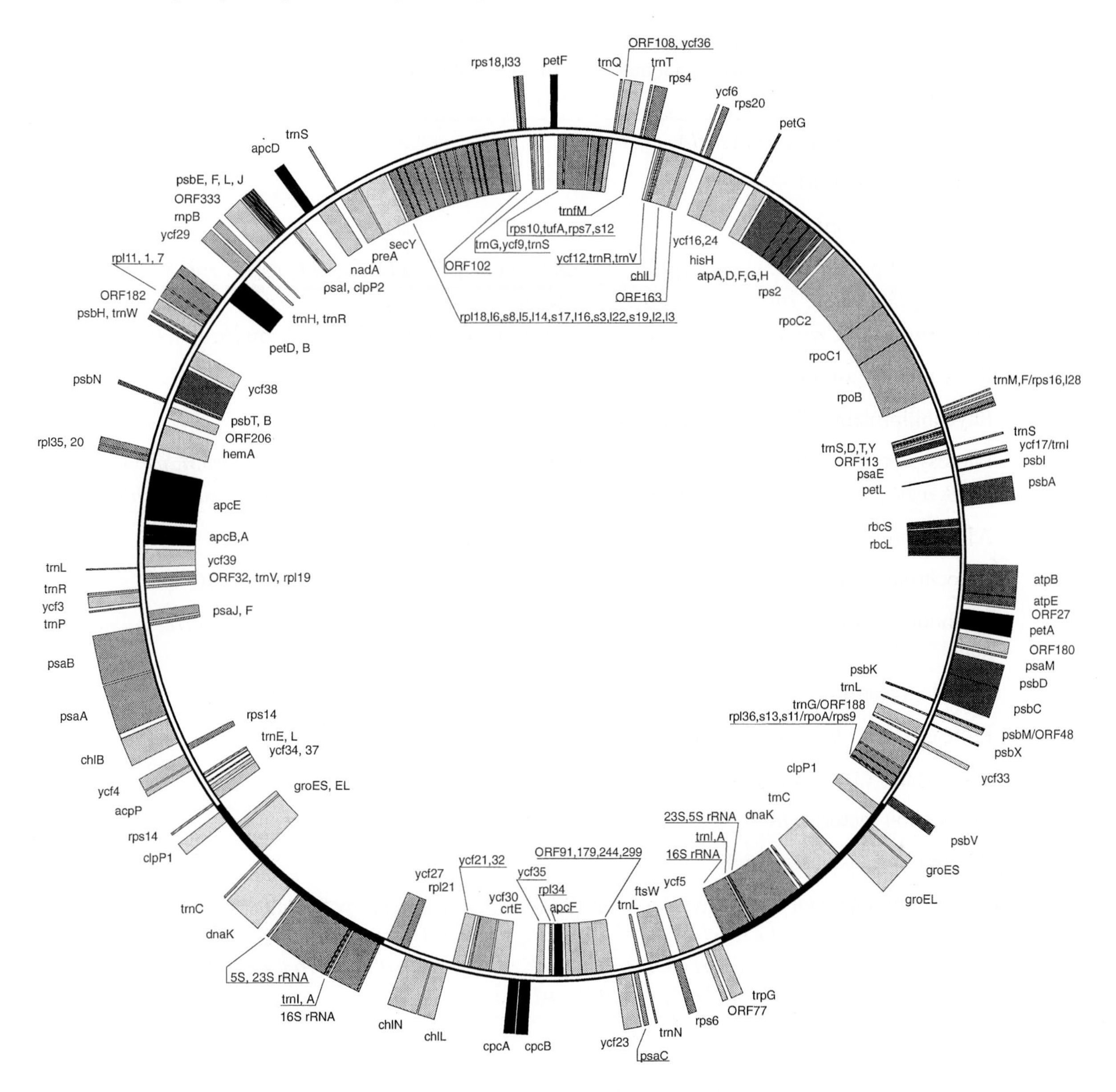

Fig. 1. Gene map of the cyanelle genome from *Cyanophora paradoxa* UTEX555. The filled region between the two circles indicates the inverted repeat. The direction of transcription is clockwise for the genes on the outer circle

MUNHOLLAND 1995). With some exceptions, e.g., the *ndh* genes (missing from all algal plastomes investigated thus far), *atpI*, or *accD*, the cyanelle genome contains the standard set of chloroplast-encoded genes. In addition the cyanelle genome encodes many more ribosomal proteins, several enzymes involved in anabolic pathways other than photosynthesis, chaperones, (putative) transcription factors, and components of ABC transporters (Table 1). In higher plants most chloroplast-

Table 1. Cyanelle genes from *Cyanophora paradoxa*. Genes marked with an asterisk are not found on other plastid genomes. Genes underlined are absent from the chloroplast genomes of higher plants

Ribosomal RNAs (3): *rrsA*, *rrlA*, *rrfA*

Transfer RNAs (36)

Other RNAs: *rnpB*

Ribosomal proteins (37): *rpl1*, *rpl2*, *rpl3*, *rpl5*, *rpl6*, *rpl7*, *rpl11*, *rpl14*, *rpl16*, *rpl18*, *rpl19*, *rpl20*, *rpl21*, *rpl22*, *rpl28*, *rpl33*, *rpl34*, *rpl35*, *rpl36*, *rps2*, *rps3*, *rps4*, *rps5*, *rps6*, *rps7*, *rps8*, *rps9*, *rps10*, *rps11*, *rps12*, *rps13*, *rps14*, *rps16*, *rps17*, *rps18*, *rps19*, *rps20*

RNA polymerase (4): *rpoA*, *rpoB*, *rpoC1*, *rpoC2*

Phycobiliproteins (7): *apcA*, *apcB*, *apcD*, *apcE*, *apcF*, *cpcA*, *cpcB*

Photosystem I and II proteins (24): *psaA*, *psaB*, *psaC*, *psaE*, *psaF*, *psaI*, *psaJ*, *psaM*, *psbA*, *psbB*, *psbC*, *psbD*, *psbE*, *psbF*, *psbH*, *psbI*, *psbJ*, *psbK*, *psbL*, *psbM*, *psbN*, *psbT*, *psbV*, *psbX*

ATPase proteins (7): *atpA*, *atpB*, *atpD*, *atpE*, *atpF*, *atpG*, *atpH*

Cytochrome and ferredoxin proteins (7): *petA*, *petB*, *petD*, *petL*, *petX*, *petF*, *petG*

Anabolic enzymes (13): *rbcL*, *rbcS*, *chlB*, *chlI*, *chlL*, *chlN*, *acpP*, *nadA**, *preA*, *crtE**, *hemA**, *hisH*, *trpG*

Peptidoglycan biosynthesis: *ftsW**

Proteases (2): *clpP1*, *clpP2*

Chaperones (3): *dnaK*, *groEL*, *groES**

Translation factor: *tufA*

Preprotein translocase: *secY*

ORFs with unknown or putative function (37): *ycf3*, *ycf4*, *ycf5*, *ycf6*, *ycf7*, *ycf9*, *ycf12*, *ycf16*, *ycf17*, *ycf21*, *ycf23*, *ycf24*, *ycf27*, *ycf29*, *ycf30*, *ycf32*, *ycf33*, *ycf34*, *ycf35*, *ycf36*, *ycf37*, *ycf38*, *ycf39*, *orf27*, *orf48*, *orf77*, *orf91*, *orf102*, *orf108*, *orf163*, *orf179*, *orf180*, *orf182*, *orf188*, *orf206*, *orf244**, *orf299**, *orf333*

localized biosynthetic enzymes are nuclear-encoded, thereby concealing the essentially prokaryotic nature of these proteins, mechanisms and pathways. This is immediately apparent when the genomes of primitive plastids like the cyanelles are surveyed (Table 2). The endosymbiotic event resulted initially in a redundancy of biosynthetic pathways. In the case of fatty acid biosynthesis and (partially) amino acid biosynthesis, the host cell seems to have abandoned its own set of enzymes. The plastid proteins show higher sequence similarity to prokaryotic counterparts than to the multiple forms eventually occurring in the cytoplasm.

Table 2. Anabolic enzymes encoded by cyanelles. *Inferred from significant sequence similarity to *ispB* (octaprenylpyrophosphate synthase) of *Escherichia coli* (Asai & al. 1994). Putative function in plastoquinone side chain biosynthesis. This gene is also present on the *Porphyra purpurea* plastome

Gene	Pathway	Function
acpP	fatty acid biosynthesis	acyl carrier protein
chlB	chlorophyll biosynthesis	protochlorophyllide reductase subunit
chlL	– ,, –	– ,, –
chlN	– ,, –	– ,, –
chlI	– ,, –	Mg-Protoporphyrin IX chelatase
hemA	chlorophyll & heme biosynthesis	glutamyl-tRNA reductase
crtE	isoprenoid pathway	geranylgeranyl pyrophosphate synthase
preA	– ,, –	nonaprenylpyrophosphate synthase*
nadA	$NAD(P)^+$ biosynthesis	quinolinate synthase
hisH	amino acid biosynthesis	N-Acetylglutamate-gamma-semialdehyde dehydrogenase
trpG	– ,, –	anthranilate synthase component II

Features of genome organization

Inverted repeats. The most conspicuous feature of the gross organization of the cyanelle genome is the 11.3 kb inverted repeat (IR), which corresponds to about half the size of higher plant chloroplast IRs. Another obvious feature is the small intergenic spacer regions between cyanelle genes. In a few cases (*orf299/orf244, ycf16/ycf24, atpD/atpF, psbD/psbC*) adjacent genes have been found to overlap by 3 to 16 bp. Moreover, only few non-coding regions that extend over several hundred bp are observed. Only a single intron has been identified: the 232 bp group I intron in the anticodon loop of $trnL^{UAA}$. These three features explain why cyanelles can encode around 50 genes more than higher plant chloroplasts which have slightly larger genomes. About half of the IR regions are occupied by the two rDNA units. In cyanelles and in the chloroplasts of *Chlamydomonas reinhardtii* the 16S rDNA 5′ ends are proximal to the small single copy region whereas in the other cases they are oriented such that they are proximal to the large single copy region. The rDNA spacer is small, as is typical for algae that lack chlorophyll *b*, and harbors *trnI* and *trnA* as in most plastids and prokaryotes (Löffelhardt & Bohnert 1994a, b).

Transfer RNAs. Cyanelles encode 36 tRNAs for all 20 standard amino acids, sufficient by far for the needs of in organello protein biosynthesis and exceeding by four to five the number usually found in higher plant chloroplasts (Sugiura1992). In most cases tRNA genes are not clustered with the exception of the putative transcription unit *trnS, trnD, trnT,* and *trnY.* The transcription of two genes, *trnH* and *trnR*, flanking *rnpB* on the opposite strand was studied and found to correspond to the organization in prokaryotes with respect to e.g., promoters, termination signals. (Baum & Schön 1996). The *rnpB* gene, also present on the *Porphyra* *purpurea* plastome, specifies the essential RNA component of RNaseP (SHEVELEV & al. 1995).

Promoters. In several cases, especially for genes or transcription units in the IR, but also for *psbA* and the *str* operon, putative promoter motifs can be observed that are similar in both sequence and spacing to the canonical sequences from *Escherichia coli* and other eubacteria. Based upon the sequence similarity of sigma factors from all eubacteria, including the nuclear-encoded but chloroplast-targeted RpoD homologs of red algae (LIU & TROXLER 1996), it has been suggested that promoter motifs in all bacteria should be quite similar in spacing and sequence to the consensus promoters of *E. coli* (GRUBER & BRYANT 1997). Although regulation at the level of transcription is known to occur in cyanobacteria (CURTIS & MARTIN 1994), it is possible that in cyanelles, as in chloroplasts, gene expression is predominantly regulated by post-transcriptional events, including translational regulation. However, it should be noted that three cyanelle ORFs (*ycf27, ycf29,* and *ycf30*), that are conserved among primitive plastid genomes, show significant sequence similarity to prokaryotic transcription regulatory factors of the OmpR and LysP classes. The occurrence of these putative regulators suggests that some transcriptional regulation occurs in cyanelles, although the targets of this regulation are not known at present. The good preservation of canonical promoters in the IR might be due to copy correction through intramolecular recombination (BOHNERT & LÖFFELHARDT 1982), a mechanism which probably operates upon non-coding regions as well. For other putative transcription units, only weak consensus promotor sequences could be recognized; rarely promoters for the prokarotic-type RNA polymerase seem to be missing altogether. In these rare instances, the corresponding genes might be transcribed by the second, imported, "eukaryotic" RNA polymerase of plastids (LINK 1996).

Shine-Dalgarno sequences. Many genes show short poly-purine stretches complementary to the 3′ end of the cyanelle 16S rRNA (-CCUCCUUU-3′OH) at a distance of 7 to 12 bases upstream of the initiation codon. Typical ribosome binding sites (RBS) are AAGG, AGGA, GGAG, GAGG. In a few cases the RBS are further upstream as observed for *psbA* (36 b) or completely missing. Within transcription units this might be indicative of a "relay race" type translation.

Transcription units. The gene arrangements observed suggest a predominance of polycistronic transcripts (LÖFFELHARDT & BOHNERT 1994a, b). Proven cases of cotranscription are given in Table 3. As in chloroplasts and cyanobacteria, processing of the primary transcripts to smaller mRNAs seems to be rather common (MICHALOWSKI & al. 1990, KRAUS & al. 1990).

Transcription units common among all plastid types and cyanobacteria. The organization of plastid ribosomal protein genes in general reflects their prokaryotic origin (SUGIURA 1992). This is even more pronounced with primitive plastids due to the higher number of genes (and transcription units) retained. The large ribosomal protein gene cluster found in all plastid types, although sometimes split into parts and with variations in gene content but never in gene order, is similar to that found in cyanobacteria such as *Synechocystis* sp. PCC 6803: there it comprises 29 genes, including 25 genes for ribosomal proteins (KANEKO & al. 1996). Compared with *Escherichia coli* (LINDAHL & ZENGEL 1986), the adjacent "S10", *spc,* and α operons are fused with the distant L13 operon and the single

Table 3. Transcription units ($5' \rightarrow 3'$) among cyanelle genes as detected by northern hybridization

Genes contained	Transcript size (bases)	Reference
atpB, atpE	2800	Löffelhardt & al. (1987)
psaA, psaB	6000	– ,, –
psbC, psbD	3200	– ,, –
petB, petD	1400	– ,, –
cpcB, cpcA	1350	Lemaux & Grossman(1985)
apcA, apcB	1200	– ,, –
rbcL, rbcS	2600	Starnes & al. (1985)
rpl33, rps18	1100	Evrard & al. (1990)
rps12, rps7, tufA, rps10	4200	Kraus & al. (1990)
rpl3, rpl2, rps19, rpl22, rps3, rpl16, rps17, rpl14, rpl5, rps8, rpl6, rpl18, rps5	7500	Michalowski & al. (1990) Evrard & al. (1990)

gene *rpl31*. Interestingly, the genes *rps4* and *rps14* found in the *E. coli* α and *spc* operons, respectively, are missing in the cyanobacterial cluster and are positioned singly elsewhere in the genome, as is the case for plastid DNAs (Fujishiro & al. 1996). In cyanelles, the large ribosomal protein gene cluster is split into two parts comprising in total 19 genes (ten genes found in the *Synechocystis* sp. PCC 6803 cluster have been transferred to the nucleus). A rearrangement, with the breakpoint near the end of the *spc*-derived portion, resulted in the organization shown in Fig. 1. In cyanobacteria, cyanelles and higher plant chloroplasts the (rudimentary) *str* operon is not contained in the large cluster. The largest r-protein gene clusters found among plastids are those of *Porphyra purpurea* (Reith & Munholland 1995) and *Odontella sinensis* (Kowallik & al. 1995) containing 30 and 29 genes, respectively, including the *str* genes. The location of *rps10* 3′ to *tufA* in the cyanelle genome is paralleled in *Synechocystis* sp. PCC 6803 and the plastid genomes of rhodophyte and chromophyte algae (Reith 1995, Douglas 1994) and differs from the arrangement found in *Escherichia coli*. The transcription unit *rpl33-rps18* is common among plastids and cyanobacteria, whereas in *E. coli* the genes *rpl28* and *rpl33* form an operon (Lindahl & Zengel 1986). Yet another case where cyanobacteria differ from the genome organization found in heterotrophic Gram-negative bacteria is the presence of "outsider" genes in the r-protein operons. Examples are *infA* (initiation factor I) and *adk* (adenylate kinase), respectively, which have also been detected in an analogous environment in the large r-protein gene cluster of *Bacillus subtilis* (Suh & al. 1996, Yasumoto & al. 1996). The *infA* gene is also present in the large cluster of higher plant chloroplasts, but is absent from the plastid genomes of *Cyanophora paradoxa*, *Porphyra purpurea*, *Odentella sinensis*, and *Euglena gracilis* (Hallick & al. 1993).

The *psaA-psaB* genes, encoding the largest subunits of the photosystem I reaction center, form a dicistronic transcription unit in all oxygenic phototrophs investigated, with the exception of the highly rearranged plastome of *Chlamydomonas reinhardtii*. The same applies for the photosystem II genes *psbD-psbC*. The genes *psbE-psbF-psbL-psbJ* constitute a further ubiquitous operon. The *orf333* uptream from cyanelle *psbE* is found is this position in cyanobacteria, too. ORF333 is the product of a nuclear gene in *Arabidopsis thaliana* and is absolutely required for the assembly of functional PS II units (P. Westhoff, pers. comm.).

Transcription units common with other primitive plastids and cyanobacteria. The *rbcL* and *rbcS* genes, encoding the large and small subunits of Rubisco, form an operon in all cyanobacteria and algae devoid of chlorophyll *b*. An exception to this rule are dinoflagellates, for which a nuclear location for *rbcL* was recently reported, but this Rubisco is a Form II enzyme which has no small subunits (Morse & al. 1995). In this phylum the situation is complicated by the occurrence of nonphotosynthetic forms that might have acquired cleptochloroplasts (Whatley 1993). Cyanelle DNA harbors transcription units encoding components of the phycobilisome core (*apcE-apcA-apcB*) and phycobilisome peripheral rods (*cpcB-cpcA*), as well as individual genes *apcF* and *apcD* encoding core components. In cyanobacteria, the *apcC* gene (allophycocyanin linker polypeptide) is typically found downstream from *apcA-apcB,* and although the *apcE* gene is often located upstream from *apcA*, this is not always the case (Bryant 1991). The prokaryotic r-protein operons *rpl11-rpl1-rpl12* and *rpl35-rpl20* are also conserved on the genomes of *Cyanophora paradoxa*, *Porphyra purpurea* and *Odontella sinensis*. A pair of cyanelle genes for ABC transporter components, *ycf16-ycf24*, shows high identity scores to the respective red algal and diatom plastid genes. The putative transcription unit *chlL-chlN*, specifying enzymes of chlorophyll biosynthesis (Table 2), also occurs on the *P. purpurea* plastome and on cyanobacterial genomes.

Transcription units shared with various other plastid types but not with cyanobacteria. The strongest indication for a common origin of all plastid types is the gene cluster shown in Fig. 2. A single gene (*rps20*) and three transcription units (*rpoBC1C2, rps2-tsf*, and *atpIHFGDA*(*C*)) that are widely separated on cyanobacterial genomes seem to have been fused together after the endosymbiotic event. This cluster is also present with some variation in gene content, but not in gene order, in cyanelles and rhodoplasts as well as in higher plant chloroplasts. While the large r-protein gene cluster could have arisen after multiple endosymbiotic events provided a cyanobacterium was the invader in each case, the existence of this "diagnostic" cluster in plastids of different evolutionary levels can only be explained when a singular primary endosymbiotic event is assumed (Kowallik 1994; Löffelhardt & Bohnert 1994a, b; Löffelhardt 1995; Löffelhardt & al. 1997; Reith 1995). In *Odontella sinensis*, this cluster is bipartite and it is com-

Cyanobacteria *rps20* 5'-*rpoBC1C2*-3' 5'-*rps2-tsf*-3' 5'-*atpIHGFDAC*-3'

P. purpurea 5'-*rps20-rpoB-rpoC1-rpoC2-rps2-tsf-atpI-atpH-atpG-atpF-atpD-atpA*-3'
atpC

C. paradoxa 5'-*rpoB-rpoC1-rpoC2-rps2-atpH-atpG-atpF-atpD-atpA*-3'
rps20 **tsf,atpI** **atpC**

Higher plants 5'-*rpoB-rpoC1-rpoC2-rps2-atpI-atpH-atpF-atpA*-3'
rps20 **tsf** **atpG** **atpD** **atpC**

Fig. 2. Gene clusters diagnostic for plastid evolution. Genes indicated in boldface below the respective plastid operons have been transferred to the nuclear genome

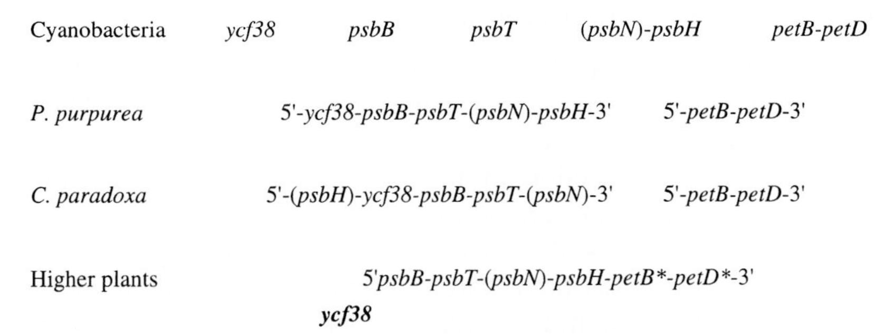

Fig. 3. Arrangement of *ycf38, psbB, psbT, psbN, psbH, petB* and *petD* genes in cyanobacteria, *Porphyra purpurea*, *Cyanophora paradoxa*, and higher plant chloroplast genomes. The parentheses indicate that *psbN*, and *psbH* for *C. paradoxa*, are transcribed from the complementary strand from the other genes. The asterisks after *petB* and *petD* genes of higher plant chloroplasts indicate the presence of introns in these genes. Genes indicated in boldface below the respective plastid operons have been transferred to the nuclear genome

pletely disintegrated in *Chlamydomonas reinhardtii*, which shows that there is no particular selection pressure to maintain or to reach this kind of gene arrangement.

Another remarkable example of a conserved gene arrangement found in many plastid genomes is the arrangement of the *psbB-psbT-(psbN)-psbH-petB-petD* genes (see Fig. 3). In higher plants, these genes form a cluster in the order written, although the *psbN* gene is transcribed from the opposite strand with respect to the other genes (indicated by parentheses in Fig. 3). In *Cyanophora paradoxa* and *Porphyra purpurea*, the *ycf38* gene is found immediately upstream from *psbB* on the same strand, but the *petB-PetD* operon is no longer associated with this gene cluster. However, the *psbH* gene of *C. paradoxa* has been transposed from downstream from *psbB-psbT-(psbN)* to a position upstream from *ycf38*, where it is now divergently transcribed from *ycf38*. In *Synechocystis* sp. PCC 6803, these genes occur as four distinct transcription units, although interestingly the *psbH* and *psbN* genes are adjacent and divergently transcribed.

A final example is the arrangement of the *rbcL* and *atpBE* genes. In both *C. paradoxa* (where *rbcS* forms a dicistronic operon with *rbcL*) and higher plant chloroplasts, these three genes are adjacent and divergently transcribed. In each case the intercistronic spacer that contains the two promoters is approximately 0.5 kb in length. These two dicistronic units are not adjacent in cyanobacterial genomes nor in the plastid genomes of most other algae, however.

Transcription units common with cyanobacteria only. There are a few cases where the cyanelle genome contains genes and transcription units that are absent from the *Porphyra purpurea* rhodoplast genome in spite of its 40% surplus in size and gene content. One of them is *groES-groEL*: the chaperonin-10 homolog is nucleus-encoded in the red alga. Another example is ORF244-ORF299 encoding two components of an ABC transporter, possibly for manganese, according to the significant sequence similarity to the cyanobacterial *mntA* and *mntB* genes (Bartsevich & Pakrasi 1995).

The nuclear genome of *Cyanophora paradoxa*

Much less is known about genes encoded in the nucleus of *C. paradoxa* than about cyanelle-encoded genes because of persistent problems in establishing good cDNA and genomic libraries. The primary interest has been nuclear-coded genes for cyanelle-targeted polypeptides. Inspection of their transit sequences and, subsequently, homologous and heterologous import experiments in vitro, was expected to reveal whether the import apparatus of cyanelles differed significantly from that of higher plant chloroplasts or whether both shared a common organization. In the latter case this would support the claim that cyanelles and all other plastid types originated from a common ancestral organelle, the result of a singular primary endosymbiotic event. This conclusion has already emerged from comparative analysis of plastid genome organization (see above). The transit sequences of three precursor proteins, for ferredoxin-NADP$^+$ oxidoreductase (FNR, Jakowitsch & al. 1993), glyceraldehyde-3-phosphate dehydrogenase (H. Brinkmann, pers. comm.), and cytochrome c_{553} (J. Steiner, unpubl.) did indeed conform to stroma targeting peptides from chloroplast precursors in both amino acid composition and hydropathy profile. Heterologous import of preFNR from *C. paradoxa* into isolated pea chloroplasts and of preFNR from a higher plant, *Mesembryanthemum crystallinum*, into isolated cyanelles was achieved (Jakowitsch & al. 1996). Using precytochrome c_{553} from *C. paradoxa*, which has a bipartite targeting signal, "conservative sorting" to the thylakoid lumen of spinach or pea chloroplasts could be demonstrated (J. Steiner & R. B. Klösgen, unpubl.). From this we conclude that the peptidoglycan wall, once present in the semiautonomous endosymbiont (i.e. the "protoplastid") and still present in the extant cyanelles, does not constitute a barrier to protein translocation. Recently, the fine structure of *C. paradoxa* cyanelle peptidoglycan was resolved (Pfanzagl & al. 1996a). A characteristic substitution, N-acetylputrescine, was also found in the cyanelle walls from other glaucocystophyceae (Pfanzagl & al. 1996b) and might have the function to increase the mesh size and thus the permeability for proteins of the murein sacculus.

Much effort has been expended on attempts to isolate genes for enzymes of peptidoglycan synthesis and metabolism from *C. paradoxa*. Surprisingly, only one gene of this type, *ftsW*, resides on the cyanelle genome. Due to the minute amounts of such gene products present in cyanelles and the low degree of sequence conservation observed even among closely related bacteria, this group of around 30 nuclear genes still awaits characterization.

During our search for nuclear-encoded cyanelle polypeptides, a clone was isolated with the aid of an antiserum directed against the total of soluble cyanelle proteins that showed 55% identity in amino acid sequence to ATP-citrate lyase (ACL) from rat liver. The truncated cDNA (2.1 kb) covers approximately 65% of the coding region including the carboxy-terminus. Verification of ACL as a cyanelle import product would be a proof for a novel anaplerotic pathway that contributes to the pool of acetyl coenzyme A in plastids. Presently this is the only case for which a genomic, albeit truncated, clone exists. Sequence analyses indicate the presence of 11 introns (40–60 bp each) with conserved GT/AG borders and consensus regions at the putative branch points (Y. Ma, unpubl.).

Further cDNA sequences from *C. paradoxa* include transketolase (Y. MA, unpubl.) and the cytoplasmic ribosomal protein L13a (J. STEINER & C. HINK, unpubl.). The latter sequence is the first from the plant kingdom. As with the homolog from rat, distinct leucine zipper and basic domain leucine zipper regions are observed (CHAN & al. 1994).

This work has primarily been supported by a grant from the U.S. Department of Agriculture (National Research Initiative – Plant Genome Program 91-37300-6338) to H.J.B. and D.A.B. and by the Austrian National Bank (Jubiläumsfondsprojekt 5436; to W.L.). Additional support was provided by the National Science Foundation (DMB-8818997 and MCB-9206851; D.A.B.) and the Arizona Agricultural Experimental Station (H.J.B.).

References

ARYEE, T. N. T., 1989: Isolation and characterization of the ribosomal RNA genes from the Pringsheim and Kiess isolates of *Cyanophora paradoxa*. – Ph. D. Thesis, University of Vienna.

ASAI, K., FUJISAKI, S., NISHIMURA, Y., NISHINO, T., OKADA, K., NAKAGAWA, T., KAWAMUKAI, M., MATSUDA, H., 1994: The identification of *Escherichia coli ispB* (*cel*) gene encoding the octaprenyl diphosphate synthase. – Biochem. Biophys. Res. Commun. **202**: 340–354.

BARTSEVICH, V. V., PAKRASI, H., 1995: Molecular identification of an ABC transporter complex for manganese. Analysis of a cyanobacterial mutant strain impaired in the photosynthetic oxygen evolution process. – EMBO J. **14**: 1845–1853.

BAUM, M., SCHÖN, A., 1996: Localization and expression of the closely linked cyanelle genes for RNase P RNA and two transfer RNAs. – FEBS Lett. **382**: 60–64.

BERENGUER, J., ROJO, F., DE PEDRO, M. A., PFANZAGL, B., LÖFFELHARDT, W., 1987: Penicillin-binding proteins in the cyanelles of *Cyanophora paradoxa*, an eukaryotic photoautotroph sensitive to β-lactam antibiotics. – FEBS Lett. **224**: 401–405.

BHATTACHARYA, D., HELMCHEN, T., BIBEAU, C., MELKONIAN, M., 1995: Comparisons of nuclear-encoded small subunit ribosomal RNAs reveal the evolutionary position of the *Glaucocystophyta*. – Molec. Biol. Evol. **12**: 415–420.

BOHNERT, H. J., LÖFFELHARDT, W., 1982: Cyanelle DNA from *Cyanophora paradoxa* exists in two forms due to intramolecular recombination. – FEBS Lett. **150**: 403–406.

BREITENEDER, H., SEISER, C., LÖFFELHARDT, W., MICHALOWSKI, C., BOHNERT, H. J., 1988: Physical map and protein gene map from the second known isolate of *Cyanophora paradoxa* (Kies-strain). – Curr. Genet. **13**: 199–206.

BRYANT, D. A., 1991: Cyanobacterial phycobilisomes: progress toward complete structural and functional analysis via molecular genetics. – In BOGORAD, L., VASIL, I. L., (Eds): Cell culture and somatic cell genetics of plants, **7B**: the photosynthetic apparatus: molecular biology and operation, pp. 257–300. – San Diego: Academic Press.

CHAN, Y.-L., OLVERA, J., GLÜCK, A., WOOL, I. G., 1994: A leucine zipper-like motif and a basic region-leucine zipper-like element in rat ribosomal protein L13a. J. Biol. Chem. **269**: 5589–5594.

COLEMAN, A. W., GOFF, L. J., 1991: DNA analysis of eukaryotic algal species. – J. Phycol. **27**: 463–473.

CURTIS, S. E., MARTIN, J. A., 1994: Transcription apparatus and transcriptional regulation. – In BRYANT, D. A., (Ed.): The molecular biology of cyanobacteria, pp. 613–639. – Dordrecht: Kluwer Academic Publishers.

DOUGLAS, S. E., 1994: Chloroplast origins and evolution. – In BRYANT, D. A., (Ed.): The molecular biology of the cyanobacteria, pp. 91–118. – Dordrecht: Kluwer Academic Publishers.

EVRARD, J.-L., JOHNSON, C., JANSSEN, I., LÖFFELHARDT, W., WEIL, J.-H., KUNTZ, M., 1990: The cyanelle genome of *Cyanophora paradoxa*, unlike the chloroplast genome, codes for the L3 protein. – Nucleic Acids Res. **18**: 1115–1119.

FUJISHIRO, T., KANEKO, T., SUGIURA, M., SUGITA, M., 1996: Organization and transcription of a putative gene cluster encoding ribosomal protein S14 and an oligopeptide permease-like protein in the cyanobacterium *Synechocystis* sp. strain PCC 6803. – DNA Res. **3**: 165–169.

GRUBER, T., BRYANT, D. A., 1997: Molecular systematic studies of eubacteria using s^{70}-type sigma factors of group 1 and group 2. – J. Bacteriol. **179**: 1734–1747.

HALLICK, R. B., 1989: Proposals for the naming of chloroplast genes. II. Update to the nomenclature of genes for thylakoid membrane polypeptides. – Pl. Molec. Biol. Reporter **7**: 266–275.

– BOTTOMLEY, W., 1983: Proposals for the naming of chloroplast gene. – Pl. Molec. Biol. Reporter **1**: 38–43.

– BAIROCH, A., 1994: Proposals for the naming of chloroplast genes. III. Nomenclature for open reading frames encoded in chloroplast genomes. – Pl. Molec. Biol. Reporter **12**: S29–30.

– HONG, L., DRAGER, R. G., FAVREAU, M. R., MONTFORT, A., ORSAT, B., SPIELMANN, A., STUTZ, E., 1993: Complete sequence of *Euglena gracilis* chloroplast DNA. – Nucleic Acids Res. **21**: 3537–3544.

HELMCHEN, T. A., BHATTACHARYA, D., MELKONIAN, M., 1995: Analyses of ribosomal RNA sequences from glaucocystophyte cyanelles provide new insights into evolutionary relationships of plastids. – J. Molec. Evol. **41**: 203–210.

JAKOWITSCH, J., BAYER, M. G., MAIER, T. L., LÜTTKE, A., GEBHART, U. B., BRANDTNER, M., HAMILTON, B., NEUMANN-SPALLART, C., MICHALOWSKI, C. B., BOHNERT, H. J., SCHENK, H. E. A., LÖFFELHARDT, W., 1993: Sequence analysis of pre-ferredoxin-NADP$^+$-reductase cDNA from *Cyanophora paradoxa* specifying a precursor for a nucleus-encoded cyanelle polypeptide. – Pl. Molec. Biol. **21**: 1023–1033.

– NEUMANN-SPALLART, C., MA, Y., STEINER, J. M., SCHENK, H. E. A., BOHNERT, H. J., LÖFFELHARDT, W., 1996: In vitro import of pre-ferredoxin- NADP$^+$-oxidoreductase from *Cyanophora paradoxa* into cyanelles and into pea chloroplasts. – FEBS Lett. **381**: 153–155.

KANEKO, T., SATO, S., KOTANI, H., TANAKA, A., ASAMIZU, E., NAKAMURA, Y., MIYAJIMA, N., HIROSAWA, M., SUGIURA, M., SAKAMOTO, S., KIMURA, T., HOSOUCHI, T., MATSUNO, A., MURAKI, A., KAKAZAKI, N., NARUO, K., OKOMURA, S., SHIMPO, S., TAKEUCHI, C., WADA, T., WATANABE, A., YAMADA, M., YASUDA, M., TABATA, S., 1996: Sequence analysis of the genome of the unicellular cyanobacterium *Synechocystis* sp. PCC 6803. II. Sequence determination of the entire genome and assignment of potential protein-coding regions. – DNA Res. **3**: 109–136.

KIES, L., 1992: *Glaucocystophyceae* and other protists harbouring prokaryotic endocytobionts. – In REISSER, W., (Ed.): Algae and symbioses, pp. 353–377. – Bristol: Biopress.

KOWALLIK, K., 1994: From endosymbionts to chloroplasts: (evidence for a single prokaryotic/eukaryotic endocytobiosis. – Endocytobiosis & Cell Res. **10**: 137–149.

– STOEBE, B., SCHAFFRAN, I., FREIER, U., 1995: The chloroplast genome of a chlorophyll $a + c$ containing alga, *Odontella sinensis*. – Pl. Molec. Biol. Reporter **13**: 336–342.

KRAUS, M., GÖTZ, M., LÖFFELHARDT, W., 1990: The cyanelle *str* operon from *Cyanophora paradoxa*: sequence analysis and phylogenetic implications. – Pl. Molec. Biol. **15**: 561–573.

LEMAUX, P. G., GROSSMAN, A. R., 1985: Major light-harvesting polypeptides encoded in polycistronic transcripts in an eukaryotic alga. – EMBO J. **4**: 1911–1919.

LINDAHL, L., ZENGEL, J. M., 1986: Ribosomal genes in *Escherichia coli*. – Annu. Rev. Genet. **20**: 297–326.

LINK, G., 1996: Green life: control of chloroplast gene transcription. – BioEssays **18**: 465–471.

LIU, C.-Y., TROXLER, R. F., 1996: Molecular characterization of a positively photoregulated nuclear gene for a chloroplast RNA polymerase sigma factor in *Cyanidium caldarium*. – Proc. Natl. Acad. Sci. USA **93**: 3313–3320.

LÖFFELHARDT, W., 1995: Molecular analysis of plastid evolution. – In JOINT, I., (Ed.): Molecular ecology of aquatic microbes, NATO ASI Series G, 138, pp. 265–278. – Berlin, Heidelberg: Springer.

– BOHNERT, H. J., 1994a: Molecular biology of cyanelles. – In BRYANT, D. A., (Ed.): The molecular biology of the cyanobacteria, pp. 65–89. – Dordrecht: Kluwer Academic Publishers.

– – 1994b: Structure and function of the cyanelle genome. – In JEON, K. W., JARVIK, J., (Eds): Int. Rev. Cytol. **151**, pp. 29–65. – Orlando: Academic Press.

– MUCKE, H., CROUSE, E. J., BOHNERT, H. J., 1983: Comparison of the cyanelle DNA from two different strains of *Cyanophora paradoxa*. – Curr. Genet. **7**: 139–144.

– BREITENEDER, H., SEISER, C., ARYEE, D. N. T., MICHALOWSKI, C., KALING, M., BOHNERT, H. J., 1987: The cyanelle genome from *Cyanophora paradoxa*: chloroplast and cyanobacterial features. – Ann. New York Acad. Sci. **503**: 550–552.

– STIREWALT, V. L., MICHALOWSKI, C. B., ANNARELLA, M., FARLEY, J. Y., SCHLUCHTER, W. M., CHUNG, S., NEUMANN-SPALLART, C., STEINER, J. M., JAKOWITSCH, J., BOHNERT, H. J., BRYANT, D. A., 1997: The complete sequence of the cyanelle genome from *Cyanophora paradoxa*: the genetic complexity of a primitive plastid. – In SCHENK, H. E. A., HERRMANN, R. G., JEON, K. W., MÜLLER, N. E., SCHWEMMLER, W., (Eds): Eukaryotism and symbiosis, pp. 40–48. – Heidelberg: Springer.

MARCO, Y., ROCHAIX, J.-D., 1981: Comparison of the ribosomal DNA organisation in *Chlamydomonas* species. – Chromosoma **81**: 629–640.

MICHALOWSKI, C. B., PFANZAGL, B., LÖFFELHARDT, W., BOHNERT, H. J., 1990: The cyanelle S10/*spc* operons from *Cyanophora paradoxa*: Molec. Gen. Genet. **224**: 222–231.

MORSE, D., SALOIS, P., MARCOVIC, P., HASTINGS, J. W., 1995: A nuclear-encoded form II RuBisCO in dinoflagellates. – Science **268**: 1622–1624.

PFANZAGL, B., ZENKER, A., PITTENAURE, E., ALLMAIER, G., MARTINEZ–TORRECUADRADA, J., SCHMID, E. R., DE PEDRO, M. A., LÖFFELHARDT, W., 1996a: Primary structure of cyanelle peptidoglycan of *Cyanophora paradoxa*: a prokaryotic cell wall as part of an organelle envelope. – J. Bacteriol. **178**: 332–339.

– ALLMAIER, G., SCHMID, E. R., DE PEDRO, M. A., LÖFFELHARDT, W., 1996b: N-Acetylputrescine as a characteristic constituent of cyanelle peptidoglycan in glaucocystophyte algae. – J. Bacteriol. **178**: 6994–6997.

REITH, M., 1995: Molecular biology of rhodophyte and chromophyte plastids. – Annu. Rev. Pl. Physiol. Pl. Molec. Biol. **46**: 549–575.

– MUNHOLLAND, J., 1995: Complete nucleotide sequence of the *Porphyra purpurea* chloroplast genome. – Pl. Molec. Biol. Reporter **13**: 332–335.

SHEVELEV, E., BRYANT, D. A., LÖFFELHARDT, W., BOHNERT, H. J., 1995: Ribonuclease-P RNA gene of the plastid chromosome from *Cyanophora paradoxa*. – DNA Res. **2**: 231–234.

STARNES, S. M., LAMBERT, D. H., MAXWELL, E. S., STEVENS, S. E., PORTER, R. D., SHIVELY, J. M., 1985: Cotranscription of the large and small subunit genes of ribulose-1,5-bisphosphate carboxylase/oxygenase in *Cyanophora paradoxa*. – FEMS Microbiol. Lett. **28**: 165–169.

STIREWALT, V. L., MICHALOWSKI, C. B., LÖFFELHARDT, W., BOHNERT, H. J., BRYANT, D. A., 1995: Nucleotide sequence of the cyanelle genome from *Cyanophora paradoxa.* – Pl. Molec. Biol. Reporter **13**: 327–332.

SUGIURA, M., 1992: The chloroplast genome. – Pl. Molec. Biol. **19**: 149–169.

SUH, J. W., BOYLAN, S. A., PRICE, C. W., 1996: Genetic and transcriptional organization of the *Bacillus subtilis spc*-alpha region. – Gene **169**: 17–23.

WHATLEY, J. M., 1993: The endosymbiotic origin of chloroplasts. – In JEON, K. W., JARVIK, J., (Eds): Int. Rev. Cytol. **144**, pp. 259–299. – Orlando: Academic Press.

YASUMOTO, K., LIU, H., JEONG, S. M., OHASHI, Y., KAKINUMA, S., TANAKA, K., KAWAMURA, F., YOSHIKAWA, H., TAKAHASHI, H., 1996: Sequence analysis of a 50 kb region between *spoOH* and *rrnH* on the *Bacillus subtilis* chromosome. – Microbiology **142**: 3039–3046.

The evolution of cryptophytes

MARTIN J. FRAUNHOLZ, JUERGEN WASTL, STEFAN ZAUNER, STEFAN A. RENSING, MARGITTA M. SCHERZINGER, and UWE-G. MAIER

Key words: Cryptophytes. – Nucleomorph, secondary endosymbiosis, genome project.

Abstract: Cryptophytes are chlorophyll a/c- and phycobiliprotein-containing algae that evolved via secondary endosymbiosis. It is shown that a intracellular symbiosis between a red alga-like organism and an enigmatic host cell led to the origin of the cryptophytes. We discuss the possible evolution of cryptophytes and describe recent projects concerning evolutionary aspects of cryptophyte biology.

Cryptophytes (Cryptomonads, *Cryptophyceae*), first reported by EHRENBERG (1831), are small, unicellular biflagellates found in marine or freshwater habitats. Most of these mastigotes are photosynthetic and motile. Typical features that can be distinguished by light microscopy are a flattened asymmetric cell and characteristic swimming motions. As it is difficult to identify cryptomonads using light microscopy, a growing number of studies have used electron microscopy as a basis for identification. These studies have also revealed intracellular components of the cryptomonad cell (reviewed in GILLOTT 1990, BRETT & al. 1994, CAVALIER-SMITH & al. 1996). Aside from a vestibulum/furrow/gullet complex, the most prominent feature is the plastid. As in other chromistan algae (CAVALIER-SMITH 1986, 1995), the plastid is surrounded by two membrane pairs. The inner pair is the plastid envelope, whereas the outer membrane pair is the plastid endoplasmatic reticulum (PER; GILLOT 1990). However, unlike other chromistan algae, the two membrane pairs diverge, thereby creating a small eukaryotic compartment, termed the periplastidal space (reviewed in MCFADDEN 1993). Some eukaryotic cell components have been identified within this compartment. These include 80S ribosomes (SITTE & BALTES 1990, MCFADDEN & al. 1994a) and a small organelle, the nucleomorph (GREENWOOD 1974, GREENWOOD & al. 1977).

The nucleomorph of Cryptophytes is a vestigial nucleus

Electron microscopy shows the nucleomorph to be an organelle surrounded by a double membrane with small pores (GILLOT & GIBBS 1980, SANTORE 1982). The hypothesis that the nucleomorph is a reduced eukaryotic nucleus (e.g., LUDWIG & GIBBS 1987) has been confirmed by the detection of both DNA (LUDWIG & GIBBS 1985, HANSMANN & al. 1986) and RNA, the latter being located in a nucleolus-like region of the nucleomorph (HANSMANN 1988). Further evidence for this hypothesis has also been obtained by the direct isolation of the nucleomorph (Hansmann & Eschbach 1990) and karyotyping of its genome (Eschbach & al. 1991). This has an average length of 600 kb and is organized into three linear chromosomes (Rensing & al. 1994). *In situ* hybridizations using 18S rRNA probes confirm its eukaryotic nature (McFadden 1990). An important aspect of these observations is that they support the hypothesis that cryptomonads have evolved through a process involving secondary endocytobiosis (e.g., Ludwig & Gibbs 1987, Gibbs 1983).

Evolution of complex plastids in secondary endocytobiosis

Such a hypothesis of secondary endocytobiosis, that is, the acquisition of a plastid from a eukaryotic rather than a prokaryotic symbiont, was proposed by Taylor (1974). It is suggested that a symbiosis between a phototrophic eukaryote in a eucyte followed by successive reduction of the eukaryotic cell structures of the endocytobiont (except the former plasma membrane) could give rise to plastids that

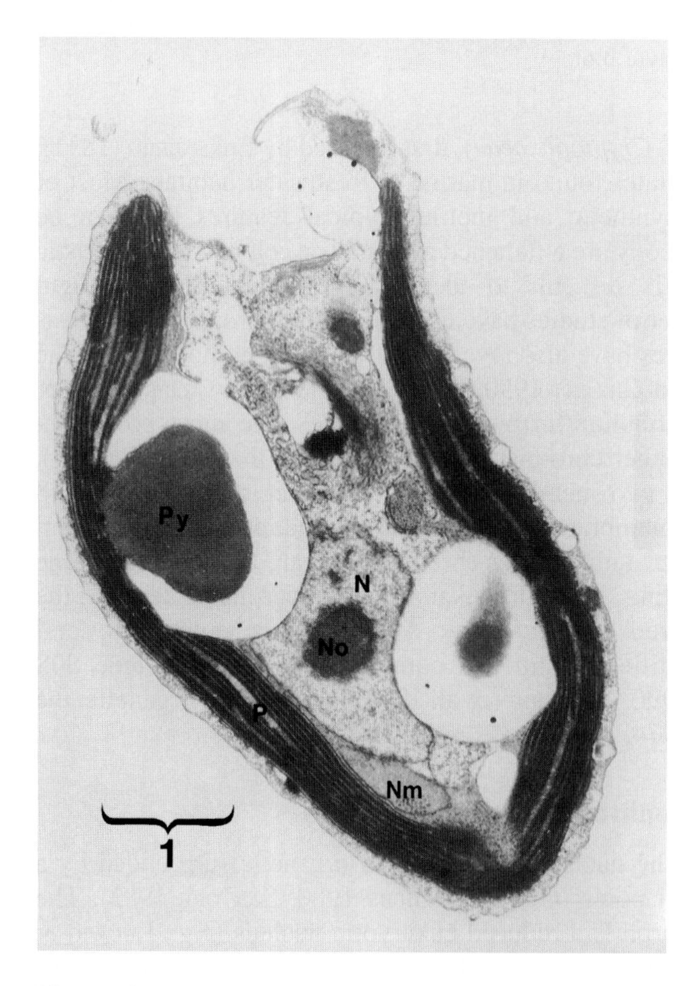

Fig. 1. Electron micrograph of *Guillardia theta*. *Py* pyrenoid, *N* nucleus, *No* nucleolus, *Nm* nucleomorph. Bar: 1 μm

are surrounded by four membranes. These are the so-called complex plastids (e.g., SITTE 1993; CAVALIER-SMITH 1986, 1995). If this scenario is correct then cryptophytes might be considered as "missing links". That is, these small flagellates may represent an intermediate stage in the evolution of complex plastids. In cryptophytes and in contrast to heterokont algae or haptophytes, the eukaryotic phototrophic endosymbiont is only partly reduced with the nucleomorph being the vestigial nucleus of the former free-living eukaryotic symbiont (LUDWIG & GIBBS 1987, reviewed in MCFADDEN & GILSON 1995).

Further evidence for the chimeric nature of cryptophytes has recently come from studies using modern molecular techniques. Characterisation of the small subunit rRNA-genes from the nucleomorph (located in the symbionts compart-

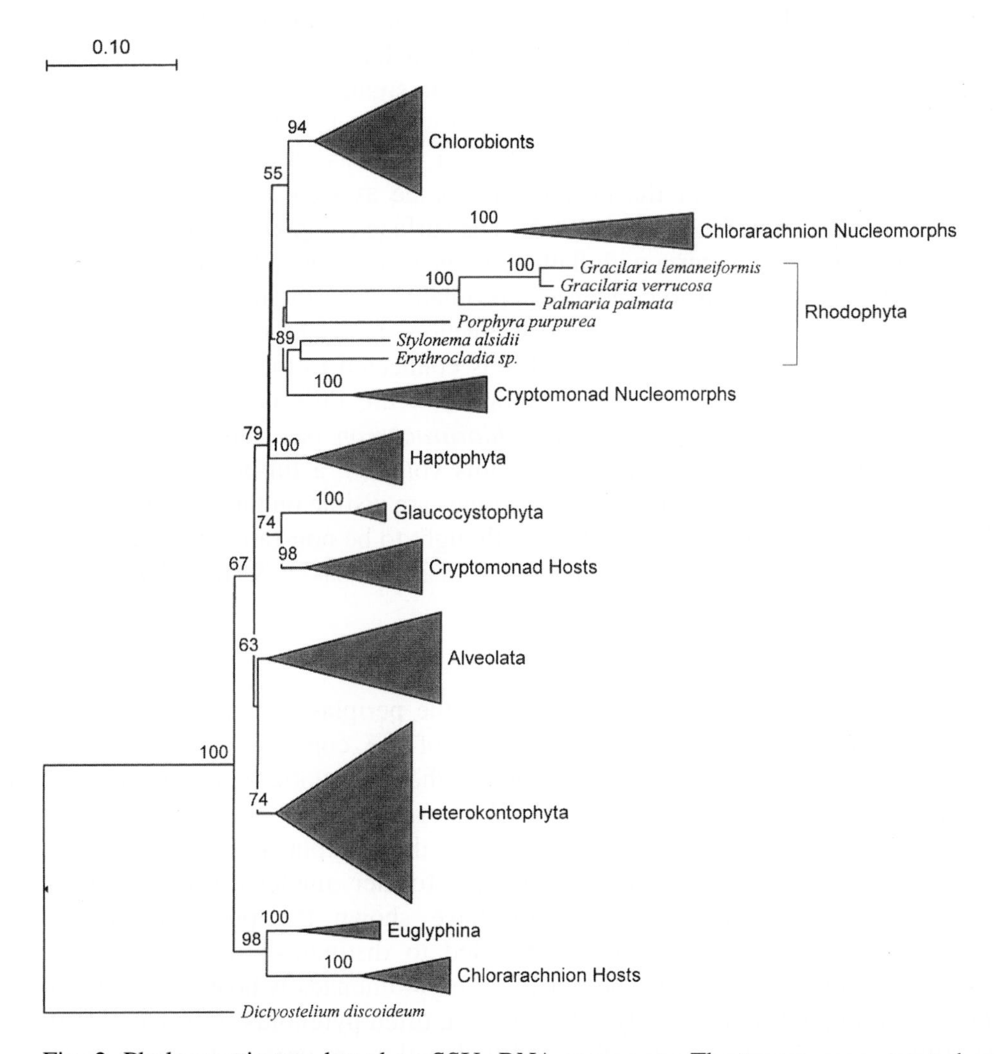

Fig. 2. Phylogenetic tree based on SSU-rRNA sequences. The tree was constructed using substitution rate calibration (VAN DE PEER & al. 1996b) and neighbor-joining. One-thousand bootstrap samples were calculated, the results being shown as numbers at the branches if over 50%. The distance scale represents 10% divergence. Monophyletic lineages are shown as triangles

ment) and from the nucleus (hosts compartment) of cryptophytes (DOUGLAS & al. 1991, MAIER & al. 1991) indicates different phylogenetic origins for sequences from these genomes. This observation supports the inference that cryptophytes indeed evolved through a process of secondary endocytobiosis (see Fig. 2 for a more recent tree).

The host

The host cell of cryptomonads possesses the typical features of a eukaryotic cell. Other features are distinctly unique and include a visible vestibulum, located at the anterior side of the cell that opens to a gullet or progresses into a furrow (MUNAWAR & BISTRICKI 1979; KLAVENESS 1981; KUGRENS & al. 1986; SANTORE 1987; HILL & WELTHERBEE 1990; HILL 1991a,b). This structure is lined by large ejectosomes. Two flagella emerge from the vestibulum and differ in length. The larger flagellum possesses two rows of mastigonemes, whereas the smaller one has only one row (HIBBERD & al. 1971). Cryptophytes have no cell wall. Instead they are surrounded by a complex periplast. This unique cell cover consists of a surface periplast component on the outside of the plasma membrane as well as an inner surface component (reviewed in BRETT & al. 1994). The surface components are thought to be proteinaceous. Mitochondria with flattened cristae are typical for cryptomonads. Mitosis of the nucleus leads to condensed chromatin, but individual chromosomes are not visible. During mitosis, the nuclear envelope partly breaks down (GILLOTT 1990). As in other eukaryotic cells, the ER is connected to the nuclear envelope and in cryptomonads, the outermost membrane of the PER is connected to the nuclear envelope. This is in contrast to *Chlorarachnion* (see Division *Chlorarachniophyta*, by MCFADDEN & al. 1997a in this volume), a further nucleomorph-containing organism. In cryptomonads, ribosomes are visible on the host side of the PER. The distribution of these ribosomes is thought to be non-random leading to a hypothesis for transport of proteins across the PER (HOFMANN & al. 1994).

The symbiont

The eukaryotic compartment of the symbiont, the periplastidal space, is located between the second and the third membrane of the complex plastid, that is, between the former cytoplasmic membrane of the eukaryotic symbiont and the outer membrane of the plastid envelope (WHATLEY & al. 1979; CAVALIER-SMITH 1982, 1986). The only organelle present in the periplastidal space is the nucleomorph which is the remnant of the former nucleus (GIBBS 1983). Examinations using electron microscopy have shown the presence of 80S ribosomes as well as starch grains, but failed to distinguish other eukaryotic features (e.g., MCFADDEN 1993). The plastid of cryptomonads is boat- or H-shaped and contains, in photosynthetically active genera, often pyrenoids (CAVALIER-SMITH & al. 1996). The pigment composition and its location are unique. Cryptomonads contain chlorophyll a and c, together with the phycobilins, phycocyanin or phycoerythrin. The phycobilins are not arranged in phycobilisomes as in the red algae or the glaucocystophytes. Moreover, they are located within the thylakoid lumen (reviewed in GIBBS 1993). Two different locations of the nucleomorph have been observed – in an invagination of the plastid envelope into a groove within the pyrenoid, or, as in most genera, free in the periplastidal space (e.g., CAVALIER-SMITH & al. 1996). The nucleomorph genome is not complexed with typical histones (HANSMANN & al. 1987, GIBBS 1993, MÜLLER & al. 1994) and condensed nucleomorph-chromatin has not been described. Division of the nucleomorph is amitotic during the preprophase stage of mitosis of the nucleus (MCKERRACHER & GIBBS 1982, MORRALL & GREENWOOD 1982).

Evolution of cryptophytes and phylogenetic inference from molecular sequences

As mentioned above, early sequence studies using SSU rRNA-genes specific for the host and the symbiont supported the phylogenetic inference that cryptophytes evolved through secondary endocytobiosis. In these studies an affiliation of the cryptophyte nucleus with green algal nuclei was shown, whereas the symbiont (nucleomorph) tended to group with the red algae (DOUGLAS & al. 1991, MAIER & al. 1991). The inclusion of *Chlorarachnion* sequences in such phylogenetic studies at first appeared to support monophyly of nucleomorph sequences for cryptophytes and chlorarachniophytes. However, recent studies suggest such a grouping was an artifact of a long branching effect (FELSENSTEIN 1978, HENDY & PENNY 1989). More robust phylogenies seem to group cryptomonad nucleomorph sequences with those of the red algae and chlorarachnion nucleomorph sequences with the green algae (CAVALIER-SMITH & al. 1996; VAN DE PEER & al. 1996a,b). BHATTACHARYA & al. (1995a) included sequences from glaucocystophytes in these data. His phylogenetic tree reconstruction found some support for a recent common ancestor of glaucocystophytes and the nuclei of cryptomonads. Although the support for this grouping was weak, the conclusion has been supported in recent studies which use new correction formulae to help retrieve phylogenetic signals from anciently diverged sequence data (VAN DE PEER & al. 1996a,b). Also, a colourless zooflagellate, *Goniomonas*, whose nuclear SSU rRNA sequence was determined and analysed by MCFADDEN & al. (1994b), appears to be related to cryptophyte nuclei. It is not yet clear if this organism is a cryptomonad, which may have lost its plastid secondarily.

Phylogenies have also been reconstructed from the symbiont plastid located SSU-RNA. These are congruent with phylogeny reconstructed for the symbiont-nucleomorph sequences. That is, cryptomonad plastid SSU rRNA sequences group with those of red algae (MCFADDEN & al. 1995; BHATTACHARYA & MEDLIN 1995; VAN DE PEER & al. 1996a,b). Furthermore, phylogenetic investigations with Rubisco have shown that cryptophytes, like red algae, also use the type I purple bacteria form of Rubisco (MARTIN & al. 1992).

Evolution of cryptophytes at the generic level

The taxonomic significance of the location of the nucleomorph in a pyrenoid structure and of the type of phycobilins has been the subject of some controversy. In three genera, *Rhodomonas, Storeatula* and *Rhinomonas*, the nucleomorph is embebbed in the pyrenoid. Recently, CAVALIER-SMITH & al. (1996) demonstrated, using phylogenetic analysis of 18S sequence data, that these genera form a natural group that was separate from cryptophytes with free-lying nucleomorphs. In their nucleomorph-encoded 18S rRNA phylogeny, a clade of cryptomonads with blue pigments (phycocyanin) forms a sister group with red-pigmented (phycoerythrin) cryptophytes. Therefore, cryptophytes could have originated monophyletically. The most parsimonious explanation of these data is that the progenitor could have had allophycocyanin, phycoerytrhin and phycocyanin. Hence after establishing the red algal-like organism, the symbiont has first lost allophycocyanin and differentially either phycoerythrin or phycocyanin. Taken together the different biochemical and molecular data strongly support the hypothesis that cryptophytes evolved through a process of secondary endocytobiosis, that is, by the uptake of a red algal-like organism by a phagotrophic unicellular eukaryotic cell. However, in respect to other chromistan algae, the explanation of the existence of a nucleomorph remains still open.

Did cryptophytes evolve in a modified scenario of secondary endocytobiosis?

It is commonly agreed that the host involved in secondary endocytobiosis was a phagotrophic eukaryotic cell. Reconstruction of algal evolution from phylogenetic analysis of 18S rRNA sequences leads to the hypothesis that chromistan algae diverged shortly after the symbiogenetic origin of plastids (CAVALIER-SMITH 1982, 1986, 1996). Given this, the reduction of the symbiont's eukaryotic structures is complete in some secondarily evolved algae like the heterokonts or haptophytes. That is, apart from the remnant of the plasma membrane, no nucleomorph including periplastidal space exists as it does in cryptophytes and *Chlorarachnion.* A modified scenario of secondary endocytobiosis, first proposed by SCHNEPF (1993) for the evolution of dinoflagellates, was hypothesised for nucleomorph-containing organisms (HÄUBER & al. 1994). In this model the host is not a heterotrophic eukaryote.

Recent phylogenetic trees have clades which contain organisms with cyanelles (primary endosymbiosis) and algae established in secondary endocytobiosis (BHATTACHARYA & al. 1995a,b; VAN DE PEER & al. 1996a,b). In the case of the cryptophytes, the host cells form a rather weak grouping with the glaucocystophytes (BHATTACHARYA & al. 1995a) whereas the hosts of (*Chlorarachnion*) group together with the *Euglyphina* (BHATTACHARYA & al. 1995b). One of the *Euglyphina, Paulinella chromatophora*, harbors a possible cyanelle, too. It is not an unlikely assumption that the host of cryptophytes (and chlorarachniophytes and other heterokont algae?) was a phototroph. This could be an explanation for the need of a nucleomorph in some algae. The scenario that horizontal gene transfer from a primary endocytobiont to the nucleus of the host occurred, thereby creating a nucleus with the typical content of former prokaryotic genes, coding for plastidal proteins, is well accepted. On the assumption that these chimeric organisms, composed as a result of horizontal gene transfer, have not lost the ability to phagocytize, it is possible that another alga was engulfed and integrated as a second endocytobiont. Consequently, an organism evolved with two kinds of plastids. If, under some environmental conditions, the secondary plastid is more efficient than the primary, the latter plastid could be eliminated. Such eukaryotic cells are then predicted to evolve a functional secondary endocytobiont, exploiting the situation that most of the protocytic plastidal genes are already located in the nucleus of the cell. The eukaryotic endocytobiont can use the functionally active, nuclear-encoded plastid proteins from the host, and only a transport system across the two outermost membranes has to evolve. The advantage of this model is that horizontal gene transfer between two eukaryotic nuclei (the symbiont's and host's nucleus) is not required.

But why is a nucleomorph maintained? If the gene content and the quality of horizontal gene transfer in primary endocytobiosis is equal or very similar in respect to the two eukaryotic partners in a secondary endocytobiosis, the eukaryotic compartments of the secondary endocytobiont could be eliminated. However, as it is thought in the case of cryptophytes and chlorarachniophytes (HÄUBER & al. 1994), if the amount of gene transfer in both partners differs, remnants, at least, of the cytoplasmic components have to be maintained. This could include an organelle which encodes the missing genes (the nucleomorph) as well as a machinery to express the functions (the periplastidal space including its 80S ribosomes). This modified scenario of secondary endocytobiosis makes no further assumptions than the traditional model and avoids the necessity of postulating genetic transfer between two eukaryotic nuclei. Furthermore, the modified model is able to explain the existence of nucleomorphs in cryptophytes and chlorarachniophytes.

Cryptophytes as symbionts

Cryptophytes are highly chimaeric cells. But organisms exist that contain cryptophyte-like organisms as endocytobionts. Dinoflagellates, known for the ability to integrate taxonomically different endocytobionts (SCHNEPF 1993), are described as the hosts for cryptophyceen-like secondary endocytobionts (e.g., WILCOX & WEDEMEYER 1984, SCHNEPF & ELBRÄCHTER 1988). In this example, cryptophytes act as solar-powered "tertiary endocytobionts" resulting in extremely chimaeric cells.

The nucleomorph and its gene content

The nucleomorph genome of cryptomonads has an average length of 600 kb. In all cryptophytes investigated so far, the nucleomorph genome comprises three chromosomes (RENSING & al. 1994, Fig. 3). Furthermore, all of these chromosomes contain rRNA-genes. The retention of the nucleomorph including its cytoplasm raises the question what role the nucleomorph plays. A possible interplay between nucleus, nucleomorph and plastid suggests that the function of the nucleomorph is the maintainance of the plastid, either by supplying essential proteins or by facilitating the passage of plastid proteins from the host cytoplasm to the plastid. In order to determine whether the nucleomorph contained functional genes other than rRNA-genes, HOFMANN & al. (1994) probed chromosome II of the cryptomonad *Rhodomonas salina* for active protein-coding genes. They were able to show that chaperone genes are located on this chromosome and that these genes are transcribed. The nucleomorph is thought therefore, to be functionally important.

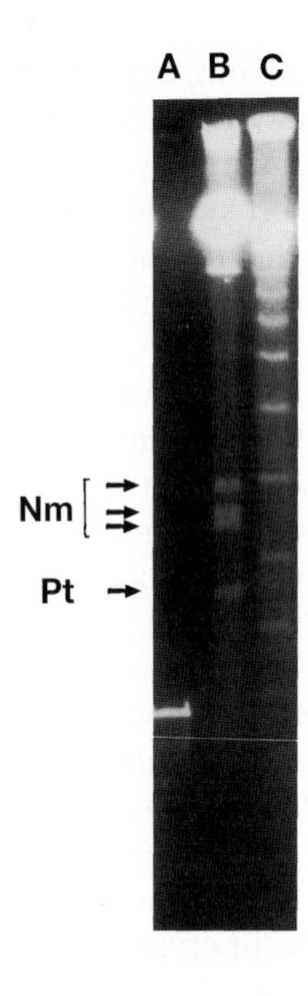

Fig. 3. Pulsed field gel of separated chromosomes of *Guillardia theta*. Lane A: lambda chromosome. Lane B: whole cell preparation of chromosomes of *Guillardia theta*. The three nucleomorph-chromosomes (Nm) as well as the nicked plastid chromosome (pt) are indicated. Lane C: lambda-chromosome concatemeres

Recently, a genome project was started to determine the entire sequence of the three nucleomorph-chromosomes of the cryptomonad, *Guillardia theta* (HILL & WETHERBEE 1990; MCFADDEN & al. 1997b). Already several rRNA- and protein-encoding genes have been identified. These include genes for house-keeping functions, e.g., transcription factors, proteins involved in DNA replication and ribosomal proteins. Additionally, genes for signal transduction and protein degradation have been identified. Genes coding for plastidal proteins have been isolated from one chromosome so far. This means that the nucleomorph is not a rudimentary organelle.

The periplastidal space as a novel compartment where multiprotein complexes are encoded by different genomes

Protein-coding genes in cryptomonads nucleomorphs are separated by less than 1 kb. Therefore, with an average length of 600 kb, only 300 genes, approximately, could be located on the nucleomorph chromosomes. This seems insufficient to maintain functional and structural control of the plastid's and endosymbiont's cytoplasm. Thus, the nucleomorph and its cytoplasm are, in addition to mitochondria and plastids, a third semiautonomous system. But in nucleomorph-containing algae, two eukaryotic nuclei could provide the subunits for a multicomponent protein complex.

Conclusions

Aside from some minor questions, the evolution of cryptophytes is thought to be clearly understood. However, the existence of a nucleomorph with its minimized genome provides a unique opportunity to understand a number of biological phenomena. Therefore, by comparison with other related and unrelated genomes, the evolutionary implications of the nucleomorph genome in the cryptophytes will in the future be further studied using molecular methods.

The authors would like to thank TRISH MCLENACHAN and PETE LOCKHART for helpful comments on the manuscript. We are supported by the Deutsche Forschungsgemeinschaft.

References

BHATTACHARYA, D., MEDLIN, L., 1995: The phylogeny of plastids: a review based on comparisons of small-subunit ribosomal RNA coding regions. – J. Phycol. **31**: 489–498.

– HELMCHEN, T., BIBEAU, C., MELKONIAN, M., 1995a: Comparison of nuclear-encoded small-subunit ribosomal RNAs reveal the evolutionary position of the *Glaucocystophyta*. – Molec. Biol. Evol. **12**: 415–420.

– – MELKONIAN, M., 1995b: Molecular evolutionary analysis of nuclear-encoded small subunit ribosomal RNA identify an independent rhizopod lineage containing the *Euglyphidae* and the *Chlorarachniophyta*. – J. Euk. Microbiol. **42**: 65–69.

BRETT, S. J., PERASSO, L., WETHERBEE, R., 1994: Structure and development of the cryptomonad periplast: a review. – Protoplasma **181**: 106–122.

CAVALIER-SMITH, T., 1982: The origins of plastids. – Biol. J. Linn. Soc. **17**: 289–306.

– 1986: The kingdom *Chromista*: origin and systematics. – In ROUND, F., CHAPMAN, D. J., (Eds): Progress in phycological research **3**, pp. 309–347. – Bristol: Biopress.

– 1995: Membrane heredity, symbiogenesis, and the multiple origins of algae. – In ARAI, R., KATO, M., DOI, Y., (Eds): Biodiversity and evolution, pp. 75–114. – Tokyo: National Science Museum.

– COUCH, J. A., THORSTEINSEN, K. E., GILSON, P., DEANE, J. A., HILL, D. R. A., MCFADDEN, G. I., 1996: Cryptomonad nuclear and nucleomorph 18S rRNA phylogeny. – Eur. J. Phycol. (in press).

DOUGLAS, S. E., MURPHY, C. A., SPENCER, D. F., GRAY, M. W., 1991: Molecular evidence that cryptomonad algae are evolutionary chimaeras of two phylogenetically distinct eukaryotes. – Nature **350**: 148–151.

EHRENBERG, C. G., 1831: Über die Entwicklung und Lebensdauer der Infusionstiere; nebst ferneren Beiträgen zu einer Vergleichung ihrer organischen Systeme. – Akademie der Wissenschaften, Berlin: 1–154.

ESCHBACH, S., HOFMANN, C. J. B., MAIER, U.-G., SITTE, P., HANSMANN, P., 1991: A eukaryotic genome of 660 kb: electrophoretic karyotype of nucleomorph and cell nucleus of the cryptomonad alga, *Pyrenomonas salina*. – Nucl. Acids Res. **19**: 1779–1781.

FELSENSTEIN, J., 1978: Cases where parsimony or compatibility methods will be positively misleading. – Syst. Zool. **27**: 401–410.

GIBBS, S. P., 1983: The cryptomonad nucleomorph: Is it the vestigal nucleus of a eukaryotic endosymbiont? – In SCHENK, H. E., SCHWEMMLER, W., (Eds): Endocytobiology II. Intracellular space as an oligogenetic system. Proc. 2nd Int. Colloquium Endocytobiology. pp. 987–992. – Berlin, New York: de Gruyter.

– 1993: The evolution of the algal chloroplast. – In LEWIN, R. A., (Ed.): Origins of plastids. Symbiogenesis, prochlorophytes and the origin of chloroplasts, pp. 107–121. – New York, London: Chapman & Hall.

GILLOT, M. A., 1990: Phylum *Cryptophyta* (Cryptomonads). – In MARGULIS, L., CORLISS, J., MELKONIAN, M., CHAPMAN, D., (Eds): Handbook of *Protoctista*, pp. 139–151. – Boston: Jones & Bartlett.

– GIBBS, S. P., 1980: The cryptomonad nucleomorph: its ultrastructure and evolutionary significance. – J. Phycol. **16:** 558–568.
Greenwood, A. D., 1974: The *Cryptophyta* in relation to phylogeny and photosynthesis. – In SANDERS, J. V., GOODCHILD, D. J., (Eds): Electron microscopy 1974, pp. 566–567. – Canberra: Australian Academy of Sciences.
– GRIFFITHS, H. B., SANTORE, U. J., 1977: Chloroplasts and cell compartments in *Cryptophyceae.* – Brit. Phycol. J. **42**: 152–160.
HANSMANN, P., 1988: Ultrastructural localization of RNA in cryptomonads. – Protoplasma **146**: 81–88.
– ESCHBACH, S., 1990: Isolation and prelimary characterization of the nucleus and the nucleomorph of a cryptomonad, *Pyrenomnas salina.* – Eur. J. Cell Biol. **52**: 373–378.
– FALK, H., SCHEER, U., SITTE, P., 1986: Ultrastructural localization of DNA in two *Cryptomonas* species by use of a monoclonal DNA antibody. – Eur J. Cell Biol. **42**: 152–160.
– MAERZ, M., SITTE, P., 1987: Investigations on genomes and nucleic acids in cryptomonads. – Endocyt. Cell Res. **4**: 289–295.
HÄEUBER, M. M., MÜLLER, S. B., SPETH, V., MAIER, U.-G., 1994: How to evolve a complex plastid? – A hypothesis. – Bot. Acta **107**: 383–386.
HENDY, M. D., PENNY, D., 1989: A framework for the quantitative study of evolutionary trees. – Syst. Zool. **38**: 297–309.
HIBBERD, D. J., GREENWOOD, A. D., GRIFFITHS, A. H., 1971: Observations on the ultrastructure of the flagella and periplast in the *Cryptophyceae*. – Brit. Phycol. J. **6**: 61–71.
HILL, D. R. A., 1991a: *Chroomonas* and other blue–green cryptomonads. – J. Phycol. **27**: 133–145.
– 1991b: A revised circumscription of *Cryptomonas* (*Cryptophyceae*) based on examination of Australian strains. – Phycologia **30**: 179–188.
– WETHERBEE, R., 1990: *Guillardia theta* gen. et spe. nov. (*Cryptophyceae*). – Canad. J. Bot. **68**: 1873–1876.
HOFMANN, C. J. B., RENSING, S. A., Häuber, M. M., MARTIN, W. F., MÜLLER, S. B., COUCH, J., MCFADDEN, G. I., IGLOI, G. L., MAIER, U.-G., 1994: The smallest eukaryotic genomes encode a protein gene: towards an understanding of nucleomorph functions. – Molec. Gen. Genet. **243**: 600–604.
KLAVENESS, D., 1981: *Rhodomonas lacustria* (PASCHNER & RUTTNER) JAVORNICKY (*Cryptomonadia*): ultrastructure of the vegetative cell. – J. Protozoo **28**: 83–90.
KUGRENS, P., LEE, R. E., ANDERSEN, R. A., 1986: Cell form and surface patterns in *Chroomonas* and *Cryptomonas* cells (*Cryptophyta*) as revealed by scanning electron microscopy. – J. Phycol. **22**: 512–522.
LUDWIG, M., GIBBS, S. P., 1985: DNA is present in the nucleomorph of cryptomonads: further evidence that the chloroplast evolved from a eukaryotic endosymbiont. – Protoplasma **127**: 9–20.
– – 1987: Are the nucleomorphs of cryptomonads and *Chlorarachnion* the vestigal nuclei of eukaryotic endosymbionts? – Ann. New York Acad. Sci. **501**: 198–211.
MAIER, U.-G., HOFMANN, C. J. B., ESCHBACH, S., WOLTERS, J., IGLOI, G. I., 1991: Demonstration of nucleomorph-encoded eukaryotic small subunit ribosomal RNA in cryptomonads. – Molec. Gen. Genet. **230**: 155–160.
MARTIN, W., SOMMERVILLE, C. C., LOISEAUX-DE GOER, S., 1992: Molecular phylogenies of plastid origins and algal evolution. – J. Molec. Evol. **35**: 385–404.
MCFADDEN, G. I., 1990: Evidence that cryptomonad chloroplasts evolved from photosynthetic eukaryotic endosymbionts. – J. Cell Sci. **95**: 303–308.
– 1993: Second-hand chloroplasts: evolution of cryptomonad algae. – Adv. Bot. Res. **19**: 189–230.

– GILSON, P. R., 1995: Something borrowed, something green: lateral transfer of chloroplasts by secondary endosymbiosis. – TREE **10**: 12–17.

– – DOUGLAS, S. E., 1994a: The photosynthetic endosymbiont in cryptomonad cells produce both chloroplast and cytoplasmic-type ribosomes. – J. Cell. Sci. **107**: 649–657.

– – HILL, D. R. A., 1994b: *Goniomonas*: rRNA sequences indicate that this phagotrophic flagellate is a close relative of the host component of cryptomonads. – Eur. J. Phycol. **29**: 29–32.

– – WALLER, R. F., 1995: Molecular phylogeny of chlorarachniophytes based on plastid rRNA and *rbc*L sequences. – Arch. Protistenk. **145**: 231–239.

– – HOFMANN, C. J. B., 1997a: Division *Chlorarachniophyta*. – Pl. Syst. Evol., [Suppl.] **11**: 175–185.

DOUGLAS, S. E., CAVALIER-SMITH, T., HOFMANN, C. J. B., MAIER, U.-G., 1997b: Bonsai genomics: sequencing the smallest eukaryotic genomes. – Trends Genet. (in press).

MCKERRACHER, L., GIBBS, S. P., 1982: Cell and nucleomorph division in the alga *Cryptomonas*. – Canad. J. Bot. **60**: 2440–2452.

MORRAL, S., GREENWOOD, A. D., 1982: Ultrastructure of nucleomorph division in species of the *Cryptophyceae* and its evolutionary implications. – J. Cell Sci. **54**: 311–318.

MÜLLER, S. B., RENSING, S. A., MAIER, U.-G., 1994: The cryptomonad histone H4-encoding gene: structure and chromosomal localization. – Gene **150**: 299–302.

MUNAWAR, M., BISTRICKI, T., 1979: Scanning electron microscopy of some nanoplankton cryptomonads. – Scanning Electron Microsec. **3**: 247–252.

RENSING, S. A., GODDEMEIER, M., HOFMANN, C. J. B., MAIER, U.-G., 1994: The presence of a nucleomorph hsp 70 gene is a common feature of *Cryptophyta* and *Chlorarachniophyta*. – Curr. Genet. **26**: 451–455.

SANTORE, U. J., 1982: The distribution of the nucleomorph in the *Cryptophyceae*. – Cell Biol. Int. Rep. **6**: 1055–1063.

– 1987: A cytological survey of the genus *Chroomonas* – with comments on the taxonomy of this natural group of the *Cryptophyceae*. – Arch. Protistenk. **134**: 83–114.

SCHNEPF, E., 1993: From prey via endosymbiont to plastid: comparative studies in dinoflagellates. – In LEWIN, R. A., (Ed.): Origins of plastids. Symbiogenesis prochlorophytes and the origin of chloroplasts, pp. 53–76. – New York, London: Chapman & Hall.

– ELBRAECHTER, M., 1988: Cryptophycean-like double membrane-bound chloroplast in the dinoflagellate, *Dinophysis* EHRENB,: evolutionary, phylogenetic and toxicological implications. – Bot. Acta **101**: 196–203.

SITTE, P., 1993: Symbiogenetic evolution of complex cells and complex plastids. – Eur. J. Protistol. **29**: 131–143.

– BALTES, S., 1990: Morphometric analysis of two cryptomonad species: Quantitative evaluation of fine-structure changes in an endocytobiotic system. – In NARDON, P., GIANINAZZI-PEARSON, V., GRENIER, A. M., MARGULIS, L., SMITH, D. C., (Eds): Endocytobiology IV, pp. 229–233. – Paris: Institut national de la Recherche Agrouoinguls

TAYLOR, F. J. R., 1974: Implications and extensions of the serial endosymbiosis theory of the origin of eukaryotes. – Taxon **23**: 229–258.

VAN DE PEER, Y., RENSING, S. A., MAIER, U.-G., DE WACHTER, R., 1996a: Substitution rate calibration of small subunit rRNA identifies chlorarachniophyte endosymbionts as remnants of green algae. – Proc. Natl. Acad. Sci. USA **93**: 7732–7736.

– VAN DER AUWERA, G., DE WACHTER, R., 1996b: The evolution of *Stramenopiles* and *Alveolata* as derived by "substitution rate calibration" of small ribosomal subunit RNA. – J. Molec. Evol. **42**: 201–210.

WHATLEY, J. M., JOHN, P., WHATLEY, F. R., 1979: From extracellular to intracellular: the establishment of mitochondria and chloroplasts. – Proc. Roy. Soc. London B **204**: 165–187.

WILCOX, L. W., WEDEMEYER, G. J., 1984: *Gymnodinium acidotum*, a dinoflagellate with an endosymbiotic cryptomonad. – J. Phycol. **20**: 236–242.

Division *Chlorarachniophyta*

GEOFFREY I. MCFADDEN, PAUL R. GILSON, and CLAUDIA J. B. HOFMANN

Key words: *Chlorarachniophyta, Chlorarachnion.* – Secondary endosymbiosis, introns, operons, nucleomorph, plastid, chloroplast.

Abstract: Chlorarachniophyte algae contain a complex, multi-membraned chloroplast derived from the endosymbiosis of a eukaryotic alga. Phylogenetic trees indicate that the host is closely related to filose amoebae and sarcomonads whereas the endosymbiont is most closely related to green algae. The endosymbiont is greatly reduced retaining only the plastid, plasmamembrane, a modicum of cytoplasm, and the nucleus. The vestigial nucleus of the endosymbiont, called the nucleomorph, contains three small linear chromosomes with a haploid genome size of 380 kb and is the smallest known eukaryotic genome. The overall gene organisation of the nucleomorph genome is extraordinarily compact making this a unique model for eukaryotic genomics.

Chlorarachnion reptans GEITLER, the type species of the division *Chlorarachniophyta*, was described by LOTHAR GEITLER (1930) from Las Palmas in the Canary Islands. The cells are amoeboid with pseudopodia interconnecting as many as 150 cells into a reticuloplasmodial continuum (GEITLER 1930). The reticulopodia entrap bacteria, flagellates, and eukaryotic algae which are transported to the cell and digested (GEITLER 1930). Currently, there are four genera and six species described (ISHIDA 1994, ISHIDA & HARA 1994, HARA & al. 1992), but one genus, *Cryptochlora*, is not yet characterised electron microscopically (CALDERON-SAENZ & SCHNETTER 1987, 1989; BEUTLICH & SCHNETTER 1993) and several undescribed species await full characterisation. Distribution is worldwide, concentrated in moderately tropical zones, but biogeographical information is incomplete. A sexual cycle is reported (ISHIDA 1994, GRELL 1990, BEUTLICH & SCHNETTER 1993) but no details of meiosis or ploidy levels for various phases have been established, so proof of true sexuality is wanting.

An extraordinary feature of *Chlorarachnion* is the presence of grass-green plastids with a pyrenoid. The combination of green plastids and reticuloplasmodial habit presented GEITLER (1930) with a systematic dilemma and he considered *Chlorarachnion* could be either a rhizopodial euglenoid or a heterokont. For the following 37 years *Chlorarachnion* languished in obscurity until RICHARD NORRIS (1967) isolated it from the Gulf of California in Mexico. Subsequently, HIBBERD & NORRIS (1984) undertook electron microscopical examination of *Chlorarachnion reptans*. On the basis of the very unusual nature of the plastids, they erected a new division – division *Chlorarachniophyta* (alternatively phylum *Chlorarachnida*,

Hibberd 1990). The plastids, which number between five and seven in the amoeboid cells of *Chlorarachnion reptans*, are surrounded by four membranes. Between the inner and outer pairs of chloroplast membranes is a small volume of cytoplasm containing particles resembling 80S-sized ribosomes. Also situated between the inner and outer pairs of membranes is a double membrane-bound organelle known as the nucleomorph. The two membranes surrounding the nucleomorph are interrupted by pores similar to those observed in nuclei (Hibberd & Norris 1984). The nucleomorph was recognised as likely representing a vestigial nucleus, and Hibberd & Norris (1984) posited that the plastid is a reduced eukaryotic endosymbiont (Fig. 1).

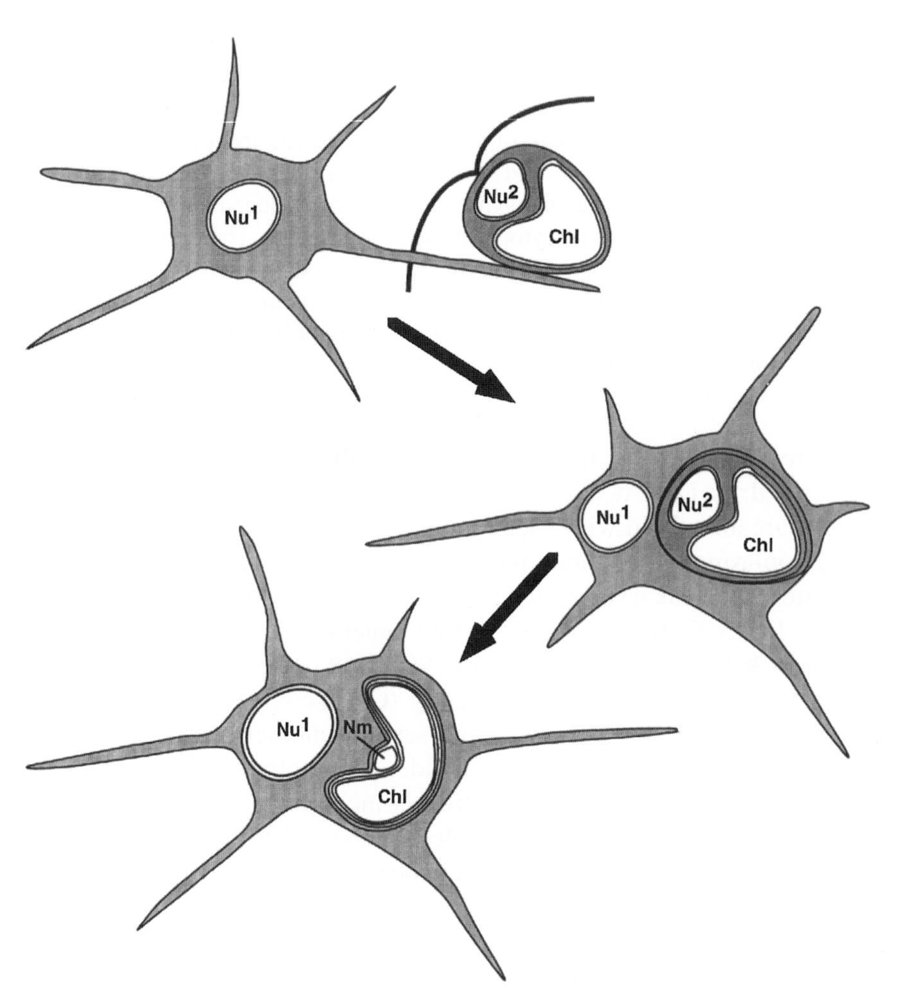

Fig. 1. Possible scenario for establishment of a secondary endosymbiosis producing chlorarachniophyte organisms. An amoeba captures a green alga (either flagellated as shown or non-flagellate) on a reticulopodium. The algal cell is internalised within a food vacuole but not digested. Photosynthate is translocated from the algal endosymbiont to the host. Genes are transferred from the endosymbiont (either from the plastid or nuclear genomes or both) to the host nucleus. The endosymbiont loses redundant structures (e.g., cell wall, cytoskeleton, mitochondria, endomembrane system) and functions

A key test of whether the nucleomorph is indeed a relict nucleus was to identify DNA within. Using the DNA-binding fluorochrome 4′-6-diamidino-2-phenylindole (DAPI), LUDWIG & GIBBS (1987, 1989) identified a modest amount of DNA within the nucleomorph. In addition to DNA, LUDWIG & GIBBS (1989) also reported the nucleomorph to contain a fibrillogranular region, proposed to represent a nucleolus. While the presence of DNA and fibrillogranular region reinforced the notion that the nucleomorph could be a vestigial nucleus, the first observations of division, which is amitotic and occurs by simple bifurcation (LUDWIG & GIBBS 1989), suggest that the nucleomorph has been highly modified during its tenure as an endosymbiont. Presence of a nucleolus-like fibrillogranular body suggested the nucleomorph might encode ribosomal RNAs that could be components of the ribosome-like particles in the surrounding periplastidal space. In situ hybridization confirmed that transcripts of rRNAs accumulate in the nucleomorph fibrillogranular region and periplastidal cytoplasm (MCFADDEN 1992).

While the early in situ hybridization experiments (MCFADDEN 1992) corroborated the hypothesis that the nucleomorph is a vestigial nucleus of an eukaryotic endosymbiont, they did not exclude the possibility that the nucleomorph arose autogenously because they used a universal probe unable to discern between nuclear and nucleomorph transcripts. To prove unequivocally that the nucleomorph is derived from an endosymbiont, it was necessary to demonstrate that nucleomorph genes were evolutionarily divergent from nuclear genes. PCR amplification of eukaryotic rRNA genes from total chlorarachniophyte DNA yielded two different genes (MCFADDEN & al. 1994), and in situ hybridization using gene-specific probes demonstrated that one gene originates from the nucleus while the other is harboured by the nucleomorph (MCFADDEN & al. 1994). The nucleomorph gene encodes 18S-like transcripts that accumulate in the periplastidal space, presumably in ribosomes therein (MCFADDEN & al. 1994). The in situ hybridization results fulfil the predictions of HIBBERD & NORRIS' (1984) hypothesis, and leave no doubt that chlorarachniophytes acquired their chloroplasts by secondary endosymbiosis of a photosynthetic eukaryote.

Organization and information content of nucleomorph genome

Using the nucleomorph-specific rRNA gene as a probe, three small, linear chromosomes (145 kb, 140 kb, 95 kb) were identified as belonging to the nucleomorph (MCFADDEN & al. 1994). Other, larger chromosomes were identified as nuclear chromosomes (MCFADDEN & al. 1994), and two small, linear chromosomes (72 kb and 36 kb) derive from the mitochondrion (GILSON & al. 1995). Since the nucleomorph chromosomes derive from the original chromosomes of a eukaryotic nucleus they were expected to carry telomeres; the post-replicatively added repeat motifs that cap all eukaryotic chromosomes (BIESSMANN & MASON 1994, SHIPPEN 1993). A terminal fragment of the 95 kb nucleomorph chromosome has been cloned and sequenced and found to carry 32 repeats of the heptanucleotide motif TCTAGGG (GILSON & MCFADDEN 1995). The repeat motif is similar to the various telomere sequence repeats of other eukaryotic chromosomes (BIESSMANN & MASON 1994, SHIPPEN 1993), particularly that of plants and the green alga *Chlorella* (TTTTGGGn) (HIGASHIYAMA & al. 1995). The nucleomorph

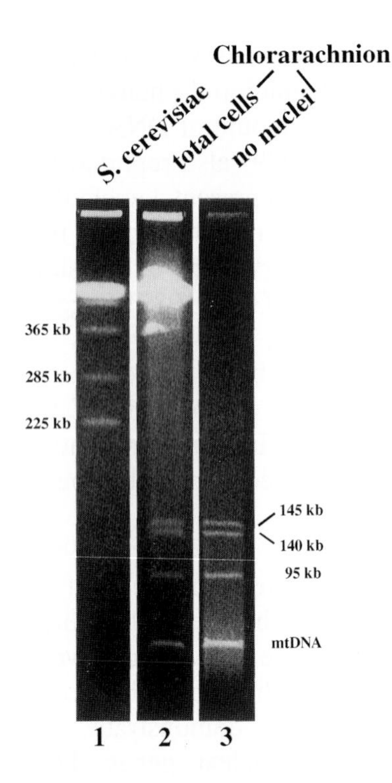

Fig. 2. Pulsed field gel electrophoresis of complete chromosomal DNAs from chlorarachniophyte strain 621 (lane 2) and the organelle DNA fraction (strain 621) in which nuclear DNA has been removed (lane 3). Molecular weight markers from *Saccharomyces cerevisiae*. are shown in lane 1. The organelle fraction contains only four linear chromosomes: the 72 kb mitochondrial chromosome (mtDNA), and the three previously identified nucleomorph chromosomes (145 kb, 140 kb, 95 kb). Plastid DNA (presumably circular) remains in the loading wells. Removal of the nuclei was accomplished by lysing the cells in 0.15 M NaCl, 0.5M EDTA, 0.1M TRIS/HCl pH 9.0. The nuclei remained intact after mild proteolysis and were sedimented prior to precipitation of organelle DNA

telomere fragment from the 95 kb chromosome was used as a probe to identify further nucleomorph chromosomes. Only two other chromosomes (the 145 kb and 140 previously identified with the nucleomorph rRNA probe) carry the telomere motif. Assuming that all other nucleomorph chromosomes carry the same telomere motif, it can be concluded that the nucleomorph harbours only three chromosomes giving a haploid genome size of only 380 kb (Gilson & McFadden 1995).

Further evidence that nucleomorphs harbour only three chromosomes comes from subtractive DNA isolation procedures (Fig. 2). When nuclei are removed during DNA isolation, only five chromosomes (145 kb, 140 kb, 95 kb, 72 kb & 36 kb) are observed in pulsed field gels (Fig. 2). These comprise the three previously identified nucleomorph chromosomes (McFadden & al. 1994, Gilson & McFadden 1995) plus the two mitochondrial chromosomes (Gilson & al. 1995). In addition to the nucleomorph and mitochondrial genomes, the nuclear-free DNA preparation also contains the chloroplast DNA, but this DNA (which is presumably circular) appears to remain in the loading wells during pulsed field gel electrophoresis (Fig. 2). Unless the chlorarachniophyte nucleomorph contains circular chromosomes that lack telomeres, only three of the endosymbiont's original complement chromosomes have persisted.

The gross architecture of the three nucleomorph chromosomes is unusual (Gilson & McFadden 1995). All chromosome termini are apparently identical with each chromosome carrying an 8.5 kb inverted repeat (Fig. 3). The repeat element comprises the telomere, about 1 kb of DNA thus far not demonstrated to encode anything, and a typical eukaryotic rRNA cistron (viz. 18S rRNA, ITS 1, 5.8S rRNA, ITS 2, 28S rRNA). It is assumed that some mechanism of gene conversion

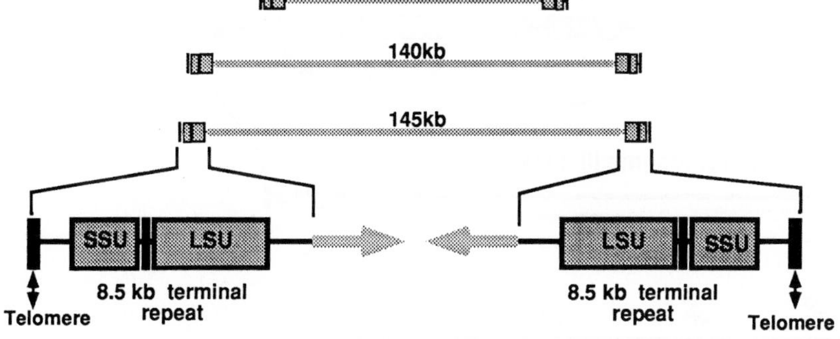

Fig. 3. Overview maps of the three nucleomorph chromosomes. Each chromosome carries two copies of a repeat comprised of reiterated telomere motifs and rDNA cistron. The repeats are arranged in inverted orientation at each end of the chromosome

operates both intra- and interchromosomally to maintain identity of the termini but these details are unknown.

Nucleomorph genes

Nucleotide sequence data shows that the modest amount of nucleomorph DNA can encode a remarkable number of genes (GILSON & MCFADDEN 1996). Packed into 30 kb of DNA examined thus far are six rRNAs, one snRNA, and eight protein genes (Fig. 4). Assuming this density is representative, we estimate the nucleomorph retains a mere 300 genes from the original cohort. With its drastically reduced genome, the nucleomorph must be dependent on the host, and we assume that numerous redundant genes must have been lost. It also seems likely that a number of genes have been transferred to the host nucleus. The nucleomorph genome has clearly been under pressure to reduce its size. The genes are separated by minimal spacer DNA (average 100 bp) and the introns (which are surprisingly numerous) are the smallest spliceosomal introns described at 18–20 nucleotides (GILSON & MCFADDEN 1996). Genome compression also appears to have resulted in the creation of eukaryotic operons where two genes are cotranscribed to produce a dicistronic messenger RNA (GILSON & MCFADDEN 1996).

The genes identified thus far fall into the 'housekeeping' category with probable roles in transcription, transcript processing, translation, and protein processing or targeting (Fig. 4). The nucleomorph apparently encodes numerous components essential to the expression of information content, but it is not yet clear what purpose this housekeeping serves. Presumably the nucleomorph encodes factors essential to the symbiotic partnership. An obvious possibility are proteins essential for plastid biogenesis and function (MCFADDEN & GILSON 1995). Since the nucleomorph was originally the nucleus of a free-living alga, it probably carried many hundreds of genes for plastid proteins. While the reduced nature of the nucleomorph suggests that a great many of these plastid protein genes must now

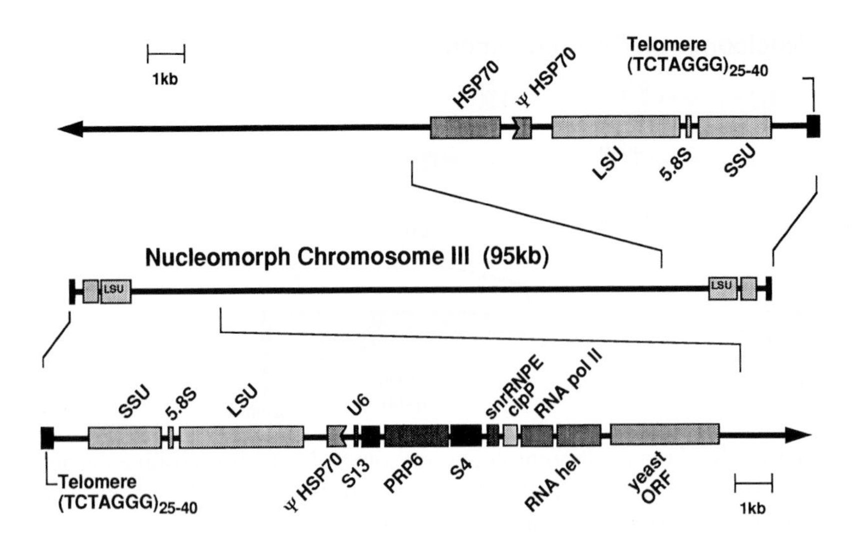

Fig. 4. Gene map for the portions of chlorarachniophyte nucleomorph chromosome III (95 kb) characterised thus far. Genes occupy both strands (the slant of the gene name indicates the direction of transcription) and are densely packed. Non-coding portions (ie 5′ and 3′ untranslated portions of the mRNA transcript) of select genes overlap the adjacent gene. Some genes (e.g., snRNP E and clpP) are contranscribed in a single message

reside in the host nucleus, the nucleomorph may contain a small number of genes encoding essential plastid components.

One might imagine that these last remaining plastid protein genes could eventually be transferred to the nucleus leaving the nucleomorph free to degenerate and perhaps disappear, as has been proposed for heterokont algae (Cavalier-Smith 1995a). However, it seems likely that nucleomorph genes have undergone drastic modifications during endosymbiosis (viz. extreme reduction of intron size, overlaps with neighbouring genes, and fusion into nouveaux operons) so perhaps these changes to their architecture prevent transfer of the final residue into the now inappropriate milieu of a standard nucleus. In other words, the high eccentricity of the nucleomorph's molecular biology may have ensured its perpetuity. Of course, it is also entirely possible that no chloroplast proteins are encoded by the nucleomorph – it could simply exist to produce components of the necessary pathway (periplastidal cytoplasm and enveloping membrane) through which the nuclear-encoded plastid proteins must travel (McFadden & Gilson 1996).

The symbiotic partners

The endosymbiont. Due to the severe reduction of the endosymbiont (all the major cellular structures excepting the nucleus, plastid and plasma membrane are gone) it was difficult to determine, on morphological grounds, what type of alga the original endosymbiont might have been (Hibberd & Norris 1984). The pigment repertoire – which includes chlorophylls a and b, β-carotene, and some unusual xanthophylls – suggested a green alga (Hatakeyama & al. 1991; Sasa & al. 1992; Hibberd & Norris 1984; Ludwig & Gibbs 1987, 1989), but the absence of starch in the chlorarachniophyte plastid is atypical of green algae rendering a biochemistry-based identification equivocal.

Fortunately, the nucleomorph retains rRNA genes that can be used to trace the evolutionary affinities of the alga originally engulfed by the chlorarachniophyte host.

Unfortunately, initial attempts (McFadden & al. 1994, Cavalier-Smith & al. 1994) to incorporate chlorarachniophyte nucleomorph sequences into trees of eukaryotic rRNAs met with difficulties, apparently because the endosymbiont genes have undergone a period of accelerated evolution that prevented commonly used tree reconstruction methods (distance, parsimony, and maximum likelihood) from identifying their evolutionary affinities with the data sets available at the time. Using a new tree inference method designed to cope with problems of long branch length, Van De Peer & al. (1996) demonstrate that the chlorarachniophyte endosymbiont is indeed a green alga, probably most closely related to the chlorophycean lineage containing *Chlamydomonas* and *Chlorella*. Use of maximum likelihood tree inference with a very large number of taxa or distance methods correcting for accelerated evolution also positions chlorarachniophyte endosymbionts within chlorophycean algae (Cavalier-Smith & al. 1996; Fig. 5).

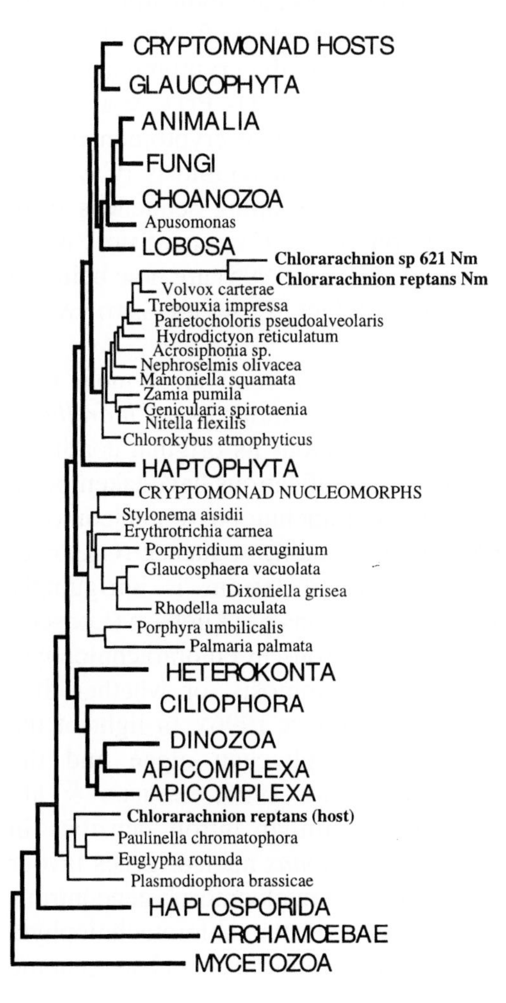

Fig. 5. Maximum likelihood tree inferred from 92 nuclear small subunit (18S) rRNA sequences using 1632 alignment positions (redrawn from Cavalier-Smith & al. 1996). Taxa in capital letters are represented by multiple sequences. The chlorarachniophyte endosymbionts (Nm) form a sister-group to *Volvox*. The chlorarachniophyte host is a sister-group to the filose amoebae (*Paulinella* and *Euglypha*). The cryptomonad endosymbionts (NUCLEOMORPHS) are positioned within the red algae as sister to *Stylonema*. The cryptomonad hosts are shown as a sister-group to the glaucophytes. Three separate fast DNAml trees were calculated using a different random order to taxon addition and allowing local rearrangements across ten branches. The tree shown had the highest log likelihood (−50397.931463). Overall, more than 350 000 trees were examined

In addition to the nucleomorph genes it is also possible to use plastid sequences to identify the evolutionary affinities of the endosymbiont. While plastid small subunit rRNA sequence initially appeared useful in that they did not display a particularly high rate of evolution, they presented another problem in that they had a bias in nucleotide composition (McFadden & al. 1995). The chlorarachniophyte plastid rRNA sequences have a relatively high AT content (54–56%) which can confound the commonly used tree inference tools (Lockhart & al. 1994). Using the LogDet transformation, we were able to compensate for this compositional bias and recover trees indicating a green algal ancestry of the endosymbiont (McFadden & al. 1995). Analysis of the plastid small subunit rRNAs using transversion distance (another method circumventing problems with nucleotide compositional bias) also identified the endosymbiont as a chlorophycean green alga (Van De Peer & al. 1996). These trees, based on nucleotide on nucleotide sequence, are also in reasonably close agreement with trees of the large subunit of Rubisco (*rbc*L) (McFadden & al. 1995). Since the *rbc*L trees were inferred from amino acid sequence, they are presumed to be largely free of the effects of nucleotide bias (McFadden & al. 1995). Further, confirmation of the green algal ancestry of chlorarachniophyte endosymbionts has come from analysis of plastid large subunit rRNAs (Lemieux & al., unpubl.) which again corroborates data indicating a green algal ancestry for the endosymbiont. The now consistent finding that the chlorarachniophyte endosymbiont was a green alga, while the cryptomonad endosymbiont was a red alga (e. g., McFadden & al. 1995, Van De Peer & al. 1996, Cavalier-Smith & al. 1996; Fig. 5) excludes the possibility that cryptomonads and chlorarachniophytes arose from a common secondary endosymbiosis (Cavalier-Smith 1993, Cavalier-Smith & al. 1994). Early trees grouping sequences from chlorarachniophyte and cryptomonad nucleomorphs (e.g., Cavalier-Smith & al. 1994, Bhattacharya & al. 1995a, Bhattacharya & Medlin 1995) are now believed to have suffered from the long-branches-attract artefact (Cavalier-Smith & al. 1996, Van De Peer & al. 1996).

The identification of the chlorarachniophyte endosymbiont is of more than academic interest. Knowing that the engulfed alga was similar to a *Chlorella* or *Chlamydomonas* would allow us to make predictions about its original condition and extrapolate what molecular and morphological modifications have taken place during its tenure as endosymbiont. For instance, chlorarachniophyte nucleomorph genes contain a staggering 5.5 introns per kilobase of coding sequence (Gilson & McFadden 1996). This extraordinarily high density of introns is intriguingly similar to that of *Chlamydomonas*, which carries 5.8 introns per kilobase (Logsdon & Palmer, pers. comm.). It is not yet known if the introns in chlorarachniophyte nucleomorph genes were present prior to the endosymbiosis, or whether they proliferated after the endosymbiosis (Gilson & McFadden 1996). In light of the trees showing a close relationship between chlorophycean algae and the chlorarachniophyte endosymbiont (Van De Peer & al. 1996; Lemieux & al., unpubl.; Cavalier-Smith & al. 1996; Fig. 5) the similar intron densities in chlorarachniophyte nucleomorph genes and *Chlamydomonas* nuclear genes favour the former scenario. Intriguingly, the chlorarachniophyte nucleomorph gene introns are markedly smaller than *Chlamydomonas* introns. If the chlorarachniophyte nucleomorph gene introns were present in the endosymbiont's nuclear genome prior to endosymbiosis, they likely underwent severe reduction in the course of subsequent evolution. Given the overall tendency for size reduction, it is not surprising that the introns should have been 'pruned'. The narrow size window of chlorarachniophyte nucleomorph gene introns (all 18, 19 or 20 bp) perhaps suggests that size reduction has converged to a limiting point. It will be interesting now to compare intron position and size from chlorarachniophyte nucleomorph genes with homologues from chlorophycean algae like *Chlamydomonas* and *Chlorella*.

The host. While much attention has focused on the identity of the endosymbiont, the host's evolutionary affinities have also been revealed recently. BHATTACHARYA & al. (1995b) have demonstrated that chlorarachniophyte hosts are sister to the filose amoebae (*Euglypha* and *Paulinella*). Neither of these amoebae contains a eukaryotic endosymbiont so the chlorarachniophyte lineage presumably diverged at a point subsequent to the establishment of the endosymbiotic relationship. Curiously, one species of *Paulinella* (*P. chromatophora*) contains cyanelle-like structures (KIES 1974). *Paulinella* and the filose amoebae/chlorarachniophyte lineage are not closely related to glaucophytes (which possess true plastid-like cyanelles) so significance of the cyanelle-like structures in *Paulinella* remains enigmatic. A second group of eukaryotes, the sarcomonads, is also closely allied to the chlorarachniophyte host and filose amoeba (CAVALIER-SMITH 1995b). These organisms, which exist as flagellates or amoeba, have no photosynthetic representatives. A third group of amoeboid organisms, the plasmodiophorans which are pathogens of plants, also groups with chlorarachniophytes/filose amoebae (CAVALIER-SMITH & al. 1996; Fig. 5) indicating that the chlorarachniophyte host arose from a large and diverse assemblage of amoeboid organisms.

Concluding remarks

The chlorarachniophytes have rocketed from biological obscurity to being forerunners as a model for understanding the principles of secondary endosymbiotic acquisition of plastids (PALMER & DELWICHE 1996). Identification of extant relatives of the two parts of this evolutionary amalgam allow us to reconstruct endosymbiotic events (either conceptually or perhaps even physically) that led to a novel lineage of organisms.

The key to this story has been the retention of the nucleomorph. Molecular dissection of this most unusual genome identifies it as an interesting and tractable model for investigating eukaryotic genome architecture. The nucleomorph also holds the promise of discovering novel genetic phenomena. The study of stripped down genomes (i.e. viruses, mitochondria, chloroplasts and parasite nuclear genome) has revealed a cornucopia of peculiar genetic processes such as RNA editing, trans splicing, overlapping genes, intron homing, eukaryotic operons, and altered genetic codes. Since the chlorarachniophyte nucleomorph is the smallest known eukaryotic genome, its modus operandi will be anything but conventional.

References

BEUTLICH, A., SCHNETTER, R., 1993: The life cycle of *Cryptochlora perforans* (*chlorarachniophyceae*). – Bot. Acta **106**: 441–447.

BHATTACHARYA, D., MEDLIN, L., 1995: The phylogeny of plastids: a review based on comparison of small-subunit ribosomal RNA coding regions – J. Phycol. **31**: 489–498.
– HELMCHEN, T., MELKONIAN, M., 1995a: Molecular evolutionary analysis of nuclear-encoded small subunit ribosomal RNA identify an independent rhizopod lineage containing the *Euglyphidae* and the *chlorarachniophyta.* – J. Euk. Microbiol. **42**: 65–69.
– – BIBEAU, C., MELKONIAN, M., 1995b: Comparisons of nuclear-encoded small-subunit ribosomal RNAs reveal the evolutionary position of the *Glaucocystophyta.* – Molec. Biol. Evol. **12**: 415–420.
BIESSMANN, H., MASON, J., 1994: Telomeric repeat sequences. – Chromosoma **103**: 154–161.
CALDERON-SAENZ, E., SCHNETTER, R., 1987: *Cryptochlora perforans*, a new genus and species of algae (*chlorarachniophyta*), capable of penetrating dead algal filaments. – Pl. Syst. Evol. **158**: 69–71.
– – 1989: Morphology, biology, and systematics of *Cryptochlora perforans* (*chlorarachniophyta*), a phagotrophic marine alga. – Pl. Syst. Evol. **163**: 165–176.
CAVALIER-SMITH, T., 1993: Kingdom *Protozoa* and its 18 phyla. – Microbiol. Rev. **57**: 953–994.
– 1995a: Membrane heredity, symbiogenesis, and the multiple origins of algae. – In ARAI, R., KATO, M., DOI, Y., (Eds): Biodiversity and evolution, pp. 75–114. – Tokyo: National Science Museum.
– 1995b: Zooflagellate phylogeny and classification – Cytology **37**: 1010–1029.
– ALLSOPP, M., CHAO, E., 1994: Chimeric conundra: are nucleomorphs and chromists monophyletic or polyphyletic? – Proc. Natl. Acad. Sci. USA **91**: 11368–11372.
– COUCH, J., THORSTEINSEN, K., GILSON, P., DEANE, J., HILL, D., McFADDEN, G., 1996: Cryptomonad nuclear and nucleomorph 18S rRNA phylogeny. – Eur. J. Phycol. **31**: 315–328.
GEITLER, L., 1930: Ein grünes Filarplasmodium und andere neue Protisten. – Arch. Protistenk. **69**: 615–635.
GILSON, P., McFADDEN, G., 1995: The chlorarachniophyte: a cell with two different nuclei and two different telomeres. – Chromosoma **103**: 635–641.
– – 1996: The miniaturised nuclear genome of a eukaryotic endosymbiont contains genes that overlap, genes that are contranscribed, and smallest known spliceosomal introns. – Proc. Natl Acad. Sci. USA **93**: 7737–7742.
– WALLER, R., McFADDEN, G., 1995: Preliminary characterization of chlorarachniophyte mitochondrial DNA – J. Euk. Microbiol. **42**: 696–701.
GRELL, K., 1990: Indications of sexual reproduction in the plasmodial protist *Chlorarachnion reptans* GEITLER. – Z. Naturforsch. **45c**: 112–114.
HARA, Y., ERATA, M., ISHIDA, K.-I., 1992: Nucleomorphs – new plant cell images of an evolutionary chimaera. – Pl. Cell Technol. **4**: 373–382.
HATAKEYAMA, N., SASA, T., WATANABE, M., TAKAICHI, S., 1991: Structure and pigment composition of *Chlorarachnion* sp. – J. Phycol. [Suppl.] **27**: 29–A154.
HIBBERD, D., 1990: Phylum *Chlorarachnida.* – In MARGULIS, L., CORLISS, J., MELKONIAN, M., CHAPMAN, D., (Eds): Handbook of *Protoctista*, pp. 288–292. – Boston: Jones & Bartlett.
– NORRIS, R., 1984: Cytology and ultrastructure of *Clorarachnion reptans* (*Chlorarachniophyta* Divisionova, *Chlorarachniophyceae* Classis Nova) – J. Phycol. **20**: 310–330.
HIGASHIYAMA, T., MAKI, S., YAMADA, T., 1995: Molecular organization of *Chlorella vulgaris* chromosome I. – Presence of telomeric repeats that are conserved in higher plants. – Molec. Gen. Genet. **246**: 29–36.

Ishida, K. -I., 1994: *Chlorarachniophyceae*. – In Hori, T., (Ed.): An illustrated atlas of the life history of algae. 3. Unicellular and flagellated algae, pp. 203–213. – Tokyo: Uchida Rokakuho.

– Hara, Y., 1994: Taxonomic studies on the *Chlorarachniophyta*. I. *Chlorarachnion* globosum sp. nov. – Phycologia **33**: 351–358.

Kies, L., 1974: Electron microscopical investigations on *Paulinella chromatophora* Lauterborn, a thecamoeba containing blue-green endosymbioints (cyanelles). – Protoplasma **80**: 69–89.

Lockhart, P., Steel, M., Hendy, M., Penny, D., 1994: Recovering evolutionary trees under a more realistic model of sequence evolution. – Molec. Biol. Evol. **11**: 605–612.

Ludwig, M., Gibbs, S., 1987: Are the nucleomorphs of cryptomonads and *Chlorarachnion* the vestigial nuclei of eukaryotic endosymbionts? – Ann. New York Acad. Sci. **503**: 198–211.

– – 1989: Evidence that nucleomorphs of Chlorarachnion reptans (*Chlorarachniophyceae*) are vestigial nuclei: morphology, division and DNA-DAPI fluorescence. – J. Phycol. **25**: 385–394.

McFadden, G., 1992: Evolution of algal plastids from eukaryotic endosymbionts. – In Harris, N., Williams, D., (Eds): In situ hybridization: application to developmental biology and medicine., pp. 143–156. – Cambridge: Cambridge University Press.

– Gilson, P., 1995: Something borrowed, something green: lateral transfer of chloroplasts by secondary endosymbiosis. – Trends Ecol. Evol. **10**: 12–17.

– – 1996: What's eating Eu? The role of eukaryote/ eukaryote endosymbioses in plastid origins. – Endocyt. Cell Res. (in press).

– Hofmann, C., Adcock, G., Maier, U.-G., 1994: Evidence that an amoeba acquired a chloroplast by retaining part of an engulfed eukaryotic alga. – Proc. Natl. Acad. Sci. USA **91**: 3690–3694.

– – Waller, R., 1995: Molecular phylogeny of chlorarachniophytes based on plastid rRNA and *rbc*L sequences. – Arch. Protistenk. **145**: 231–239.

Norris, R., 1967: Micro-algae in enrichment cultures from Puerto Penasco, Sonora, Mexico. – Bull. South. Cal. Acad. Sci. **66**: 233–250.

Palmer, J., Delwiche, C., 1996: Second-hand chloroplasts and the case of the disappearing nucleus. – Proc. Natl. Acad. Sci. USA **93**: 7432–7435.

Sasa, T., Takaichi, S., Hatakeyama, M., Watanabe, M., 1992: A novel carotenoid ester, loroxanthin dodecanoate, from *Pyramimonas parkeae* (*Prasinophyceae*) and a chlorarachniophycean alga. – Pl. Cell Physiol. **33**: 921–925.

Shippen, D., 1993: Telomeres and telomerases. – Curr. Opin. Genet. Devel. **3**: 759–763.

Van De Peer, Y., Rensing, S., Maier, U.-G., De Wachter, R., 1996: Substitution rate calibration of small subunit rRNA identifies chlorarachniophyte endosymbionts as remnants of green algae. – Proc. Natl Acad. Sci. USA **93**: 7732–7736.

Phylogenetic relationships of the 'golden algae' (haptophytes, heterokont chromophytes) and their plastids

Linda K. Medlin, Wiebe H. C. F. Kooistra, Daniel Potter, Gary W. Saunders, and Robert A. Andersen

Key words: *Chrysophyceae, Haptophyceae.* – Actin, chloroplast, chromophyte, diatoms, heterokont, plastid, phylogeny, *rbc*L, stramenopiles, SSU rRNA, *tuf*A.

Abstract: The phylogenetic relationships of the "golden algae", like all algae, were rarely addressed before the advent of electron microscopy because, based upon light microscopy, each group was so distinct that shared characters were not apparent. Electron microscopy has provided many new characters that have initiated phylogenetic discussions about the relationships among the "golden algae". Consequently, new taxa have been described or old ones revised, many of which now include non-algal protists and fungi. The haptophytes were first placed in the class *Chrysophyceae* but ultrastructural data have provided evidence to classify them separately. Molecular studies have greatly enhanced phylogenetic analyses based on morphology and have led to the description of additional new taxa. We took available nucleotide sequence data for the nuclear-encoded SSU rRNA, fucoxanthin/chlorophyll photosystem I/II, and actin genes and the plastid-encoded SSU rRNA, *tuf*A, and *rbc*L genes and analysed these to evaluate phylogenetic relationships among the "golden algae", viz., the *Haptophyceae* (= *Prymnesiophyceae*) and the heterokont chromophytes (also known as chromophytes, heterokont algae, autotrophic stramenopiles). Using molecular clock calculations, we estimated the average and earliest probable time of origin of these two groups and their plastids. The origin of the haptophyte host-cell lineages appears to be more ancient than the origin of its plastid, suggesting that an endosymbiotic origin of plastids occurred late in the evolutionary history of this group. The pigmented heterokonts (heterokont chromophytes) also arose later, following an endosymbiotic event that led to the transfer of photosynthetic capacity to their heterotrophic ancestors. Photosynthetic haptophytes and heterokont chromophytes both appear to have arisen at or shortly before the Permian-Triassic boundary. Our data support the hypothesis that the haptophyte and heterokont chromophyte plastids have independent origins (i.e., two separate secondary endosymbioses) even though their plastids are similar in structure and pigmentation. Present evidence is insufficient to evaluate conclusively the possible monophyletic relationship of the haptophyte and heterokont protist host cells, even though haptophytes lack tripartite flagellar hairs. The molecular data, albeit weak, consistently fail to present the heterokont chromophytes and haptophytes as monophyletic. Phylogenetic resolution among all classes of heterokont chromophytes remains elusive even though molecular evidence has established the phylogenetic alliance of some classes (e.g., *Phaeophyceae* and *Xanthophyceae*).

The "golden" algae are today commonly referred to as chromophyte algae, heterokont algae or autotrophic stramenopiles, and historically have included the haptophyte algae. They range in size from minute picoplankton (1–2 μm) to the large kelps or brown seaweeds (40 m). Brown seaweeds were utilised by humans before recorded history, and almost certainly humans have cursed diatoms innumerable times as they slipped on rocks in streams. LINNAEUS (1753) described several genera, and of these, *Fucus* is still retained as a valid genus of brown seaweeds. Shortly after LINNAEUS' seminal publication, additional macroscopic algae were described (e.g., STACKHOUSE 1809, LAMOUROUX 1813, AGARDH 1820) that are now classified in the *Phaeophyceae*. LAMOUROUX (1813) and HARVEY (1836) made a major contribution to the classification of algae when they introduced the concept of colour, or pigmentation, as an important taxonomic feature for distinguishing major groups (viz., green, brown and red algae). The microscopic "golden" algae were discovered and reported by an entirely different group of workers, the early microscopists who were studying the *Infusoria* (microscopic organisms). For example, MÜLLER (1786), EHRENBERG (1838), RABENHORST (1853) and STEIN (1878) described microscopic organisms that are today considered relatives of the brown seaweeds. Unlike the macroalgae, which were clearly viewed as plants, the microalgae were frequently placed in the kingdom *Animalia* because they are often motile.

The evolutionary relationships among these algae have been controversial as documented in their long and complicated taxonomic history. The recognition of the relationship between the large, plant-like brown seaweeds and the small golden, yellow or brown microalgae first began with the works of KLEBS (1893) and BLACKMAN (1900). BLACKMAN (1900), who is best known for his ideas on the volvocine, tetrasporine and other green algal lineages, believed green algae gave rise to green plants via an ever-increasing degree of complexity and size, and proposed a similar scheme for "golden" algae. BLACKMAN placed the simple flagellate *Chromulina* at the base of the brown lineage and, with increasing size and complexity, the evolutionary lineage culminated with the brown seaweeds (see Chapter 12). In a separate but parallel scheme, he proposed an evolutionary tree for the yellow-green algae, which he considered to be distinct from, but related to, the brown lineage (BLACKMAN 1900). Beginning near the turn of the century, PASCHER began working extensively on "golden" algae, and he proposed a number of taxonomic changes and phylogenetic hypotheses that are relevant to this chapter. PASCHER (1913) combined the chrysomonads of KLEBS (1893), the heterokonts of LUTHER (1899) and the diatoms (e.g., KÜTZING 1834, RABENHORST 1853) into the single division *Chrysophyta*. The *Chrysophyta* stood as an equal taxonomic group to the division *Phaeophyta*, although it was still implicit, if not explicit, that the *Chrysophyta* gave rise to the evolutionarily advanced *Phaeophyta*. PASCHER (1910) also made another far-reaching taxonomic decision when he placed the haptophycean family *Isochrysidaceae* with the other chrysophytes having two equal flagella. This family united the organisms currently placed in the *Haptophyceae* (e.g., *Hymenomonas*) with certain other golden algae (e.g., *Synura, Syncrypta*). As more genera and species of haptophytes were described (e.g., LOHMANN 1913, LACKEY 1939), taxonomists followed PASCHER's classification and placed these haptophytes in the class *Chrysophyceae* (see BOURRELLY 1957).

Although the first half of the twentieth century brought the description of many species, the higher level taxonomic groups of algae were usually not treated in a phylogenetic sense. For example, FRITSCH (1945) states "The *Phaeophyceae* present no obvious affinities with any other class and are indeed in most respects so sharply circumscribed that little opportunity is afforded for speculations on their relationships.... On present evidence this class must be regarded as an altogether distinct evolutionary line (*Phaeophyta*)." Similarly, evolutionary relationships among the green and red algae, cryptomonads, dinoflagellates, etc., were rarely discussed during this time because shared characters were not obvious.

At the midpoint of the twentieth century, CHADEFAUD (1950) published a seminal paper in which he erected a new group, the *Chromophyceae*, based upon similarities of the flagella. This paper not only established the chromophytes sensu lato, but it also marked the first of many papers in the second half of the century that would address issues of algal phylogeny. CHADEFAUD combined the euglenoids, dinoflagellates, cryptophytes, chrysophytes, raphidophytes, brown algae and certain protozoans into a large group that was equal in stature to the "Blue-Green Algae", "Red Algae" and "Green Algae". Probably all modern workers exclude some of the organisms included by CHADEFAUD (viz., euglenoids, choanoflagellates), and many other workers exclude the cryptomonads and the dinoflagellates; however, the concept of a "chromophyte" group still exists. The group has been modified or renamed by several workers. For example, CHRISTENSEN (1962, 1989) proposed the division *Chromophyta* for algae lacking chlorophyll *b*; CAVALIER-SMITH (1986) proposed the kingdom *Chromista* for organisms having chlorophyll *c*, chloroplast endoplasmic reticulum (CER) and tripartite tubular hairs; PATTERSON (1989) proposed the stramenopiles for organisms having tripartite tubular hairs; and both VAN DEN HOEK (1978) and MOESTRUP (1992) expanded LUTHER'S name *Heterokonatae* to include not only the yellow-green algae (including freshwater raphidophytes) (sensu LUTHER 1899) but also all algae with tripartite tubular hairs. Conversely, CAVALIER-SMITH (1986) suggests that the haptophytes, heterokonts and cryptomonads are a monophyletic group (kingdom *Chromista*) that excludes the dinoflagellates.

Even today, there is no consensus on which organisms belong within this group, and no single name is in use. The rapid accumulation of ultrastructural, plastid pigment and molecular data have resulted in both the re-definition of old names and the creation of new names for the "golden algae." There are two reasons for this: (1) the group contains both pigmented and non-pigmented organisms so that "algae", "fungi" and "protozoa" must be contended with, and (2) there has been no unequivocal evidence that supports a single phylogenetic hypothesis. The two most widely used names, "chromophyte" and "heterokont", have changed in opposite ways: the chromophytes have become more restrictive by the removal of taxa from CHADEFAUD'S (1950) original definition, and the heterokonts have become more expansive by the addition of taxa to LUTHER'S (1899) original definition. CAVALIER-SMITH & CHAO (1996) have detailed much of the taxonomic nomenclature relating to these and other names. Thus, the "golden" algae discussed in this chapter are referred to in the literature by simple names, such as the chromophytes, chromists, heterokonts and stramenopiles, as well as by compound names, such as the heterokont chromophytes or the pigmented stramenopiles. We use the name "heterokont chromophyte" because it acknowledges the two most commonly used names, and it indicates that most flagellate cells have "heterokont" flagella (= tripartite hairs, not heterodynamic flagellar beating) as well as "chromophyte" pigmentation (light-harvesting carotenoids, most with chlorophyll $a + c$). The haptophycean algae are included within the broad definition of the "golden algae" for this chapter, but the dinoflagellates and the cryptomonads are not.

Haptophyceae. The haptophyte algae were initially recognised as distinct from the *Chrysophyceae, Phaeophyceae* or *Xanthophyceae* with respect to ultrastructural featuree (PARKE & al. 1955, 1956, 1958). Ultimately, the class *Haptophyceae* was erected (CHRISTENSEN 1962) and its members were considered separate from, but related to, the *Chrysophyceae*. This decision was not met with universal support, as BOURRELLY (1968), STARMACH (1985) and others continued to place the haptophytes within the *Chrysophyceae*. However, after HIBBERD (1976) summarised the similarities and differences, he found little evidence for retaining the haptophyte taxa in the *Chrysophyceae*. Subsequently, most workers have considered the *Haptophyceae* to be distinct from the *Chrysophyceae* but often with some close, but undescribed, evolutionary relationship between the two classes. The unique or distinctive characters that separate the haptophytes from the *Chrysophyceae* and other heterokont chromophytes are: (1) haptonema, (2) flagellar transitional region and microtubular roots, (3) mitosis, (4) calcium carbonate biomineralisation (e.g., coccoliths in some representatives), (5) absence of tripartite flagellar hairs and (6) no plastid girdle lamellae (HIBBERD 1976, GREEN & al. 1989, GREEN & LEADBEATER 1994). Gene sequence data, which have been reported during the past few years, also suggest that the haptophytes are distantly related to the *Chrysophyceae* as well as to any other heterokont chromophytes (BHATTACHARYA & al. 1992, LEIPE & al. 1994, BHATTACHARYA & MEDLIN 1995, SAUNDERS & al. 1995, CARON & al. 1996, CAVALIER-SMITH & al. 1996, GREEN & DUNFORD 1996, MEDLIN & al. 1996a).

Heterokont chromophytes. The heterokont chromophytes belong to a larger group of heterokont organisms (= stramenopiles) that can be characterised as follows: (1) two flagella are typically present and they are usually of distinctly different lengths, (2) the two flagella have different patterns of motion (stiff sinusoidal beat vs. irregular undulations), (3) one flagellum typically bears two rows of tripartite flagellar hairs (VLK 1938, DODGE 1975), (4) the flagellar hairs provide a reverse thrust to the flagellar beat and therefore pull the cell rather than push it (SLEIGH 1989), (5) presence of a girdle lamella (except in the *Eustigmatophyceae*) and (6) silica biomineralisation when mineralisation is present (viz., diatoms, silica-scaled chrysophytes and synurophytes, silicoflagellates). Three features, the β-1,3-linked glucan carbohydrate storage product (CRAIGIE 1974, WANG & BARTNICKI-GARCÍA 1974), the chloroplast endoplasmic recticulum (see review in GIBBS 1993) and tubular mitochondrial cristae (TAYLOR 1976, STEWART & MATTOX 1970), are shared with the *Haptophyta.*

Electron microscopic studies have contributed substantially to our understanding of this group, providing a suite of putatively homologous characters for algal systematics. These new observations have led to the description of several new classes of heterokont chromophytes, viz., the *Eustigmatophyceae* (Hibberd & Leedale 1971), the *Dictyochophyceae* (now including the *Pedinellophyceae*) (Silva 1980), the *Synurophyceae* (Andersen 1987), the *Coscinodiscophyceae* and *Fragilariophyceae* (Round & al. 1990) and the *Pelagophyceae* (Andersen & al. 1993). However, ultrastructural data alone have been unable to resolve the phylogenetic relationships of the heterokont chromophytes (e.g., Andersen 1991, Williams 1991). In contrast, molecular data have resolved a number of phylogenetic relationships in this group. For example, SSU rRNA data have shown a relationship between the *Phaeophyceae* and the *Xanthophyceae* (Ariztia & al. 1991, Potter & al. 1997), between the *Chrysophyceae* and *Synurophyceae* (Ariztia & al. 1991, Bhattacharya & al. 1992), between the *Dictyochophyceae* and *Pelagophyceae* (Saunders & al. 1995) and between the *Sarcinochrysidales* sensu stricto and the *Pelagophyceae* (Saunders & al. 1997b). Nonetheless, these studies and others have failed to resolve unequivocally the relationships among the deeper branching heterokont chromophytes.

Hypotheses

Today, many questions remain unresolved regarding the phylogeny of the "golden" algae, but we will address two major issues in this chapter. The first is whether or not the haptophyte algae have any close evolutionary relationship with the heterokont chromophytes. The second question concerns the phylogenetic relationships among the heterokont chromophytes themselves. We have examined these relationships using new and/or existing molecular and morphological data. We proposed several hypotheses relative to these relationships and evaluated the data to determine if support for one hypothesis over another can be found. The hypotheses are outlined below.

Relationships between the haptophyte and heterokont algae. Relationships between these two groups are confounded by the questions of the monophyly of the host cells and the number of endosymbioses giving rise to their plastids. Thus, we can formulate four scenarios/hypotheses (1a–d) to explain their evolution.

Hypothesis 1a. The heterokonts and haptophytes form a monophyletic group that gained their plastids as the result of single secondary endosymbiotic event. That is, the host cells and the plastids from the two groups will have similar phylogenies because they share the same evolutionary history. Thus in the molecular analyses, both the two host cells and the two plastids should be each others' sister group, respectively.

Hypothesis 1b. The heterokonts and the haptophytes are a monophyletic group, but after their divergences, each acquired its plastid through independent endosymbioses. Thus, in the molecular analyses, the host cells are each others' sister group but their plastids are not.

Hypothesis 1c. The heterokonts and the haptophytes are not a monophyletic group. However, they both engulfed and retained a similar eukaryotic cell as their plastid. Thus in the molecular analyses, the host cells are not each others' sister group but their plastids are.

Hypothesis 1d. The heterokonts and the haptophytes are not a monophyletic group. Each gained their plastids from separate secondary endosymbiotic events. Therefore, neither the host cells nor the plastids for the two groups will have similar phylogenies because each has had an independent evolutionary history.

Hypotheses 1a, 1c, and 1d allow the possibility of plastid gain and loss in the heterokont lineage of major groups, e.g., the oomycetes.

Relationships between non-photosynthetic and photosynthetic heterokonts.

Hypothesis 2a. The heterokont chromophytes form a monophyletic group that does not include the major non-photosynthetic heterokont groups (i.e., oomycetes, hyphochytrids, thraustrochytrids).

Hypothesis 2b. The heterokont chromophytes are not a monophyletic group, i.e., one or more of the major non-photosynthetic heterokont lineages is included within the clade of heterokont chromophytes.

Relationships within the heterokont chromophytes.

Hypothesis 3a. The heterokont chromophytes contain two major monophyletic lineages: one with a well-developed flagellar apparatus and one with an often highly reduced flagellar apparatus. Those with a well-developed flagellar apparatus have microtubular roots, and the flagellar apparatus is typically distant from the nucleus, the two components often being connected via a striated rhizoplast. Those with a highly reduced flagellar apparatus typically lack microtubular roots, and the flagellar apparatus is often closely associated or directly in contact with the nuclear envelope.

Hypothesis 3b. The flagellar apparatus of heterokont chromophytes has been reduced two or more times independently, and the two lineages are not monophyletic.

Hypothesis 4a. The heterokont chromophytes contain two major monophyletic lineages: one has a diatoxanthin/diadinoxanthin-containing light-harvesting complex and the other has a violaxanthin-containing light-harvesting complex.

Hypothesis 4b. The pigmentation of heterokont chromophytes has evolved two or more times independently, and the two pigment groups are polyphyletic.

To evaluate these hypotheses, we have inferred phylogenies from both nuclear and plastid genes and compared these to other published phylogenies, where applicable. In each case we have focused on: (1) relationships between the major groups (first set of hypotheses) and (2) relationships within the heterokont group (second, third and fourth sets of hypotheses).

Our phylogenetic trees were constructed from nuclear and plastid ribosomal and protein-coding genes using the neighbor-joining method (SAITOU & NEI 1987, except for Figs. 1 & 2). Interpretations of bootstrap support for these trees is based on the analysis of bootstrap accuracy and repeatability by HILLIS & BULL (1993). With symmetrical phylogenies having an internodal change of $< 20\%$ and approximately equal rates of change, we interpret a bootstrap proportion of $\geqslant 70\%$ to indicate a $\geqslant 95\%$ probability that the recovered clade represents a true clade. We qualify our interpretation because readers may disagree with our interpretation of bootstrap support.

Nuclear genome

Among the nuclear genes we have selected for our analyses are the small subunit of the ribosomal cistron (e.g., BHATTACHARYA & al. 1992; LEIPE & al. 1994; SAUNDERS & al. 1995; CAVALIER-SMITH & al. 1996; MEDLIN & al. 1996b, c) and the multi-gene families of the fucoxanthin/chlorophyll photosystem I & II binding proteins (FCP) and the actin protein (LA ROCHE & al. 1994, CARON & al. 1996, BHATTACHARYA & EHLTING 1995, GREEN & DURNFORD 1996) (Figs. 1–3). The ribosomal genes occur in many, perhaps hundreds of copies per cell and are generally believed to evolve in a concerted fashion such that all copies are homogenised and should encode the same coding region. This effectively ensures that the ribosomal phylogenies more accurately represent species trees rather than gene trees (HILLIS & al. 1996). The phylogenies of the other two genes more likely reflect those of gene trees. It is generally believed that as more genes are compared, a better picture of the evolution of the group can be achieved as the phylogenies may converge upon one another.

In each of the three Figures presented for our analysis using nuclear genes, the heterokonts and the haptophytes are shown as separate, monophyletic groups and are never found as sister taxa. There is, however, an association of the heterokonts with the alveolates, which includes the dinoflagellates as their autotrophic members, in both the SSU rRNA analysis (Fig. 1) and in the FCP binding proteins (Fig. 2). Bootstrap support for this association is $< 50\%$ in the rRNA phylogeny (Fig. 1), but it is 100% in the FCP photosystem protein phylogeny (Fig. 2, taken from CARON & al. 1996; see also GREEN & DURNFORD 1996). Preliminary results using a combined SSU and LSU data set in a neighbor joining analysis show a strong association of the alveolates (*Prorocentrum, Toxoplasma, Tetrahymena*) with non-pigmented heterokonts (*Phytophthora*, hyphochytrids, bootstrap = 100%) (VAN DER AUWERA & DE WACHTER 1996). Similarly, the alpha tubulin gene shows a relationship between the alveolates and heterokonts (bootstrap = 50%, KEELING & DOOLITTLE 1996), with the haptophytes occupying a more distant relationship.

In the actin tree, the single haptophyte representative (*Emiliania huxleyi*) is separated from the heterokonts (Fig. 3). The short branch lengths with only moderate bootstrap support among the major groups indicate that the actin gene is unable to resolve the branching order of the eukaryotes. No dinoflagellate taxa, and only two heterokont chromophyte taxa, are included.

The molecular data do not support an affiliation of the haptophytes with the heterokonts or with any other eukaryotic group (see the absence of bootstrap support for the association of the haptophytes with any other lineage in the rRNA trees) (Fig. 1). This lack of clear bootstrap support for the haptophytes' nearest neighbor in the rRNA (and likely the actin) phylogeny has generally been assumed to be related to a very rapid evolution that occurred during the major radiation of the eukaryotic lineages, known as the crown radiation (KNOLL 1992, WAINRIGHT & al. 1993). The FCP binding protein phylogeny may help to resolve relationships among eukaryotes during this time frame; however only photosynthetic organisms can be compared, which may be misleading. Nevertheless, the most significant point to be gathered from the rRNA and FCP trees and other phylogenies is the recurrent association of the dinoflagellates with the heterokonts (with high

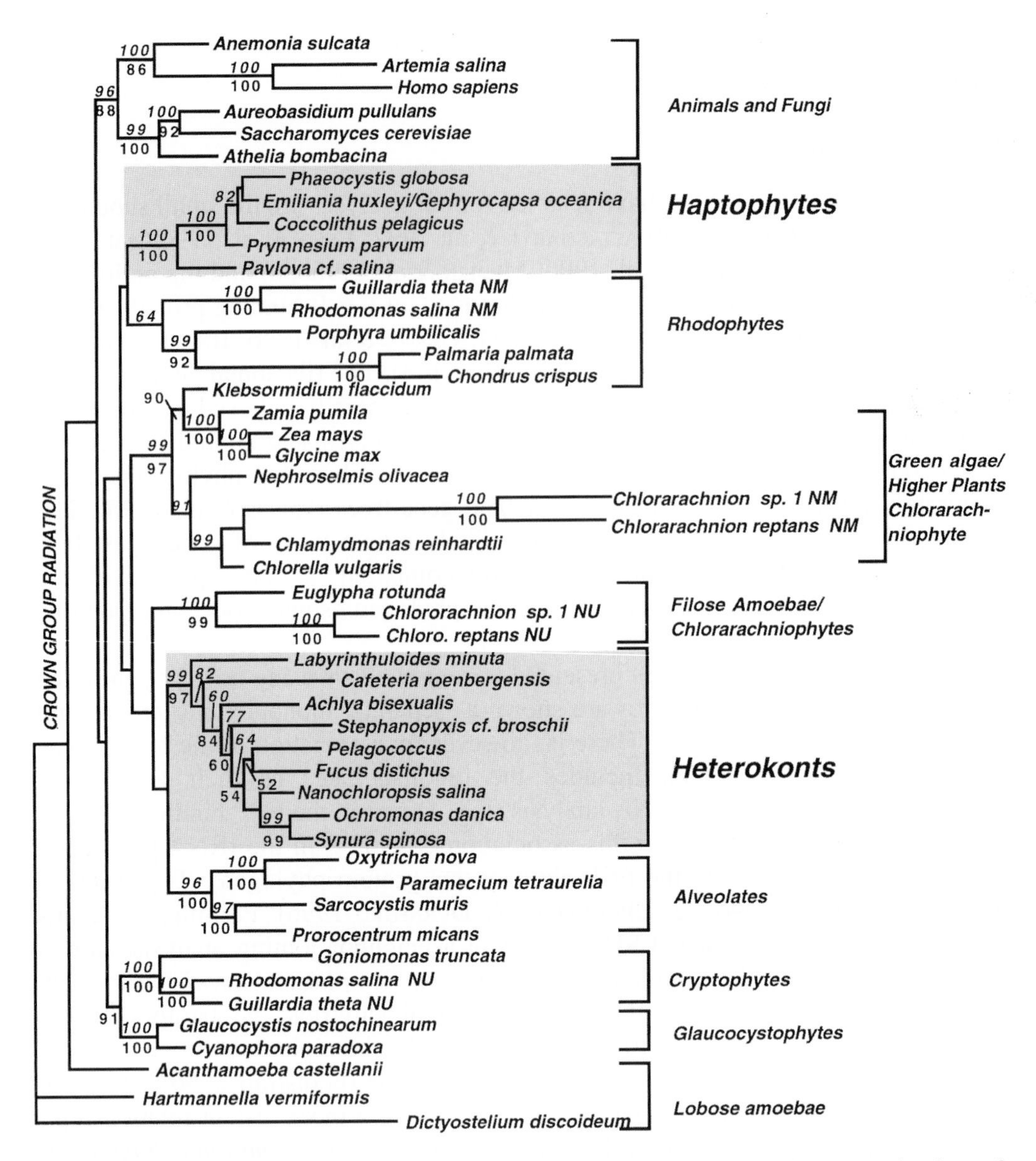

Fig. 1. Phylogenetic analysis of nuclear-encoded SSU rRNA genes from a selection of organisms belonging to the major radiation of eukaryotes using the maximum likelihood method. Representatives of all major eukaryotic groups, especially the algae, are included. Bootstrap values > 50% (100 replications, FELSENSTEIN 1985) from a neighbor-joining analysis (SAITOU & NEI 1987) of a KIMURA (1980) distance matrix and a PAUP (SWOFFORD 1993) weighted maximum parsimony analysis (MEDLIN & al. 1996b) are shown above and below the internal nodes, respectively. The positions of the haptophytes and the heterokonts are highlighted. NM refers to the gene from the nucleomorph or vestigial nucleus within the plastid; NU refers to the gene from the nucleus of the same organism. The tree is rooted within the branch leading to *Dictyostelium*. In addition to the plastid-containing groups named and bracketed in this Figure, two groups within the alveolates also contain plastids. These are the dinoflagellates (here represented by *Prorocentrum micans*) and the apicomplexans (here represented by *Sarcocystis muris*), which are thought to have acquired their plastids independently from one another (KOHLER & al. 1997, DELWICHE & PALMER 1996; see Chapter 3). Two plastid-containing groups that are not shown on this tree are the Euglenophytes, which branch well below the crown-group radiation in rRNA trees, and the enigmatic *Paulinell a chromatophora*, which branches with the filose amoeba *Euglypha rotunda* (BHATTACHARYA & al. 1995)

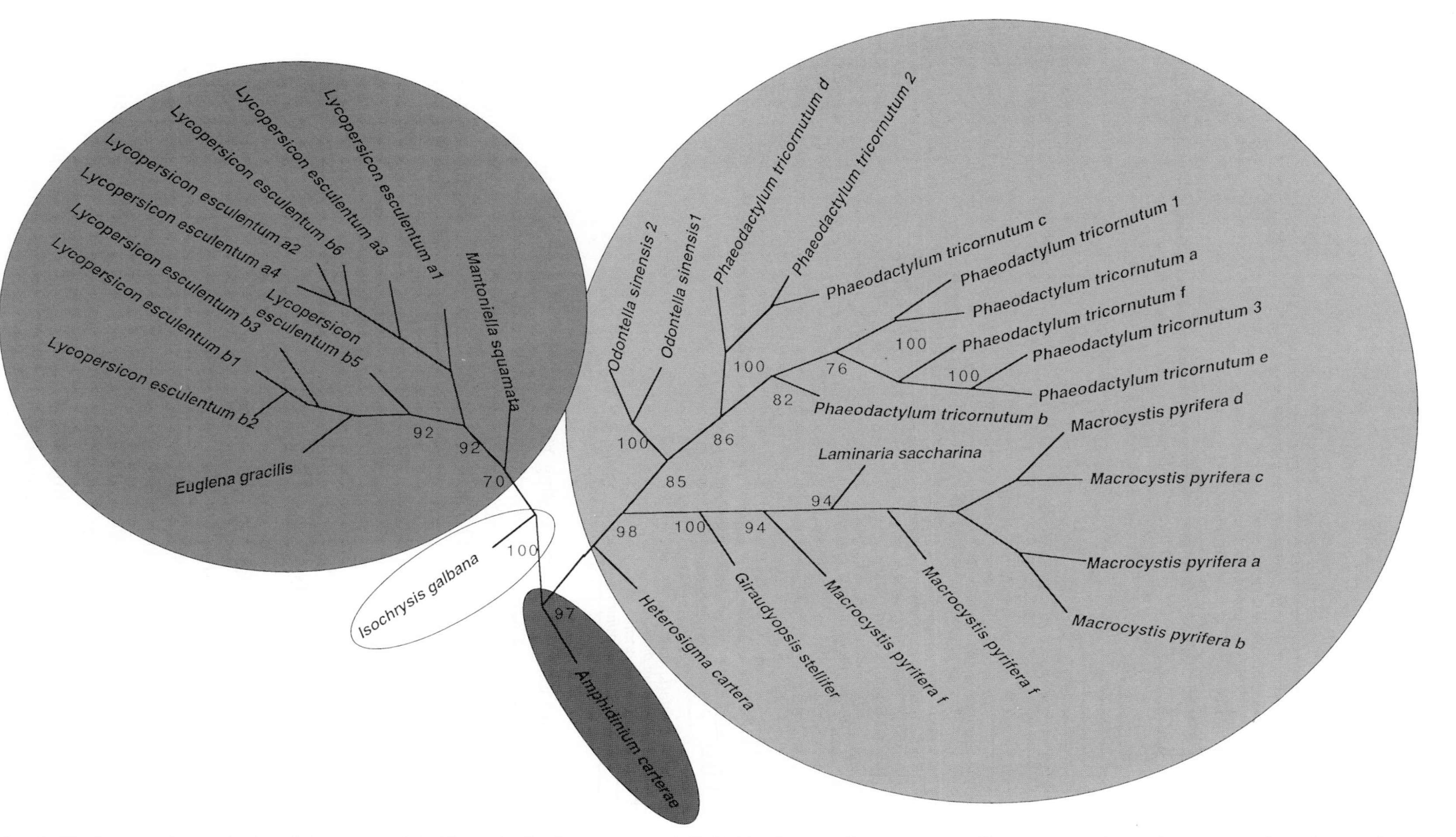

Fig. 2. Phylogenetic analysis of fucoxanthin/chlorophyll photosystem I & II-binding coding regions. This unrooted parsimony tree is redrawn from CARON & al. (1996). Bootstrap values > 90% (50 replications) are shown above the internal nodes. Major groups of algae/higher plants are highlighted as follows: clear haptophytes, light grey heterokonts, medium grey green algae and higher plants, dark grey dinoflagellates

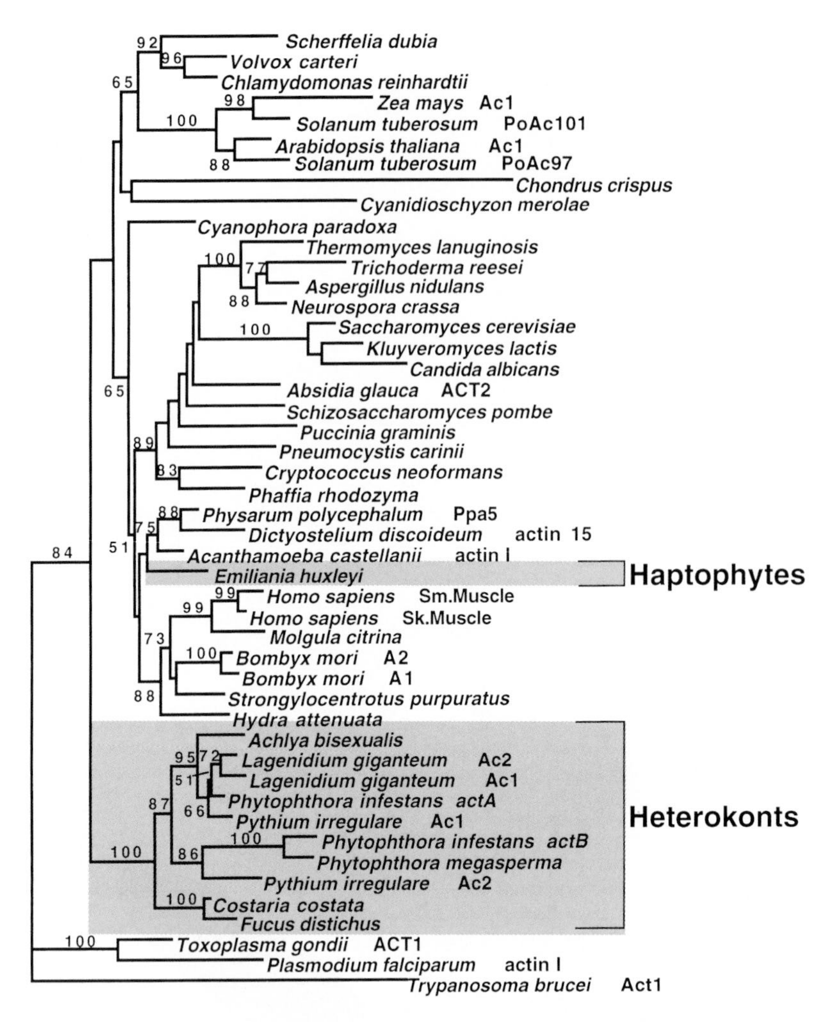

Fig. 3. Phylogenetic analysis of actin coding regions. The neighbor-joining method was used to infer the tree from a KIMURA (1980) distance matrix. Only first and second positions of actin codons were included in the analysis (702 nt). Bootstrap values (100 replications) are shown above the internal nodes. The positions of the haptophytes and the heterokonts are highlighted. The root of this phylogeny lies on the branch leading to the *Trypanosoma brucei* actin sequence

bootstrap support in some analyses) to the exclusion of the haptophytes. These data are consistent with hypotheses 1c and 1d.

Only in the large subunit rRNA phylogeny for major algal groups published to date (PERASSO & al. 1989), do the haptophytes group with the heterokonts, suggesting that hypotheses 1a and b are supported. However, taxon sampling is very low (*Ochromonas, Synura, Vacuolaria, Prymnesium, Cricosphaera*), and this may misrepresent the relationships among these and other groups. A similar relationship between the haptophytes and the heterokonts can be obtained with the SSU rRNA data set if taxon sampling is limited to only a few groups [see relationships of the *Haptophyta* to the *Heterokonta* in CAVALIER-SMITH & al. (1995), which is in contrast to their relationships in CAVALIER-SMITH & al. (1996) and CAVALIER-SMITH & CHAO (1996)]. A more extensive analysis using the entire LSU rRNA molecule for a variety of algal groups is presently being undertaken (G. VAN DER AUWERA, pers. comm.).

The cryptomonads, which according to CAVALIER-SMITH (1986) should be ancestral to the heterokonts and haptophytes in the monophyletic kingdom *Chromista*, are also phylogenetically removed from these two chromist groups. Cryptophytes show a strong relationship with the glaucocystophytes in our rRNA tree (bootstrap = 100%, Fig. 1), and cryptophytes are an independent clade (bootstrap = 99/100%) in another recent study (CAVALIER-SMITH & al. 1994, 1996). Further evidence from the GAPDH gene (LIAUD & al. 1997, bootstrap = 66%), the *sec*Y gene (VOGEL & al. 1996, bootstrap = 74%) and the stress-70 protein gene (RENSING & al. 1996, bootstrap =< 50%) also places the cryptophytes distant from the heterokonts.

Thus, the kingdom *Chromista* does not appear to be monophyletic, i.e., descended from a single endosymbiotic event that transformed its heterotrophic ancestors into "algae". In addition, the haptophytes appear to be a unique lineage with no clear sister taxon revealed. There is a recurrent association of the heterokonts with the alveolates.

Relationships within the *Haptophyta*. A moderate data set is now available for the haptophytes from the SSU rRNA genes (Fig. 4A, MEDLIN & al. 1996a and MEDLIN unpubl.). The haptophyte lineage is undifferentiated for some time after its origin (see point A on Fig. 4B) before it diverges into two groups, which correspond well with the two haptophyte subclasses, the *Pavlovophycidae* and the *Prymnesiophycidae* (JORDAN & GREEN 1994). This divergence is well supported in a bootstrap analysis, which is entirely consistent with the clear morphological differences between the subclasses (JORDAN & GREEN 1994). The *Pavlovophycidae* have unequal flagella with small tubular hairs and lack organic body scales. In contrast the *Prymnesiophycidae* have nearly equal flagella with no flagellar hairs but have organic body scales. In the taxonomic treatment put forth by JORDAN and GREEN, only a single order is retained in each subclass. The *Pavlovophycidae* contains extant species that can be traced back to earlier divergences in the rRNA tree than those in the other subclass. It contains both flagellate organisms plus an undescribed coccoid organism whose taxonomic affinities were only recognised through sequence analysis (POTTER & al. 1996). The remaining haptophytes are divided among three clades. Monophyletic groups within these three clades appear to reflect family level relationships in the *Haptophyta*. However, the genus *Chrysochromulina* is paraphyletic (see also FUJIWARA & al. 1995). It is clear that some key haptophytes (e.g., the *Isochrysidaceae*) are missing from the rRNA tree, and relationships within the tree are likely to change as more taxa are added (compare Fig. 4 with the interpretation of relationships within the haptophytes in rRNA tree in CAVALIER-SMITH & al. 1996). Significantly, all of the coccolithophorids form a monophyletic group with the family *Noelaerhabdaceae* (*Emiliania* and *Gephyrocapsa*) sister to the remainder of the lineage.

Relationships within the *Heterokonta*. Within the heterokont organisms we will discuss the rRNA-generated phylogenies, because these data are the most extensive. All recent rRNA analyses have shown the non-photosynthetic lineages

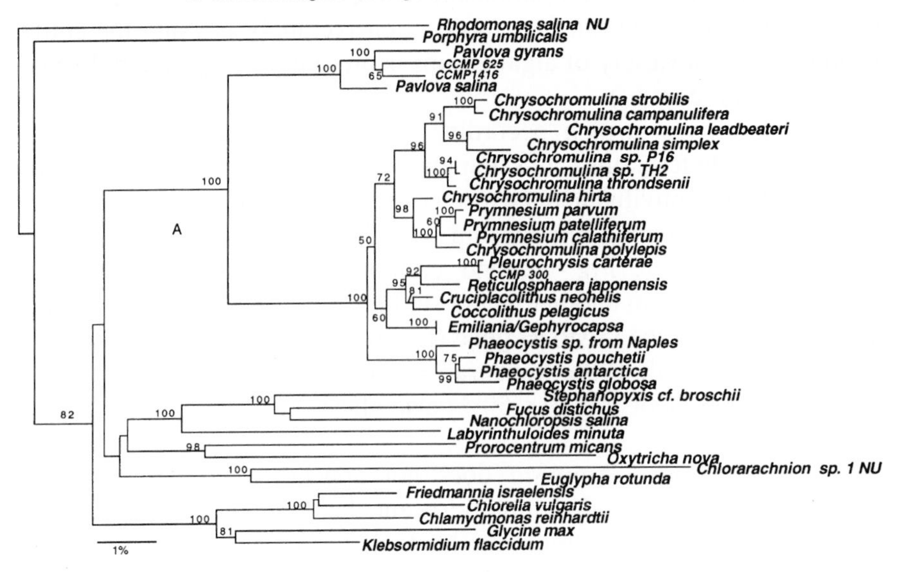
A. 18s rRNA Neighbor-Joining Tree
Rhodomonas salina NU
Porphyra umbilicalis
Pavlova gyrans
CCMP 625
CCMP1416
Pavlova salina
Chrysochromulina strobilis
Chrysochromulina campanulifera
Chrysochromulina leadbeateri
Chrysochromulina simplex
Chrysochromulina sp. P16
Chrysochromulina sp. TH2
Chrysochromulina throndsenii
Chrysochromulina hirta
Prymnesium parvum
Prymnesium patelliferum
Prymnesium calathiferum
Chrysochromulina polylepis
Pleurochrysis carterae
CCMP 300
Reticulosphaera japonensis
Cruciplacolithus neohelis
Coccolithus pelagicus
Emiliania/Gephyrocapsa
Phaeocystis sp. from Naples
Phaeocystis pouchetii
Phaeocystis antarctica
Phaeocystis globosa
Stephanopyxis cf. broschii
Fucus distichus
Nanochloropsis salina
Labyrinthuloides minuta
Prorocentrum micans
Oxytricha nova
Chlorarachnion sp. 1 NU
Euglypha rotunda
Friedmannia israelensis
Chlorella vulgaris
Chlamydmonas reinhardtii
Glycine max
Klebsormidium flaccidum
A
1%

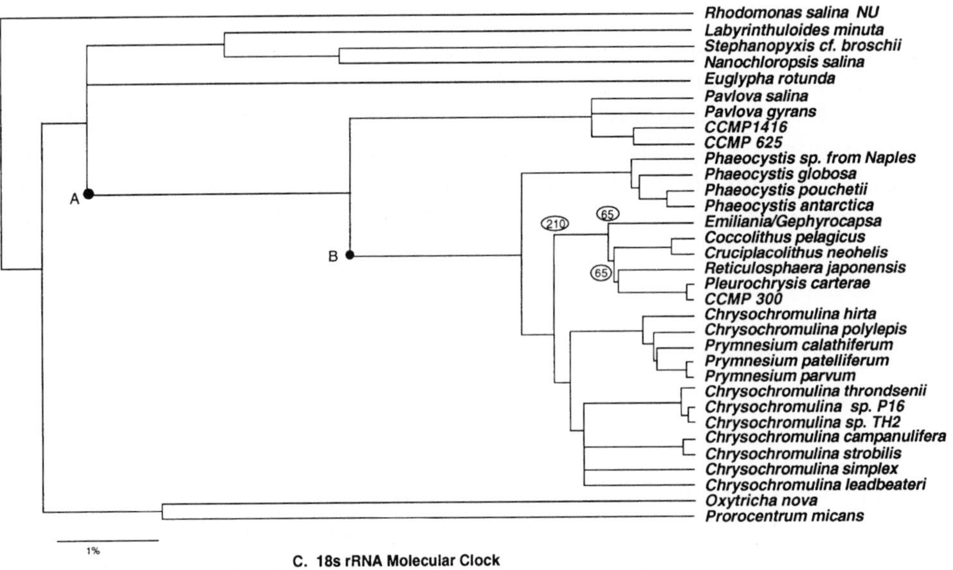
B. Linearized 18s rRNA Neighbor-Joining Tree
Rhodomonas salina NU
Labyrinthuloides minuta
Stephanopyxis cf. broschii
Nanochloropsis salina
Euglypha rotunda
Pavlova salina
Pavlova gyrans
CCMP1416
CCMP 625
Phaeocystis sp. from Naples
Phaeocystis globosa
Phaeocystis pouchetii
Phaeocystis antarctica
Emiliania/Gephyrocapsa
Coccolithus pelagicus
Cruciplacolithus neohelis
Reticulosphaera japonensis
Pleurochrysis carterae
CCMP 300
Chrysochromulina hirta
Chrysochromulina polylepis
Prymnesium calathiferum
Prymnesium patelliferum
Prymnesium parvum
Chrysochromulina throndsenii
Chrysochromulina sp. P16
Chrysochromulina sp. TH2
Chrysochromulina campanulifera
Chrysochromulina strobilis
Chrysochromulina simplex
Chrysochromulina leadbeateri
Oxytricha nova
Prorocentrum micans
A
B
210
65
65
1%

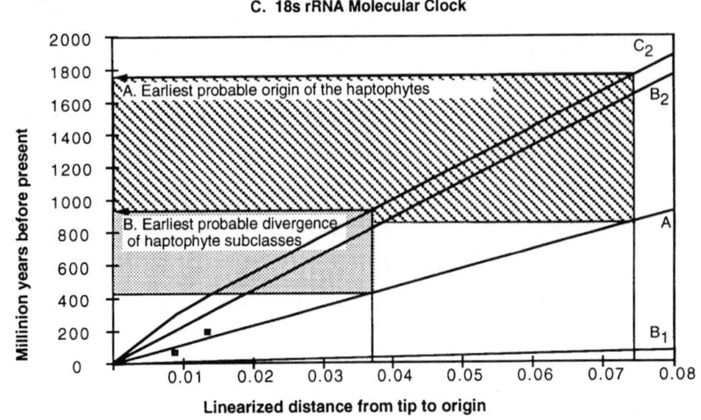
C. 18s rRNA Molecular Clock
A. Earliest probable origin of the haptophytes
B. Earliest probable divergence of haptophyte subclasses
Millinion years before present
Linearized distance from tip to origin
C2
B2
A
B1
0
200
400
600
800
1000
1200
1400
1600
1800
2000
0.01
0.02
0.03
0.04
0.05
0.06
0.07
0.08

as early divergences, whereas the photosynthetic algae emerge later as a monophyletic group. The presence of the oomycetes either outside or inside the autotrophic clade of the heterokonts has been controversial (CAVALIER-SMITH 1989, 1993; WILLIAMS 1991; LEIPE & al. 1994; CAVALIER-SMITH & al. 1996). This has fuelled speculations that the oomycetes, as well as all other heterotrophic heterokonts, may have secondarily lost their plastids. However, the monophyly for all the pigmented heterokonts has received stronger support as more pigmented taxa are included in the analyses (compare increases in bootstrap support in BHATTACHARYA & al. (1992), LEIPE & al. (1994), SAUNDERS & al. (1995), CAVALIER-SMITH & CHAO (1996), MEDLIN & al. (1997). Bootstrap support is generally lowered if many distant outgroups are included in the analyses. However, none of the recent analyses fail to present these algae as monophyletic. These data increasingly support hypothesis 2a.

Branching order within the heterokont algal groups is one of the most controversial issues concerning their phylogenetic analyses. Differences in the branching order within the pigmented heterokonts likely reflect differences in sequence alignments, analytical techniques, number of taxa and the subjective choice of the number of unambiguous nucleotides included in the analyses. In early studies, when using molecular data alone from six of the 12 or more classes of heterokont algae, the first divergence was between the diatoms and all remaining heterokont algae (BHATTACHARYA & al. 1992, LEIPE & al. 1994). However, a later cladistic analysis of 14 morphological/biochemical characters recovered a larger group within the heterokonts, which included the diatoms (SAUNDERS & al. 1995). This group could be defined morphologically as those algae containing a reduced flagellar apparatus; combined molecular and traditional data analysis further strengthened support for this clade (SAUNDERS & al. 1995).

Fig. 4. *A–C* Phylogenetic analysis of nuclear-encoded SSU rRNA genes from the *Haptophyta*. *A* Neighbor-joining tree inferred from a KIMURA (1980) distance matrix. Representatives of all subclasses/and or orders of the *Haptophyta* and the closely related groups are included. The tree was rooted on the branch leading to *Rhodomonas*. Bootstrap values (100 replications) are shown above the internal nodes. "A" marks the period of time before the *Haptophyta* diverge into their two subclasses. *B* Linearisation of the neighbor-joining tree in A according to TAKEZAKI & al. (1995) so that all rate variation in the molecule is eliminated. All significantly faster evolving taxa were excluded from the analysis. First appearances of coccolithophorid taxa from the fossil record are encircled and placed at the node where the taxa to the right are believed to have their first appearance. Point "A" marks the origin of the Haptophyta; point "B" is the divergence of its two subclasses. *C* Molecular clock constructed from *B*. Branch lengths from taxa in *B* with a fossil record were regressed against first appearance dates according to the molecular clock model in HILLIS & al. (1996). "A" is the regression line, constrained through the origin. Lines B_1 and B_2 are the 95% confidence limits around the regression line. Lines C_1 and C_2 are the 95% confidence limits for a new predicted value of time given the length of an undated node. Lower confidence limits, below zero, are reset at zero. C_1 is below the x axis and not shown. Blocks of time are shown for each group whose origin has been estimated from the molecular clock. The block spans the time of the average age of the group (from A) to the earliest probable time of origin based on the upper 95% confidence limit (C_2) of an undated node

The "reduced flagellar apparatus" group consists of the diatoms as a sister taxon to an assemblage containing the *Pelagophyceae* and other microalgae, which historically were loosely termed the "marine chrysophytes." The group is characterised by a flagellar transition region with two transitional plates and a small transitional helix below the major plate, a flagellar apparatus that lacks microtubular roots (see *Sarcinochrysidales* however, SAUNDERS & al. 1997b) and basal bodies positioned on or very near the nucleus. A paraxonemal rod, similar to that of dinoflagellates, is common in some members (paraxonemal rods of this type are absent in other heterokont chromophytes). Furthermore, there appears to be a tendency for a "sinking spindle" at the onset of mitosis (VESK & JEFFREY 1987, GREEN 1989, PICKETT-HEAPS & al. 1990), although few organisms other than diatoms have been examined in detail. The carotenoid pigments of this group are restricted to the diatoxanthin and diadinoxanthin types as well as fucoxanthin, 19′-butanoyloxyfucoxanthin and 19′-hexanoyloxyfucoxanthin (BJORNLAND & LIAAEN-JENSEN 1989); violaxanthin, anteraxanthin, zeaxanthin, heteroxanthin, vaucherioxanthin, etc. are not found in this group.

This "reduced flagellar apparatus clade" is sister taxon to a clade containing the chrysophytes/synurophytes, the eustigmatophytes, the xanthophytes and the phaeophytes. If more taxa, such as the *Sarcinochrysidales* and the *Chrysomeridales* (SAUNDERS & al. 1997b) and the *Raphidophyceae* plus additional *Xanthophyceae* (POTTER & al. 1997), are added, then the diatoms emerge before all pigmented heterokonts in molecular phylogenies (Fig. 5). The reduced flagellar apparatus lineage appears intact only in combined molecular and morphological data sets (not shown), suggesting that insufficient data exist to place the diatom branch unequivocally either within the reduced flagellar apparatus clade or outside it. Therefore, we are unable to find conclusive support for either hypothesis 3a or 3b.

The remaining pigmented heterokonts diverge into three (possibly two) clades in both the molecular only and the combined analyses. One clade contains the *Xanthophyceae* and its sister group, the *Phaeophyceae*; the *Chrysomeridales* are sister to the remainder of this clade (SAUNDERS & al. 1997b and in Fig. 5A). There

Fig. 5 *A–C*. Phylogenetic analysis of nuclear-encoded SSU rRNA genes from the *Heterokonta*. *A* Neighbor-joining tree inferred from a KIMURA (1980) distance matrix. Representatives of all classes/and or orders of the pigmented *Heterokonta* and the oomycetes are labelled on the tree, which was rooted on the branch leading to *Ulkenia*. Bootstrap values (100 replications) are shown above the internal nodes. The two light grey blocks contain algae with the diatoxanthin/diadinoxanthin-containing light-harvesting complex. The darker grey block contains algae with the violaxanthin-containing light-harvesting complex. Within this dark grey block are taxa highlighted in white that have a vaucherioxanthin (*Vacuolaria*) or heteroxanthin-containing light-harvesting complex (*Xanthophyceae*). *B* Linearisation of the neighbor-joining tree in *A* as in Fig. 4B. First appearances of diatom taxa from the fossil record are encircled and placed at the node where the taxa to the right are believed to have their first appearance. A hypothesis, which predates certain diatom taxa to have their origin before a major gap in the fossil record, i.e., at 125 Ma was used to predate three extant taxa. Point "A" marks the origin of the pigmented heterokonts; "B" is the origin of the diatoms; and "C" is the origin of the brown algae. *C* Molecular clock constructed from *B* as in Fig. 4C

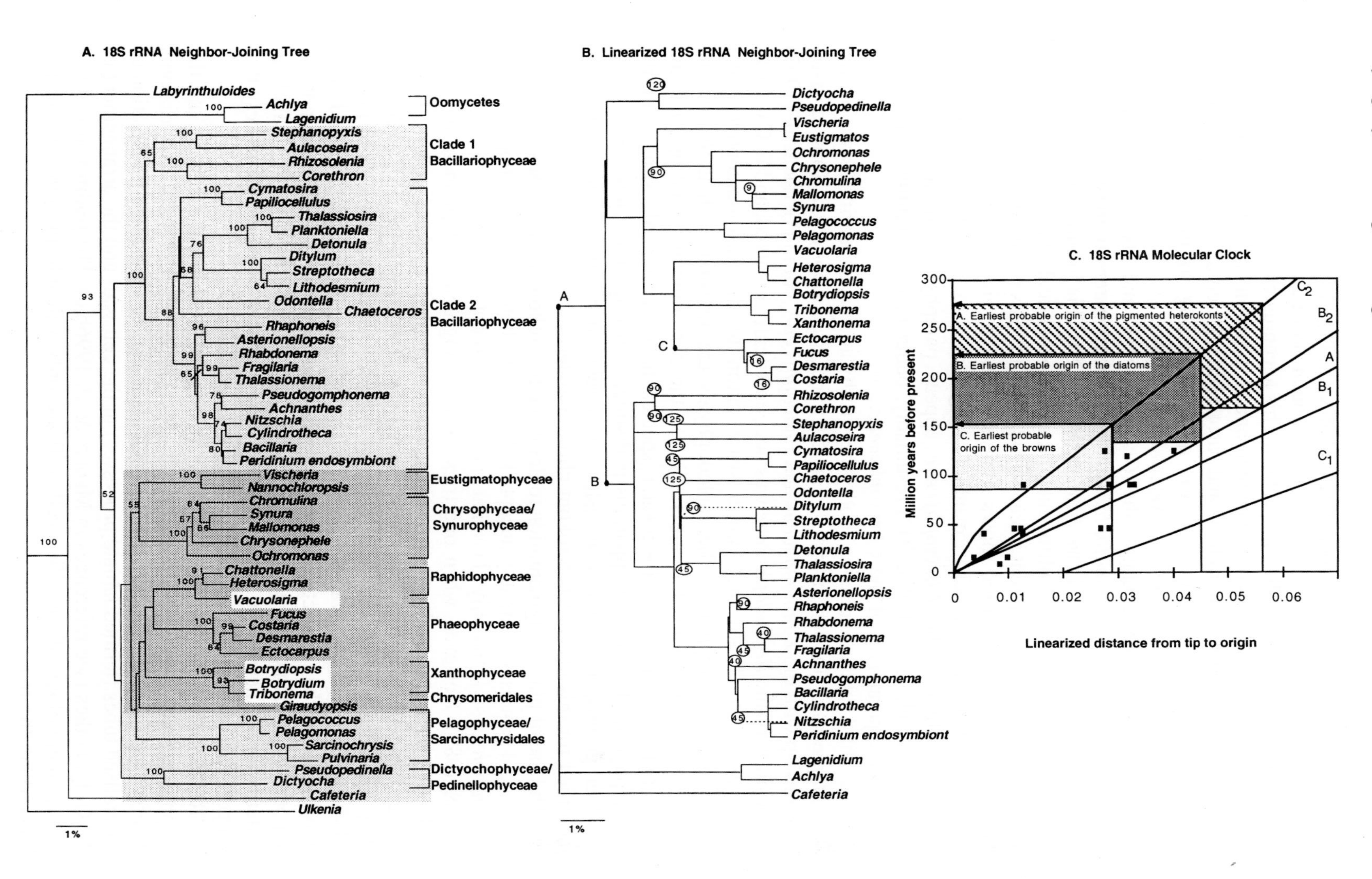
A. 18S rRNA Neighbor-Joining Tree
Labyrinthuloides
Achlya
Lagenidium
Stephanopyxis
Aulacoseira
Rhizosolenia
Corethron
Cymatosira
Papiliocellulus
Thalassiosira
Planktoniella
Detonula
Ditylum
Streptotheca
Lithodesmium
Odontella
Chaetoceros
Rhaphoneis
Asterionellopsis
Rhabdonema
Fragilaria
Thalassionema
Pseudogomphonema
Achnanthes
Nitzschia
Cylindrotheca
Bacillaria
Peridinium endosymbiont
Vischeria
Nannochloropsis
Chromulina
Synura
Mallomonas
Chrysonephele
Ochromonas
Chattonella
Heterosigma
Vacuolaria
Fucus
Costaria
Desmarestia
Ectocarpus
Botrydiopsis
Botrydium
Tribonema
Giraudyopsis
Pelagococcus
Pelagomonas
Sarcinochrysis
Pulvinaria
Pseudopedinella
Dictyocha
Cafeteria
Ulkenia
Oomycetes
Clade 1 Bacillariophyceae
Clade 2 Bacillariophyceae
Eustigmatophyceae
Chrysophyceae/ Synurophyceae
Raphidophyceae
Phaeophyceae
Xanthophyceae
Chrysomeridales
Pelagophyceae/ Sarcinochrysidales
Dictyochophyceae/ Pedinellophyceae
1%
B. Linearized 18S rRNA Neighbor-Joining Tree
Dictyocha
Pseudopedinella
Vischeria
Eustigmatos
Ochromonas
Chrysonephele
Chromulina
Mallomonas
Synura
Pelagococcus
Pelagomonas
Vacuolaria
Heterosigma
Chattonella
Botrydiopsis
Tribonema
Xanthonema
Ectocarpus
Fucus
Desmarestia
Costaria
Rhizosolenia
Corethron
Stephanopyxis
Aulacoseira
Cymatosira
Papiliocellulus
Chaetoceros
Odontella
Ditylum
Streptotheca
Lithodesmium
Detonula
Thalassiosira
Planktoniella
Asterionellopsis
Rhaphoneis
Rhabdonema
Thalassionema
Fragilaria
Achnanthes
Pseudogomphonema
Bacillaria
Cylindrotheca
Nitzschia
Peridinium endosymbiont
Lagenidium
Achlya
Cafeteria
1%
C. 18S rRNA Molecular Clock
A. Earliest probable origin of the pigmented heterokonts
B. Earliest probable origin of the diatoms
C. Earliest probable origin of the browns
Million years before present
Linearized distance from tip to origin

are few morphological and biochemical characters that unite this group. However, the zoospores of all taxa have an R_1 microtubular root that extends to the anterior of the cell, forming a beak-like projection. This accentuated anterior end places the flagellar insertion in a lateral, rather than apical, position. Other features, such as flagellar transitional region, mitosis and carotenoid pigmentation show no common thread, and for these reasons it was quite surprising to find molecular data supporting this relationship (e.g., ARIZTIA & al. 1991).

The second clade contains the *Chrysophyceae* and its sister group, the *Synurophyceae*; the *Eustigmatophyceae*, in turn is sister to these two classes (BHATTACHARYA & al. 1992, LEIPE & al. 1994; Fig. 5). This group is also difficult to characterise on the basis of ultrastructural and biochemical features. The flagellar apparatus is distinctly different in each group, and mitosis is variable where known. The carotenoid pigmentation shows some similarities (viz., violaxanthin), but the eustigmatophytes have vaucherioxanthin, a pigment not reported for chrysophytes and synurophytes. The *Chrysophyceae* appear paraphyletic in most analyses (SAUNDERS & al. 1995, 1997a, b). Presumably, better taxon sampling will resolve the possible paraphyly of the *Chrysophyceae.*

The third clade, contains both the freshwater and marine taxa of the *Raphidophyceae*. In our analysis (Fig. 5A) the *Raphidophyceae* form a sister relationship with the *Xanthophyceae* and *Phaeophyceae*, but in POTTER & al. (1997) the position of the *Raphidophyceae* was not consistently resolved, and in CAVALIER-SMITH & CHAO (1996) they are a sister taxon to the eustigmatophytes and chrysophytes/synurophytes. The raphidophytes are unusual in that the marine species have carotenoids that are similar to the chrysophytes, synurophytes and phaeophytes, whereas the freshwater species have carotenoids similar to the xanthophytes. The flagellar apparatus is distinct, showing no obvious relationship to other groups. Thus, no clear sister taxon relationship has been conclusively identified for this class of heterokont chromophytes.

The use of plastid pigmentation to delineate heterokont algal classes, as well as other algal groups, is generally accepted. Among the heterokont chromophytes, the *Raphidophyceae* is the only class that is a glaring exception (BJØRNLAND & LIAAEN-JENSEN 1989). However, when one tries to find congruence of pigment data and molecular data, the results are less clear. The xanthophyll-cycle pigments and the reduced flagellar apparatus characters have been plotted onto the SSU rRNA tree (Fig. 5A). There appears to be a tendency for early diverging lineages to possess 19′-fucoxanthin-like derivatives, and to a lesser degree, for diatoxanthin and diadinoxanthin to be restricted to early diverging lineages, weakly supporting hypothesis 4a. However, in part due to the lack of resolution in the branching patterns for the heterokont chromophytes, we are unable to state conclusively that support can be found for either hypothesis 4a or 4b.

Relationships within the diatoms. Among the diatoms, the centric and the araphid pennate forms are paraphyletic (MEDLIN & al. 1996b,c). The diatoms diverge into two clades (clade 1 + 2). Each clade can be defined by the position of specialised tubes (termed labiate processes) in the cell wall and the arrangement of the Golgi bodies (see references in MEDLIN & al. 1996b,c). The traditional features of the morphology of the silica cell wall are only valuable in defining younger branches in the tree.

Plastid genome

Endosymbiosis. The endosymbiotic hypothesis of plastid evolution maintains that plastids were acquired by primitive eukaryotic heterotrophs through the engulfment and maintenance of photosynthetic prokaryotes (SCHIMPER 1883, MERESCHKOWSKY 1905, RAVEN 1970, MARGULIS 1981). This hypothesis was once opposed by those who argued that the plastid arose directly without endosymbiosis during the evolution of the first eukaryotes (KLEIN & CRONQUIST 1967, CAVALIER-SMITH 1975), but this view is no longer supported by evolutionary biologists.

The plastids of the rhodophyte, chlorophyte and glaucocystophyte algae and the higher plants have only two membraned-plastids and are assumed to have resulted from a primary endosymbiotic event in which a eukaryotic host engulfed a prokaryotic cell. The host organisms associated with the primary endosymbiosis appear to arise as independent plastid-bearing lineages within the crown group radiation of the eukaryotes (Fig. 1, BHATTACHARYA & MEDLIN 1995).

The algae with 3–4 membraned plastids are hypothesised to have arisen through a secondary endosymbiotic event(s) in which a heterotrophic eukaryote host engulfed and reduced a photosynthetic eukaryote cell to a plastid. The additional membranes surrounding the plastid are remnants of the endosymbiosis (i.e., the host cell vacuole and the plasmalemma of the endosymbiont, see review in GIBBS 1993). Algae resulting from the secondary endosymbiosis include the euglenophytes and the chlorarachniophytes, which contain chlorophyll $a + b$, as well as the heterokont chromophytes, haptophytes, dinoflagellates and cryptophytes, most of which contain chlorophyll $a + c$ (GIBBS 1978, 1981; CAVALIER-SMITH 1989; JEFFREY 1989; ROWAN 1989; KOWALLIK 1992; VALENTIN & al. 1992).

Whereas current evidence from molecular and morphological/biochemical data suggests that the primary endosymbiotic event occurred only once, the secondary endosymbiotic event may have occurred several times (see review in BHATTACHARYA & MEDLIN 1995 and DELWICHE & PALMER 1996). The host organisms associated with the secondary endosymbioses, (viz, the euglenoids, cryptomonads, chlorarachniophytes, dinoflagellates haptophytes, and heterokont chromophytes, with 3–4 membraned plastids,) do not share a common ancestry, and thus a more likely hypothesis for their emergence as pigmented lineages is that each lineage has acquired its plastid through an independent secondary endosymbiosis, i.e. multiple secondary endosymbioses rather than through a single event (see CAVALIER-SMITH 1982). The identification of the vestigial nucleus (nucleomorph) in the plastids of the cryptophytes and chlorarachniophytes as being associated with the red algae and the green algae, respectively (Fig. 1), provides direct evidence using the nuclear genome of the endosymbiont that multiple secondary endosymbioses have occurred.

Many of the host lineages believed to have arisen from secondary endosymbiosis event(s) also have heterotrophic taxa as sister groups or as early divergences in their lineages (viz., heterokonts, cryptomonads, euglenoids, alveolates, chlorarachniophytes, see Fig. 1 and BHATTACHARYA & MEDLIN 1995). Either these lineages were originally photosynthetic and these heterotrophic taxa lost their plastids, or the lineages gained their plastids through secondary endosymbioses later in their evolution. The phylogenies of the plastids arising from

A. tuf A Neighbor-Joining Tree

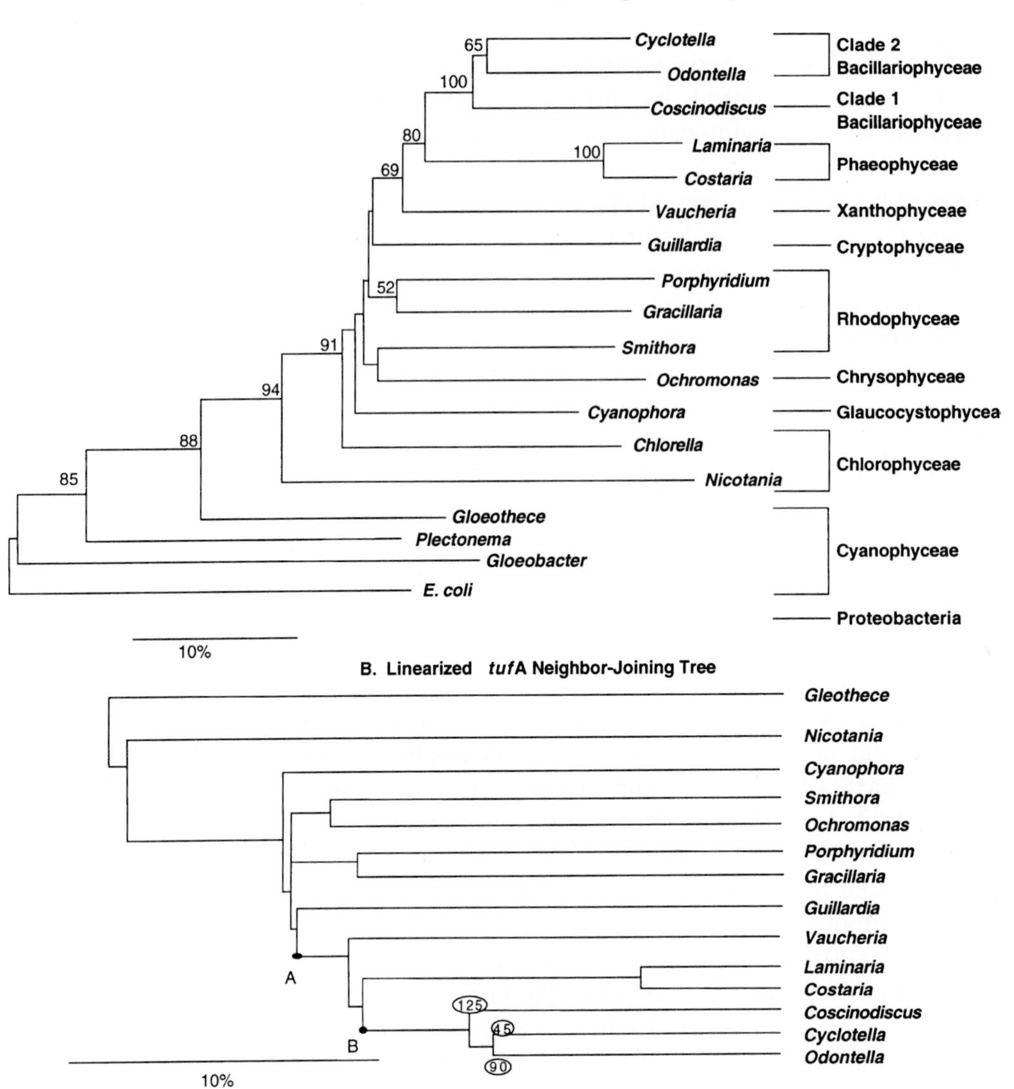
65
100
80
69
52
91
94
88
85
100
Cyclotella
Odontella
Coscinodiscus
Laminaria
Costaria
Vaucheria
Guillardia
Porphyridium
Gracillaria
Smithora
Ochromonas
Cyanophora
Chlorella
Nicotania
Gloeothece
Plectonema
Gloeobacter
E. coli
Clade 2 Bacillariophyceae
Clade 1 Bacillariophyceae
Phaeophyceae
Xanthophyceae
Cryptophyceae
Rhodophyceae
Chrysophyceae
Glaucocystophycea
Chlorophyceae
Cyanophyceae
Proteobacteria
10%
B. Linearized tufA Neighbor-Joining Tree
Gleothece
Nicotania
Cyanophora
Smithora
Ochromonas
Porphyridium
Gracillaria
Guillardia
Vaucheria
Laminaria
Costaria
Coscinodiscus
Cyclotella
Odontella
A
B
125
45
90
10%

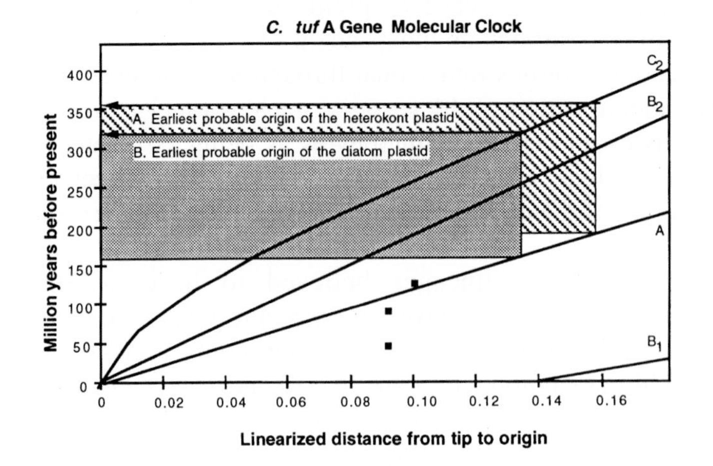
C. tuf A Gene Molecular Clock
Million years before present
0
50
100
150
200
250
300
350
400
A. Earliest probable origin of the heterokont plastid
B. Earliest probable origin of the diatom plastid
C_2
B_2
A
B_1
0
0.02
0.04
0.06
0.08
0.10
0.12
0.14
0.16
Linearized distance from tip to origin

secondary endosymbioses must be examined and compared with those of their host cells to infer the likely source of the photosynthetic eukaryote that was transformed into the plastids.

There is a moderate to extensive heterokont chromophyte, haptophyte and cryptophyte data set available for several plastid genes: the *tuf*A gene (DELWICHE & al. 1995), the large subunit of the RUBISCO operon (FUJIWARA & al. 1995, CHESNICK & al. 1996) and the small subunit of the ribosomal operon (BHATTACHARYA & MEDLIN 1995, MEDLIN & al. 1995 and unpubl.) (Figs. 6–8). In each case, we will compare the plastid phylogenies with those of the host lineages to infer the likely source of the taxa transformed into a plastid.

In the *tuf*A phylogeny (Fig. 6A), we have selected representatives from the larger data set published by DELWICHE & al. (1995) for our analysis. There are no haptophyte sequences available; however the heterokonts are well represented, and there is one cryptophyte in the tree. The *tuf*A gene suggests that plastids are monophyletic and originate from the cyanobacteria. The *tuf*A gene of 2-membraned plastids of the green algae/higher plants, the glaucocystophytes and the red algae are the first divergences from the cyanobacterial *tuf*A gene. The anomalous position of the chrysophyte *Ochromonas* among the red algae in the *tuf*A phylogeny has been discussed by DELWICHE & al. (1995) as being either a contaminant or an evolutionary novelty. The *tuf*A gene of 4-membraned plastids, represented in this tree by the heterokonts and cryptomonads, are later divergences and, fall within the red algae. Among the diatoms, *Coscinodiscus*, a centric diatom belonging to clade 1 diatoms as inferred from the nuclear-encoded SSU rRNA tree, is sister to two other centric diatoms of clade 2. The position of the cryptophyte *Guillardia* is not supported in the *tuf*A phylogeny and likely represents a problem of taxon sampling.

Although there is no bootstrap support for the clade containing the red and the chromophyte algae in the *tuf*A phylogeny, the separation of a "green lineage" from a "red plus golden lineage" is congruent with that found in the phylogenetic reconstructions from other plastid genes (see below). The lack of bootstrap support for the lineages in this phylogeny may in part reflect the asymmetry of the tree and internodal differences closer to 20% (see HILLIS & BULL 1993).

A phylogeny of the SSU rRNA gene has been constructed with the LogDet transformation [to avoid base compositional bias that can distort the relationships in this gene (LOCKHART & al. 1994)] and with the neighbor-joining analysis. The branching order of the major lineages are identical, and we present the neighbor-joining tree (Fig. 7A), which we will use below for our molecular clock

Fig. 6 *A–C*. Phylogenetic analysis of *tuf*A coding regions. *A* Neighbor-joining tree inferred from a gamma-weighted distance matrix (MEGA, KUMAR & al. 1993, a= 2) using all three codon positions. Bootstrap values (100 replications) are shown above the internal nodes. The tree was rooted on the branch leading to *E. coli*. *B* Linearisation of the neighbor-joining tree in *A* as in Fig. 4B. First appearances of diatom taxa from the fossil record are encircled and placed at the node where the taxa to the right are believed to have their first appearance. Point "A" marks the origin of the heterokont plastid; "B" is the origin of the diatom plastid. *C* Molecular clock constructed from *B* as in Fig. 4C. C_1 is below is the x axis and not shown

A. 16s rRNA Neighbor-Joining Tree

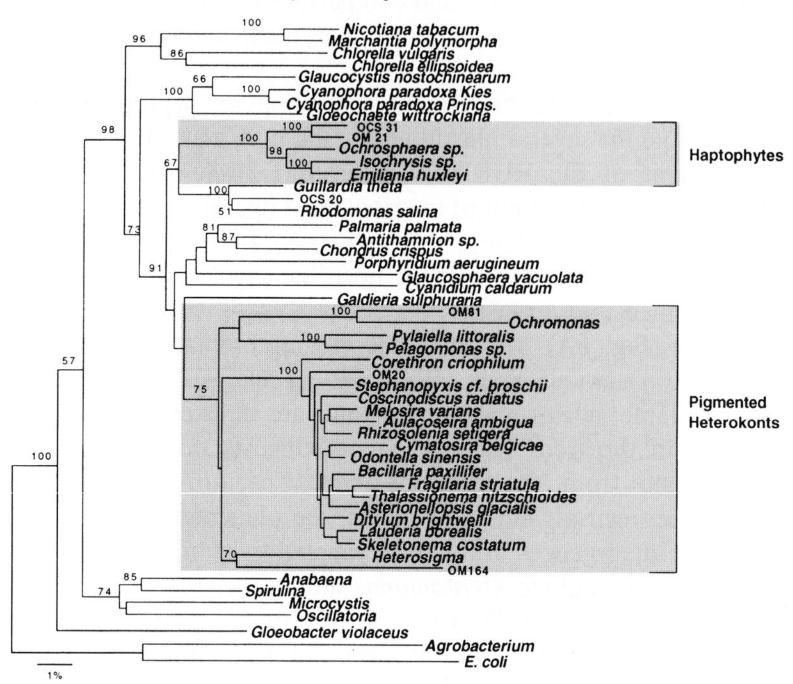

B. Linearized 16s rRNA Neighbor-Joining Tree

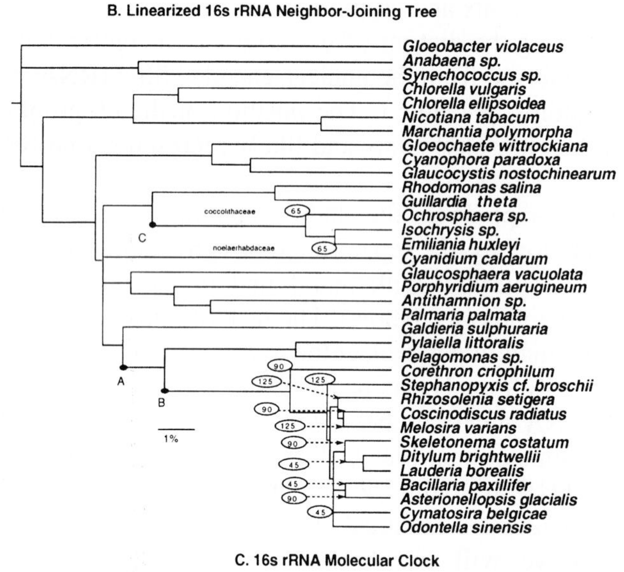

C. 16s rRNA Molecular Clock

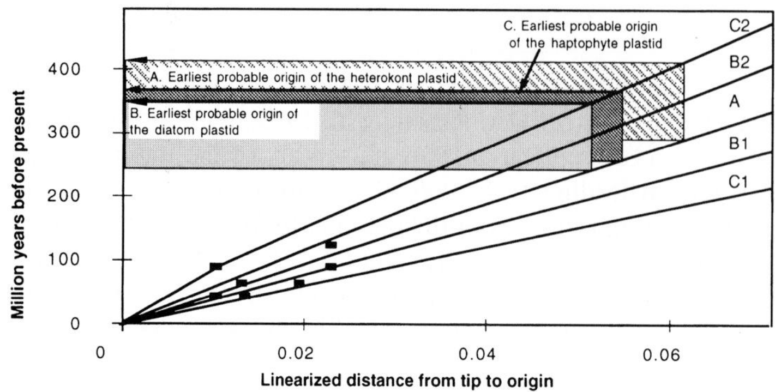

calculations. Branch lengths in the LogDet transformation are not linear using small data sets (< 2000 aa) and cannot be used for clock calculations unless they are corrected (Gu & Li 1996).

The 16S rRNA phylogeny also suggests that plastids are monophyletic (bootstrap = 98%) and originate from the cyanobacteria. The 2-membraned plastids of the green algal/higher plants, the glaucocystophytes and the red algae diverge as independent monophyletic lineages. The 4-membraned plastids of heterokont chromophytes, haptophytes and cryptophytes share a recent evolutionary history with the red algae (bootstrap = 91%). However, their hosts do not (Fig. 1), and this is supportive evidence that these lineages obtained their plastids from a red algae or a red algal-like ancestor via endosymbiosis. The haptophytes and cryptomonads are moderately supported (bootstrap = 67%) and are sister taxa to a largely unresolved assemblage of red algae and the heterokont chromophytes. The primitive unicellular reds contribute substantially to the problems of unresolved branching order within this lineage. The advanced red algae are well supported (boostrap = 81%) and are later divergences in the tree. The heterokont chromophytes are well supported (bootstrap = 75%) and branch from within the red algal lineage. Because the haptophytes and cryptomonads are held outside the true red algal lineage by moderate bootstrap support, one possible interpretation of these data is that the heterotrophic ancestors of the haptophytes and cryptomonads engulfed and retained a red algal-like ancestor, whereas the heterotrophic heterokonts are more likely to have engulfed a primitive red algae. Further taxon sampling among the primitive reds may help to improve support for the branching order among these taxa. The LogDet tree also supports the position of the haptophytes and cryptomonads outside the red algal lineage (tree not shown, but see Medlin & al. 1995).

Within the heterokonts, several lineages are recovered. These correspond to the major classes of the heterokont algae (except for the clade comprising the *Pelagophyceae* and *Phaeophyceae*), but support for the branching order is not strong. Within the diatoms, clade 1 diatoms are broken into separate lineages, however clade 2 diatoms remain intact. There are not enough identified taxa sampled in the haptophyte and cryptophyte lineages to comment on their branching order, but each is a monophyletic lineage (bootstrap = 100%).

Fig. 7 *A–C*. Phylogenetic analysis of plastid-encoded SSU rRNA genes. *A* Neighbor-joining tree inferred from a Kimura (1980) distance matrix. Representatives from all of the algae are included except for the *Chlorarachniophyta* and the *Euglenophyta*. Bootstrap values (100 replications) are shown above the internal nodes. The *Haptophyta* and the *Heterokonta* are labelled on the tree. The tree was rooted on the branch leading to *E. coli*. Terminal taxa in the tree represented by codes (e.g., OM81) are unidentified sequences from a 16S rRNA clone library provided courtesy of Dr. M. Rappé. *B* Linearisation of the neighbor-joining tree in *A* as in Fig. 4B. First appearances of diatom and coccolithophorid taxa from the fossil record are encircled and placed at the node where the taxa to the right are believed to have their first appearance. Point "A" marks the origin of the heterokont plastid; "B" is the origin of the diatom plastid; "C" is the origin of the haptophyte plastid. *C* Molecular clock constructed from *B* as in Fig. 4C

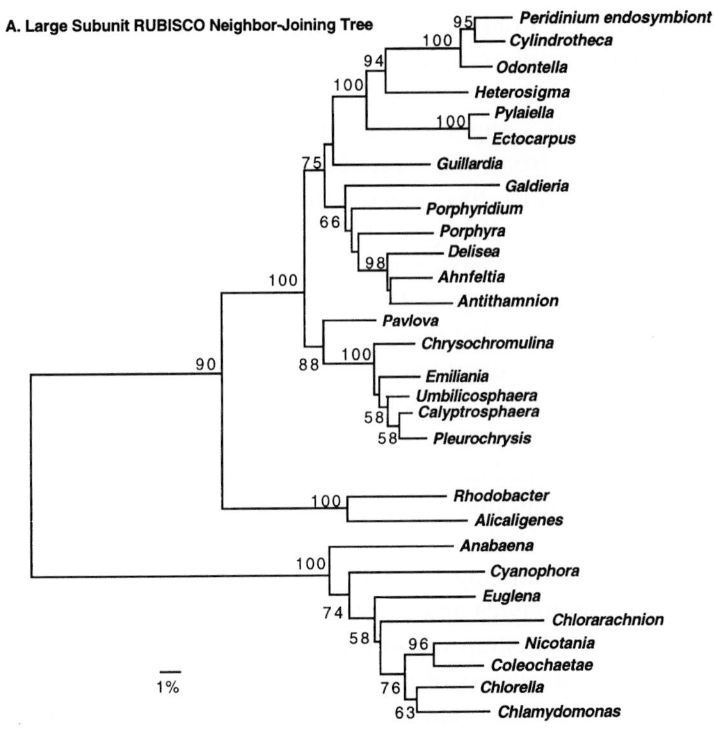

B. Linearized Large Subunit RUBISCO Neighbor-Joining Tree

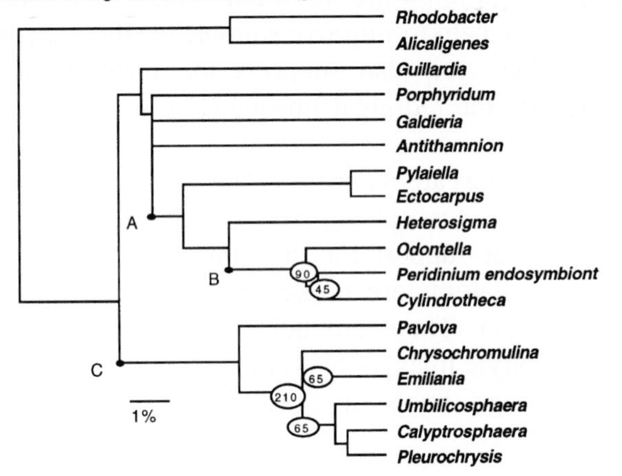

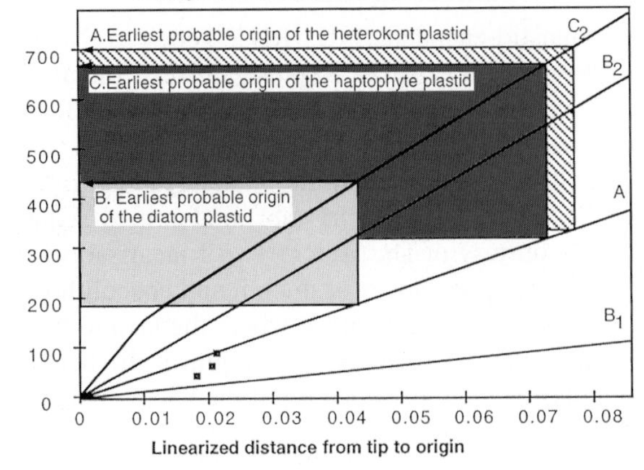

The RUBISCO gene is the only single-copy gene whose molecular analysis indicates a polyphyletic origin for the plastids (see review in PALMER 1993, LOISEAUX DE GOËR 1994). The phylogenetic relationships reconstructed from the RUBISCO tree are far more complicated than those presented here in Fig. 8 (see DELWICHE & PALMER 1996). We show only the major lineages containing the "red" and "green" type of Form I of the large subunit of RUBISCO. The two lineages are so divergent that it is not incorrect to root one lineage with the other as we have done in Fig. 8A (see DELWICHE & PALMER 1996).

The RUBISCO genes from the chlorophyll $a + b$ lineage, containing the green algae/higher plants, glaucocystophytes, chlorarachniophytes and euglenoids, originate from cyanobacterial genes (Fig. 8A). In contrast, the non-green algal lineages (excluding the dinoflagellates, which have an entirely different RUBISCO origin, see PALMER 1993) originate from *Proteobacteria* (Fig. 8A). Historically, the branching order among the major groups has changed as more taxa have been added to the phylogenetic analyses (compare LOISEAUX DE GOËR 1994, FUJIWARA & al. 1995, MCFADDEN & al. 1995, and Fig. 8A). However, it is clear that the RUBISCO gene of the heterokont chromophytes and cryptophytes in this lineage (all of which apparently obtained their plastids through a secondary endosymbiosis) share a common ancestry with the red algal lineage (bootstrap = 100%). Our tree would support the hypothesis that the heterokont chromophytes, haptophytes and cryptophytes obtained their plastids by engulfing a red algae or an ancestor that gave rise to the red algae. As with the other plastid genes, the host organism genes of the heterokont chromophytes, haptophytes and cryptophytes do not share a recent evolutionary history with the red algae.

As in the 16S rRNA tree, the haptophytes (a monophyletic group in 88% of bootstrap replicates) are recognised as a discrete sister lineage (bootstrap = 75%) to the red algae plus other chlorophyll $a + c$ algae. However, if fewer primitive red algae are included in the analysis, the cryptophytes will group with the haptophytes (bootstrap = 64%, data not shown) as sister group to the red algae plus heterokont chromophytes. In no cases are the haptophytes and heterokont chromophytes sister taxa. Lineages corresponding to the diatoms, raphidophytes and phaeophytes are recovered within the heterokont chromophytes, but taxon sampling is too low to comment on their branching order. Among the haptophytes, the divergence of the two subclasses are confirmed. The family *Noelaerhabdaceae* (*Emilinia huxleyi*)

Fig. 8 *A–C*. Phylogenetic analysis of the large subunit of RUBISCO coding regions. *A* Neighbor-joining tree inferred from a gamma-weighted distance matrix (MEGA, KUMAR & al. 1993, a = 1) using all three codon positions. Bootstrap values (100 replications) are shown above the internal nodes. The tree was rooted on the branch leading to the green algae/higher plants. *B* Linearisation of the neighbor-joining tree in *A* as in Fig. 4B. The entire green algal/higher plant lineage was evolving too fast and was eliminated from the linearisation. First appearances of diatom and coccolithophorid taxa from the fossil record are encircled and placed at the node where the taxa to the right are believed to have their first appearance. Point "A" marks the origin of the heterokont plastid; "B" is the origin of the diatom plastid; "C" is the origin of the haptophyte plastid. *C* Molecular clock constructed from *B* as in Fig. 4C. C_1 is below the x axis and not shown

is sister to the other coccolithophorids, which is congruent with the host tree (Fig. 5A).

Plastid genes from haptophytes and heterokont chromophytes were never sister taxa in any of our analyses, and the heterokont chromophyte plastid genes were always embedded within the red algal plastid lineage. The convergence of the 16S rRNA and RUBISCO plastid gene phylogenies suggests that the heterokont chromophytes likely engulfed a primitive red algae, whereas the cryptophytes and the haptophytes are more likely to have engulfed an ancestor of the red algae. This suggests that the *Chromista* are not monophyletic and that haptophyte, heterokont chromophytes and cryptophyte plastids arose from separate endosymbiotic events. These data support hypotheses 1c and 1d.

Molecular clock calculations

Molecular data are normally used to reconstruct the phylogenetic history of extant organisms. Ideally, as organisms diverge, their genomes accumulate base substitutions in a stochastic, but clock-like manner. It is now widely recognised that a universal molecular clock does not exist and that the base substitution rate varies within lineages and genes. Nevertheless, if potential errors are identified with a relative rate test and corrected by eliminating the significantly fast and slow taxa and by linearising the rate of evolution, it is then possible to use molecular data to estimate divergence times.

Using the method of HILLIS & MORITZ (1990), we have estimated from the nuclear and plastid genes the time of origin (1) for the diatoms, (2) for the heterokont chromophytes, (3) for the haptophytes and the divergence of their two subclasses and (4) for the timing of the secondary endosymbiotic events for the pigmented heterokonts and the haptophytes. We initially calculated a relative rate or branch length test in which the evolutionary rate of all pair-wise combinations of taxa was compared to several outgroups (WU & LI 1985, TAKEZAKI & al. 1995). In this manner we identified taxa not evolving within a stochastic model of base substitution. We then selected a range of taxa with varying degrees of distance from one another to be used for the construction of a linearised neighbor-joining tree in which rate variation between the taxa was assumed to be eliminated (TAKEZAKI & al. 1995). A linearised neighbor joining (NJ) tree was constructed from the nuclear-encoded SSU rRNA genes and from the plastid-encoded SSU rRNA, *rbc*L and the *tuf*A genes (Figs. 4–8B). First appearance dates of diatom and coccolithophorid taxa with a fossil record were regressed against estimated branch lengths of lineages in each tree to construct a molecular clock for each gene or group of organisms (Figs. 4–8C). First appearance dates of taxa immediately after a gap in the diatom fossil record were predated to the middle of the gap or before, it, if potential ancestors of the extant taxa could be identified in well-preserved diatom deposits before the gap (GERSONDE & HARWOOD 1990).

For each linearised tree, we estimated an average age of the clade and its earliest probable age ($p = 95\%$). The average age of any undated node was determined by multiplying the length of its median or average lineage by the regression coefficient. The earliest probable age for any undated node was taken from the upper 95% confidence limit around the age estimate given the length of its median or average lineage.

Fossil dates may seriously underestimate the first appearance date of any lineage (WRAY & al. 1996). Thus, time estimates based on the average age for the lineage, given that the fossil dates may be later than first appearances, are also likely to underestimate the origin of groups. The HILLIS & MORITZ (1990) model for the calculation of a molecular clock provides an estimate for an upper and lower 95% confidence limit for the origin of any undated node. Thus, using this calculation, the actual time of origin of any undated node in the tree, should realistically lie somewhere between the average age determined from the regression line and the earliest probable age determined from the upper 95% confidence limit. Presumed dates of origin between the lower 95% confidence limit and the regression line would be nullified by fossil taxa present during this time. With few fossil dating points, the 95% confidence interval can be quite broad, pushing the earliest probable age farther back in time.

From the SSU rRNA clock calculated for the pigmented heterokonts, we have estimated the average age of the brown algae, the diatoms and the pigmented heterokonts (see also KOOISTRA & MEDLIN 1996). The recent appearance of the brown algae is well in agreement with other molecular, morphological and biogeographic evidence (see SAUNDERS & DRUEHL 1992), but contrasts with some earlier putative brown algal fossils (TAGGART & PARKER 1976). The average age of the diatoms is very close to their first fossil record (MEDLIN & al. 1996c). Using our average and earliest probable dates for the origin of the pigmented heterokonts (170–270 Ma), we conclude that this group is unlikely to have existed much before the Permian-Triassic boundary.

In contrast, the SSU rRNA molecular clock for the haptophyte lineage indicates that they are a much older group. Their average age is 850 Ma; their earliest probable age is c. 1800 Ma. These dates may be greatly overestimated because the haptophyte SSU rRNA clock is based only on three divergence times from the coccolithophorid fossil record. Nevertheless, rate variation in the SSU rRNA gene for this group is minimal (data not shown), so we feel that our predictions of time of divergences are reasonable. Interestingly, the mean of the average age of the divergence of the two subclasses of the haptophytes (299 Ma) is closer to the average age of the pigmented heterokonts (248 Ma). These figures are determined by averaging both the nuclear and plastid age estimates.

By constructing a molecular clock from our plastid gene sequence data, we can date the timing of the endosymbiotic event leading to the transfer of photosynthetic capacity to the heterokont and haptophyte lineages. Thus, the ages derived from plastid genes can be compared to those of the origins of their host cells. If hypothesis 1a is true, then the dates for the divergence of the haptophyte and heterokont algae should be younger that the dates for the endosymbiotic event leading to the transfer of photosynthetic capacity to the haptophyte and heterokont lineages.

Hypotheses 1a, 1c and 1d can support the possibility that the early heterotrophic divergences in the heterokont lineage are the result of plastid loss. If that is true, then the endosymbioses should predate the origin of the hosts. If the converse is true, then the origin of the heterokont algae should coincide with the timing of

Table 1. Estimated average time of origin (in millions of years) of the host cells and their plastids from the heterokont chromophytes, diatoms and haptophytes

Algal groups	Host cells 18S rRNA	16S rRNA	Plastids *tuf*A	RUBISCO	Plastids Mean
Heterokont chromophytes	170	293	190	337	274
Diatoms	135	249	160	190	200
Haptophytes					
Group origin	850				
Subclass divergence	420			177	
Plastid origin		263		322	293

the secondary endosymbiotic event leading to the transfer of photosynthetic capacity to heterokont organisms. If the haptophytes obtained their plastids at the origin of their lineage, then the timing of their symbiosis should be widely disparate from that of the pigmented heterokonts given no plastid loss.

The average date for the origin of the pigmented heterokonts calculated from the SSU rRNA gene (170 Ma) is close to the mean of the average date for a secondary endosymbiotic origin of the heterokont plastid estimated from three plastid genes (274 Ma) (Figs. 6–8C, Table 1). This provides support for hypotheses 1c and 1d over the remaining two hypotheses. The consistent separation of the haptophytes from the heterokonts in all of the plastid phylogenies provides evidence to support hypothesis 1d over 1c.

The estimated average date for the origin of the haptophyte plastid is considerably younger than the origin of the host lineage. This would suggest that early members of the haptophytes were not photosynthetic and that the endosymbiosis occurred somewhere along the internode leading to the diversification of the haptophytes. Interestingly, the divergence of the two subclasses of the haptophytes, as estimated from the SSU rRNA and RUBISCO genes, is very close to the origin of the haptophyte plastid (Table 1). We hypothesise that the endosymbiotic event in the haptophyte lineage occurred just prior to the divergence of the two subclasses. Therefore, it follows that all early ancestors in the haptophyte lineage were heterotrophic and are extinct, or are undersampled.

Interestingly, the estimated times for the two secondary endosymbiotic events in haptophytes and heterokonts, respectively, are remarkably close. The mean divergence time of the two groups estimated from three genes is 281 Ma (n = 5, Table 1). Thus, the transfer of the photosynthetic capacity to these lineages occurred approximately the same time at or before the Permian-Triassic boundary (250 Ma). MEDLIN & al. (1997) have presented evidence for the correlation of the Permian-Triassic mass extinction with the re-radiation of the modern phytoplankton following this event. Both the pigmented heterokonts and the haptophytes comprise the bulk of today's eukaryotic phytoplankton in the oceans. Our molecular clock calculations would support a theory that the Permian-Triassic extinction opened many niches in the world's oceans and those organisms capable of engulfing and maintaining a photo-autotroph had an adaptive advantage. This suggests that multiple secondary endosymbioses could have occurred at a similar time.

Summary

Although remarkable progress has been made during the past 40 years, a satisfactory understanding of the phylogenetic relationships among the "golden algae" remains elusive. From the data presented in Figs. 1–3 and from other published phylogenies, no clear sister relationship has been demonstrated for the haptophyte and heterokont host cells. This is the minimum that must be presented to invoke a monophyletic origin for the two groups. Instead, a relationship between the alveolates and the heterokonts consistently reoccurs with up to 100% bootstrap support in some phylogenies. The *Haptophyceae*, once part of the *Chrysophyceae*, have been shown to be a distinct taxonomic group, however, their closest living relative remains unresolved. Further work should be undertaken from other genes to resolve conclusively the relationship between heterokonts and haptophytes: nevertheless existing date do not support a monophyletic origin for the two groups.

Both ultrastructural (two extra membranes around the plastid) and molecular data suggest the plastids of the haptophytes and heterokont chromophytes are the result of secondary endosymbiosis (probably a red algae or red alga-like organism). The plastid-encoded SSU rRNA (bootstrap = 67%) and *rbc*L data (bootstrap = 75%) support separate endosymbioses for the plastids of haptophytes and heterokont chromophytes because other groups occupy branches in phylogenetic trees between them (cryptophytes and red algae in SSU rRNA – Fig. 7; cryptophytes and red algae in *rbc*L – Fig. 8). Although bootstrap support of $\geqslant 70\%$ can indicate a probability $\geqslant 95\%$ that the recovered clade is real (Hillis & Bull 1993), we cannot state conclusively that the haptophyte and heterokont plastids are a monophyletic group (hypothesis 1a or 1c) or are two distinctly different lineages (hypotheses 1b and d). However, the consistent separation of the haptophyte and heterokont plastids in the phylogenetic analyses, taken in combination with our molecular clock calculations, favour hypothesis 1d that the two lineages are not monophyletic.

Data presented above (nuclear-encoded SSU rRNA, actin) suggest that the heterokont chromophytes are a monophyletic assemblage, which supports hypothesis 2a. Bootstrap support for the molecular data is too weak to determine the phylogenetic branching pattern among this assemblage. Thus, we cannot support either hypothesis regarding the monophyly of taxa with reduced flagellar apparatuses (hypotheses 3a, 3b). The SSU rRNA data (Fig. 5) suggest that the two carotenoid types arose independently more than once, which lends some support to hypothesis 4a.

Molecular clock-calculated dates suggest that the haptophyte host cell lineage is relatively ancient (Proterozoic-Paleozoic) but that the haptophyte plastid was acquired more recently (Mesozoic). The molecular clock-calculated dates for origin of the heterokont chromophytes are more recent (Mesozoic), both with respect to the host cell and the plastid. Perhaps coincidentally, the estimated dates for the origin of plastids in both haptophytes and heterokont chromophytes are nearly identical. Data suggest both groups first became photosynthetic at, or shortly before the Permian-Triassic boundary. This would support a hypothesis that secondary endosymbioses, which represent a major evolutionary step in the advancement of the algae, may be associated with the major climatic changes at the end Permian and the mass extinctions that followed (ERWIN 1994).

We gratefully acknowledge the technical support of U. WELLBROCK, S. WRIEDEN, and NATHALIE SIMON. Dr M. RAPPÉ kindly provided access to unpublished 16S rRNA sequence data from the diatoms, haptophytes and cryptophytes and Dr D. BHATTACHARYA provided Fig. 3. This research was supported in part by grants from the BMBF (03F0161B) and DFG (ME1480/1-2) of Germany to LKM, from the NSERC of Canada to GWS, and from the NSF (BRS94-19498) and ONR (N001492J1717) of the USA to RAA. This is AWI contribution no. 1290.

References

AGARDH, C. A., 1820: Species Algarum. **1**. – Lund.

ANDERSEN, R. A., 1987: *Synurophyceae* classis nov., a new class of algae. – Amer. J. Bot. **74**: 337–353.

– 1991: The cytoskeleton of chromophyte alga. – Protoplasma **164**: 143–159.

– SAUNDERS, G. W., PASKIND, M. P., SEXTON, J. P., 1993: The ultrastructure and 18S rRNA gene sequence for *Pelagomonas calceolata* gen. & sp. nov., and the description of a new algal class, the *Pelagophyceae* classis nov. – J. Phycol. **29**: 701–715.

ARIZTIA, E. V., ANDERSEN, R. A., SOGIN, M. L., 1991: A new phylogeny for chromophyte algae using 16S-like rRNA sequences from *Mallomonas papillosa* (*Synurophyceae*) and *Tribonema aequale* (*Xanthophyceae*). – J. Phycol. **27**: 428–436.

BHATTACHARYA, D., MEDLIN, L., 1995: The phylogeny of plastids: A review based on comparison of small subunit ribosomal RNA coding regions. – J. Phycol **31**: 489–498.

– EHLTING, J., 1995: Actin coding regions: gene family evolution and use as a phylogenetic marker. – Arch. Protistenk. **145**: 155–164.

– MEDLIN, L., WAINRIGHT, P. O., ARIZTIA, E. V., BIBEAU, C., STICKEL, S. K., SOGIN, M. L., 1992: Algae containing chlorophylls *a* and *c* are paraphyletic: molecular evolutionary analysis of the *Chromophyta*. – Evolution **46**: 1801–1817.

– HELMCHEN, T., MELKONIAN, M., 1995: Molecular evolutionary analyses of nuclear-encoded small subunit ribosomal RNA identify an independent rhizopod lineage containing the *Euglyphidae* and the *Chlorarachniophyta*. – J. Eukaryote Microbiol. **42**: 65–69.

BJØRNLAND, T., LIAAEN-JENSEN, S., 1989: Distribution patterns of carotenoids in relation to chromophyte phylogeny and systematics. – In GREEN, J. C., LEADBEATER, B. S. C., DIVER, W. L., (Eds): The chromophyte algae: problems and prospectives, pp. 37–60. – Syst. Ass. Spec. Vol. **38**. – Oxford: Clarendon Press.

BLACKMAN, F. F., 1900: The primitive algae and the flagellata. An account of modern work bearing on the evolution of algae. – Ann. Bot. **14**: 647–688.

BOURRELLY, P., 1957: Recherches sur les Chrysophycées: morphologie, phylogénie, systématique. – Rev. Algol. Mém. Hors-Séri. **1**: 1–412.

– 1968: Les Algues d'eau douce. 2. Algues jaunes & brunes. – Paris: Boubée.

CARON, L., DOUADY, D., QUINET-SZELY, DE GOËR, S., BERKALOFF, C., 1996: Gene structure of a chlorophyll a/c binding protein from a brown alga: Presence of an intron and phylogenetic implications. – J. Molec. Evol. **43**: 270–280.

CAVALIER-SMITH, T., 1975: The origin of nuclei and of eukaryotic cells. – Nature **256**: 463–468.

– 1982: The origins of plastids. – Biol. J. Linn. Soc. **17**: 289–306.

– 1986: The kingdom *Chromista*: Origin and systematics. – In ROUND F. E., CHAPMAN D. J., (Eds): Progress in phycological research, **4**, pp. 319–358. – Bristol: Biopress.

– 1989: The kingdom *Chromista*. – In Green, J. C., Leadbeater, B. S. C., Diver, W. L., (Eds): The chromophytic algae: problems and perspectives, pp. 381–407. – Oxford: Clarendon Press.
– 1993: Kingdom *Protozoa* and its 18 phyla. – Microbiol. Rev. **57**: 953–994.
– Chao, E. E., 1996: 18S rRNA sequence of *Heterosigma carterae* (*Raphidophyceae*), and the phylogeny of the heterokont algae (*Ochrophyta*). – Phycologia **35**: 500–510.
– Allsopp, M. T. E. P., Chao, E. E., 1994: Chimeric conundra: Are nucleomorphs and chromists monophyletic or polyphyletic? – Proc. Natl. Acad. Sci. USA **91**: 11368–11372.
– Chao, E. E., Allsopp, M. T. E. P., 1995: Ribosomal RNA evidence for chloroplast loss within *Heterokonta*: Pedinellid relationships and a revised classification of Ochristan algae. – Arch. Protistenk. **145**: 209–220.
– Allsopp, M. T. E. P., Haueber, M. M., Gothe, G., Chao, E. E., Couch, J. A., Maier, U.-G., 1996: Chromobiote phylogeny: the enigmatic alga *Reticulosphaera japonensis* is an aberrant haptophyte, not a heterokont. – Eur. J. Phycol. **31**: 255–264.
Chadefaud, M., 1950: Les cellules nageuses des algues dans l'embranchment des Chromophycées. – Compt. Rend. Hebd. Séances Sci. **231**: 788–790.
Chesnick, J. M., Morden, C. W., Schmieg, A. M., 1996: Identity of the endosymbiont of *Peridinium foliaceum* (*Pyrrhophyta*): Analysis of the *rbc*LS operon. – J. Phycol. **32**: 850–857.
Christensen, T., 1962: Alger. – In Böcher, T. W., Lange, M., Sørensen, T., (Ed.): Botanik, **2**/2, pp. 1–178. – Copenhagen: Munksgaard.
– 1989: The *Chromophyta*, past and present. – In Green, J. C., Leadbeater, B. S. C., Diver, W. L., (Eds): The chromophyte algae: problems and perspectives, pp. 1–12. – Oxford: Clarendon Press.
Craigie, J. S., 1974: Storage Products. – In Stewart, W. D. P., (Ed.): Algal physiology and biochemistry, pp. 206–235. – Berkeley, CA: University of California Press.
Delwiche, C. F., Palmer, J. D., 1996: Rampant horizontal transfer and duplication of Rubisco genes in *Eubacteria* and plastids. – Molec. Biol. Evol. **13**: 873–882.
– Kuhsel, M., Palmer, J. D., 1995: Phylogenetic analysis of *tuf*A sequences indicates a cyanobacterial origin of all plastids. – Molec. Phylogenet. Evol. **4**: 110–128.
Dodge, J. D., 1975: The fine structure of algal cells. – New York: Academic Press.
Ehrenberg, C. G., 1838: Die Infusionsthierchen als vollkommene Organismen. – Leipzig: Voss.
Erwin, D. H., 1994: The Permo-Triassic extinction. – Nature **367**: 231–236.
Felsenstein, J., 1985: Confidence limits on phylogenies: an approach using the bootstrap. – Evolution **39**: 783–791.
Fritsch, F. E., 1945: The structure and reproduction of the algae. **2**. – Cambridge: Cambridge University Press.
Fujiwara, S., Sawada, M. H., Someya, J., Minaka, N., Nishikawa, S., 1995: Molecular phylogenetic analysis of the *rbc*L in *Prymnesiophyta*. – J. Phycol. **30**: 863–871.
Gersonde, R., Harwood, D. M., 1990: Lower Cretaceous diatoms from ODP Leg 113 site 693 (Weddell Sea). I: Vegetative cells. – In Barker, P. F., Kennett, J. P., et al. (Eds): Proceedings of the Ocean Drilling Program, scientific results, **113**, pp. 365–402. – College Station, TX: Ocean Drilling Program.
Gibbs, S. P., 1978: The chloroplasts of *Euglena* may have evolved from symbiotic green algae. – Canad. J. Bot. **56**: 2882–2889.
– 1981: The chloroplasts of some algal groups may have evolved from endosymbiotic eukaryotic algae. – Ann. New York. Acad. Sci. **361**: 193–208.
– 1993: The origin of algal chloroplasts. – In Lewin, R. A., (Ed.): Origins of plastids, pp 107–121. – New York: Chapman and Hall.

GREEN, B. R., DURNFORD, D. G., 1996: The chlorophyll-carotenoid proteins of oxygenic photosynthesis. – Annual. Rev. Pl. Physiol. Pl. Mol. Biol. **47**: 685–714.

GREEN, J. C., 1989: Relationships between the chromophyte algae: the evidence from studies of mitosis. – In GREEN, J. C., LEADBEATER, B. S. C., DIVER, W. L., (Eds): The chromophyte algae: problems and perspectives, pp. 189–206. – Oxford: Clarendon Press.

– LEADBEATER, B. S. C., (Eds), 1994: The haptophyte algae. – Oxford: Clarendon Press.

– – DIVER, W. L., (Eds), 1989: The chromophyte algae: problems and perspectives. – Oxford: Clarendon Press.

GU, X., LI, W.-H., 1996: Bias-corrected paralinear and LogDet distances and tests of molecular clocks and phylogenies under nonstationary nucleotide frequencies. – Molec. Biol. Evol. **13**: 1375–1383.

HARVEY, W. H., 1836: *Algae*. – In MACKAY, J. T., (Ed.): Flora Hibernica, pp. 157–254. – Dublin.

HIBBERD, D. J., 1976: The ultrastructure and taxonomy of the *Chrysophyceae* and *Prymnesiophyceae* (*Haptophyceae*): a survey with some new observations on the ultrastructure of the *Chrysophyceae*. – Bot. J. Linn. Soc. **72**: 55–80.

– LEEDALE, G. F., 1971: A new algal class – the *Eustigmatophyceae*. – Taxon **20**: 523–525.

HILLIS, D. M., BULL, J. J., 1993: An empirical test of bootstrapping as a method for assessing confidence in phylogenetic analysis. – Syst. Biol. **42**: 182–192.

– MORITZ, C., 1990: An overview of applications of molecular systematics. – In HILLIS, D. M., MORITZ, C., (Eds): Molecular systematics, pp. 502–515. – Sunderland, MA: Sinauer.

– – MABLE, B. K., 1996: Molecular systematics. – Sunderland, MA: Sinauer.

JEFFREY, S. W., 1989: Chlorophyll *c* pigments and their distribution in the chromophyte algae. – In GREEN, J. C., LEADBEATER, B. S. C., DIVER, W. L., (Eds): The chromophyte algae: problems and perspectives, pp. 13–36. – Oxford: Clarendon Press.

JORDAN, R., GREEN, J. C., 1994: A check-list of the extant *Haptophyta* of the world. – J. Mar. Biol. Assoc. U.K. **74**: 149–174.

KEELING, P. J., DOOLITTLE, W. F., 1996: Alpha-tubulin from early-diverging eukaryotic lineages and the evolution of the tubulin family. – Molec. Biol. Evol. **13**: 1297–1305.

KIMURA, M., 1980: A simple method for estimating evolutionary rates of base substitution through comparative studies of sequence evolution. – J. Molec. Evol. **16**: 111–120.

KLEBS, G., 1893: Flagellatenstudien. II. – Z. Wiss. Zool. **55**: 353–445.

KLEIN, R. M., CRONQUIST, A., 1967: A consideration of the evolutionary and taxonomic significance of some biochemical, micromorphological, and physiological characters in the thallophytes. – Quart. Rev. Biol. **42**: 105–296.

KNOLL, A. H., 1992: The early evolution of eukaryotes: a geological perspective. – Science **256**: 622–627.

KOHLER, S., DELWICHE, C. F., DENNY, P. W., TILNEY, L. G., WEBSTER, P., WILSON, R. J., PALMER, J. D., ROOS, D. S., 1997: A plastid of probable green algal origin in apicomplexan parasites. – Science **275**: 1485–1489.

KOOISTRA, W. H. C. F., MEDLIN, L. K., 1996: Evolution of the diatoms (*Bacillariophyta*): IV. A reconstruction of their age from small subunit rRNA coding regions and the fossil record. – Molec. Phylogenet. Evol. **6**: 391–407.

KOWALLIK, K. V., 1992: Origin and evolution of plastids from chlorophyll-$a + c$-containing algae: suggested ancestral relationships to red and green algal plastids. – In LEWIN, R. A., (Ed.): Origins of plastids, pp. 223–263. – New York: Chapman and Hall.

KUMAR, S., TAMURA, K., NEI, M., 1993: MEGA: molecular evolutionary genetics analysis, version 1.0. – University Park, PA: Institute of Molecular Evolutionary Genetics, Pennsylvania State University.

KÜTZING, F. T., 1834: Synopsis diatomearum oder Versuch einer systematischen Zusammenstellung der Diatomeen. – Halle.
LACKEY, J. B., 1939: Notes on plankton flagellates from the Scioto River. – Lloydia **2**: 128–143.
LAMOUROUX, J. V. F., 1813: Essai sur les genres de la famille des thalassiophytes non articulées. – Ann. Mus. Hist. Nat. **20**: 21–47, 115–139, 267–293.
LAROCHE, J., HENRY, D., WYMAN, K., SUKENIK, A., FALKOWSKI, P., 1994: Cloning and nucleotide sequence of a cDNA encoding a major fucoxanthin-chlorophyll *a/c*-containing protein from the chrysophyte *Isochrysis galbana*: implications for the evolution of the *cab* gene family. – Pl. Molec. Biol. **25**: 355–368.
LEIPE, D. D., WAINRIGHT, M. L., GUNDERSON, J. H., PORTER, D., PATTERSON, D. J., VALOIS, F., HIMMERICH, S., SOGIN, M. L., 1994: The stramenopiles from a molecular perspective: 16S-like rRNA sequences from *Labyrintuloides minuta* and *Cafeteria roenbergensis*. – Phycologia **33**: 369–377.
LIAUD, M.-F., BANDT, U., SCHERZINGER, M., CERFF, R., 1997: Evolutionary origin of cryptomonad microalgae: Two novel chloroplast/cytosol-specific GAPDH genes as potential markers of ancestral endosymbiont and host cell components. – J. Molec. Evol. **44**: [Suppl 1]: S28–S37.
LINNEAUS, C., 1753: Species plantarum. – Stockholm.
LOCKHART, P. J., HOWE, C. J., BRYANT, D. A., BEANLAND, M. D., PENNY, D., 1994: Substitutional bias confounds inference of cyanelle origins from sequence data. – J. Molec. Evol. **34**: 153–162.
LOHMANN, H., 1913: Über Coccolithophoriden. – Verh. Deutsch. Zool Ges. **23**: 143–164.
LOISEAUX DE GOËR, S., 1994: Plastid lineages. – Progr. Phycol. Res. **10**: 137–177.
LUTHER, A., 1899: Über *Chlorosaccus* eine neue Gattung der Süsswasseralgen. – Beih. Kongl. Svenska Vetensk. – Akad. Handl. **24** (III, 13): 1–22.
MCFADDEN, G. I., GILSON, P. R., WALLER, R. F., 1995: Molecular phylogeny of chlorarachniophytes based on plastid rRNA and *rbc*L sequences. – Arch. Protistenk. **145**: 231–239.
MARGULIS, L., 1981: Symbiosis in cell evolution. – San Francisco: Freeman.
MEDLIN, L. K., COOPER, A., HILL, C., WRIEDEN-PRIGGE, S., WELLBROCK, U., 1995: Phylogenetic position of the *Chromista* plastids from 16S rDNA coding regions. – Curr. Genet. **28**: 560–565.
– BARKER, G. L. A., CAMPBELL, L., GREEN, J. C., HAYES, P. K., MARIE, D., WRIEDEN, S., VAULOT, D., 1996a: Genetic characterization of *Emiliania huxleyi* (*Haptophyta*). – J. Mar. Syst. **9**: 13–31.
– GERSONDE, R., KOOISTRA, W. H. C. F., WELLBROCK, U., 1996b: Evolution of the diatoms (*Bacillariophyta*): II. Nuclear-encoded small-subunit rRNA sequence comparisons confirm a paraphyletic origin for the centric diatoms. – Molec. Biol. Evol. **13**: 67–75.
– – – SIMS, P. A., WELLBROCK, U., 1996c: Evolution of the diatoms (*Bacillariophyta*): III. The age of the *Thalassiosirales*. – Beih. Nova Hedwigia **11**: 221–234.
– – – – – 1997: Is the origin of diatoms related to the end-Permian mass extinction. – In JAHN, R., MEYER, B., PREISIG, N. R. (Eds): Nova Hedwigia Festschrift für U. GEISSLER pp. 1–13. – Stuttgart: J. Cramer.
MERESCHKOWSKY, C., 1905: Über Natur und Ursprung der Chromotaphoren im Pflanzenreiche. – Biol. Centralbl. **25**: 593–604.
MÜLLER, O. M. F., 1786: Animacula infusoria fluviatilia et marina. – Copenhagen: Moller.
MOESTRUP, Ø., 1992: Taxonomy and phylogeny of the *Heterokontophyta*. – In STABENAU, H., (Ed.): Phylogenetic changes in peroxisomes of algae, phylogeny, of plant peroxisomes, pp. 383–399. – Oldenburg: University of Oldenburg.
PALMER, J. D., 1993: A genetic rainbow of plastids. – Nature **364**: 762–763.

PARKE, M., MANTON, I., CLARKE, B., 1955: Studies on marine flagellates. II. Three new species of *Chrysochromulina*. – J. Mar. Biol. Assoc. U.K. **34**: 579–609.

– – – 1956: Studies on marine flagellates. III. Three further species of *Chrysochromulina*. – J. Mar. Biol. Assoc. U.K. **35**: 387–414.

– – – 1958: Studies on the marine flagellates. IV. Morphology and microanatomy of a new species of *Chrysochromulina*. – J. Mar. Biol. Assoc. U.K. **37**: 209–228.

PASCHER, A., 1910: Chrysomonaden aus dem Hirschberger Großteiche. Untersuchungen über die Flora des Hirschberger Großteiches. I. Teil. – Monogr. Abh. Int. Rev. Gesamten Hydrobiol. Hydrogr. **1**: 1–66.

– 1913: *Chrysomonadinae*. – In PASCHER, A., (Ed.): Süsswasser-Flora Deutschlands, Österreichs und der Schweiz, **2**, pp. 7–15.

PATTERSON, D. J., 1989: Stramenopiles: chromophytes from a protistan perspective. – In LEADBEATER, B. S. C., DIVER, W. L., (Eds): The chromophyte algae, pp. 357–379. – Oxford: Clarendon Press.

PÉRASSO, R., BAROIN, A., QU, L. H., BACHELLERIE, J. P., ADOUTTE, A., 1989: Origin of the algae. – Nature **339**: 142–144.

PICKETT-HEAPS, J., SCHMID, A-M. M., EDGAR, L. A., 1990: The cell biology of diatom valve formation. – In ROUND, F. E., CHAPMAN, D. J., (Eds): Progress in phycological research, pp., 1–168. – Bristol: Biopress.

POTTER, D., LAJEUNESSE, T. C., SAUNDERS, G. W., ANDERSEN, R. A., 1996: Convergent evolution masks extensive biodiversity among marine coccoid picoplankton. – Biodiversity conservation. **6**: 99–107.

– SAUNDERS, G. W., ANDERSEN, R. A., 1997: Phylogenetic relationships of the *Raphidophyceae* and *Xanthophyceae* as inferred from nucleotide sequences of the 18S ribosomal RNA gene. – Amer. J. Bot. **84**: 966–972.

RABENHORST, L., 1853: Süsswasser-Diatomeen (Bacillarien), für Freunde der Mikroskopie bearbeitet. – Leipzig.

RAVEN, P. H., 1970: A multiple origin for plastids and mitochondria. – Science **169**: 641–646.

RENSING, S. A., OBRDLIK, P., ROBER-KLEBER, N., MÜLLER, S. B., HOFMANN, C. J. B., MAIER, U.-G., 1996: Molecular phylogeny of the stress-70 protein family with certain emphasis on algal relationships. – In: 1st European Phycological Congress, Abstracts, p. 16.

ROUND, F. E., CRAWFORD, R. M., MANN, D. G., 1990: The diatoms: Biology and morphology of the genera. – Cambridge: Cambridge University Press.

ROWAN, K. S., 1989: Photosynthetic pigments of algae. – Cambridge: Cambridge University Press.

SAITOU, N., NEI, M., 1987: The neighbor-joining method: a new method for reconstructing phylogenetic trees. – Molec. Biol. Evol. **4**: 406–425.

SAUNDERS, G. W., DRUEHL, D., 1992: Nucleotide sequences of the small-subunit ribosomal RNA genes from selected *Laminariales* (*Phaeophyta*): implications for kelp evolution. – J. Phycol. **28**: 544–549.

– POTTER, D., PASKIND, M. P., ANDERSEN, R. A., 1995: Cladistic analyses of combined traditional and molecular data sets reveal an algal lineage. – Proc. Natl. Acad. Sci. USA **92**: 244–248.

– HILL, D. R. A., TYLER, P. A., 1997a: Phylogenetic affinities of *Chrysonephele palustris* (*Chrysophyceae*) based on inferred nuclear small-subunit ribosomal RNA sequence. – J. Phycol. **33**: 132–134.

– POTTER, D., ANDERSEN, R. A., 1997b: Phylogenetic affinities of the *Sarcinochrysidales* and *Chrysomeridales* (*Heterokonta*) based on analyses of molecular and combined data. – J. Phycol. **33**: 310–318.

SCHIMPER, A. F. W., 1883: Über die Entwicklung der Chlorophyllkörner und Farbkörper. – Bot. Zeitung (Berlin) **41**: 105–112.

SILVA, P. C., 1980: Names of classes and families of living algae. – Regnum Veg. **103**: 1–156.

SLEIGH, M. A., 1989: Protozoa and other protists. – London: Arnold.

STACKHOUSE, J., 1809: Tentamen marino-cryptogamicum, ordinem novum, in genera et species distributum, in Classe XXIVta Linnaei sistens. – Mém. Soc. Imp. Naturalistes Moscou **2**: 50–97.

STARMACH, K., 1985: *Chrysophyceae* und *Haptophyceae*. – In ETTL, H., GERLOFF, J., HEYNIG, H., MOLLENHAUER, D., (Eds): Süsswasserflora von Mitteleuropa **1**. – Stuttgart: G. Fischer.

STEIN, F. VON, 1878: Der Organismus der Infusionsthiere. **3**. – Leipzig.

STEWART, K. D., MATTOX, K., 1970: Phylogeny of phytoflagellates. – In COX, E. R., (Ed.): Phytoflagellates, pp. 433–462. – New York: Elsevier/North-Holland.

SWOFFORD, D. L., 1993: PAUP, Phylogenetic analysis using parsimony, version 3.1, program and documentation. – Champaign, IL: Illinois Natural History Survey, University of Illinois.

TAGGART, R. E., PARKER, L. R., 1976: A new fossil alga from the Silurian of Michigan. – Amer. J. Bot. **63**: 1390–1392.

TAKEZAKI, N., RZHETSKY, A., NEI, M., 1995: Phylogenetic test of the molecular clock and linearized trees. – Molec. Biol. Evol. **12**: 823–833.

TAYLOR, F. J. R., 1976: Flagellate phylogeny: a study in conflicts. – J. Protozool. **23**: 28–40.

VALENTIN, K., CATTOLICO, R. R., ZETZCHE, K., 1992: Phylogenetic origin of the plastids. – In LEWIN, R. A., (Ed.): Origins of plastids, pp. 193–221. – New York: Chapman and Hall.

VAN DEN HOEK, C., 1978: Algen: Einführung in die Phykologie. – Stuttgart: Thieme.

VAN DER AUWERA, G., DE WACHTER, R., 1996: Large-subunit rRNA sequence of the chytridiomycete *Blastocladiella emersonii*, and implication for the evolution of zoosporic fungi. – J. Molec. Evol. **43**: 476–483.

VESK, M., JEFFREY, S. W., 1987: Ultrastructure and pigments of two strains of the picoplanktonic alga *Pelagococcus subviridis* (*Chrysophyceae*). – J. Phycol. **23**: 322–336.

VLK, W., 1938: Über den Bau der Geissel. – Arch. Protistenk. **90**: 448–488.

VOGEL, H., FISCHER, S., VALENTIN, K., 1996: A model for the evolution of the plastid sec apparatus inferred from *sec*Y gene phylogeny. – Pl. Molec. Biol. **32**: 685–692.

WAINRIGHT, P. O., HICKLE, G., SOGIN, M. L., STICKEL, S. K., 1993: Monophyletic origins of the *Metazoa*: an evolutionary link with the fungi. – Science **260**: 340–342.

WANG, M. C., BARTNICIA GARCIA, S., 1974: Mycolaminarins: Storage $(1 \rightarrow 3) - \beta$-D-glucans from the cytoplasm of the fungus *Phytophthora palmivora*. – Carbohydrate Res. **37**: 331–338.

WILLIAMS, D. M., 1991: Phylogenetic relationships among the *Chromista*: a review and preliminary analysis. – Cladistics **7**: 141–156.

WRAY, G. A., LEVINGTON, J. S., SHAPIRO, L. H., 1996: Molecular evidence for deep Precambrian divergences among metazoan phyla. – Science **274**: 568–573.

WU, C.-I., LI, W.-H., 1985: Evidence for higher rates of nucleotide substitution in rodents than in man. – Proc. Natl. Acad. Sci. USA **8**: 1741–1745.

Molecular and morphological phylogenies of kelp and associated brown algae

L. D. Druehl, C. Mayes, I. H. Tan, and G. W. Saunders

Key words: *Alariaceae, Laminariaceae, Lessoniaceae, Phaeophyceae.* – Kelp, molecular evolution, phylogenetics, rDNA.

Abstract: Morphological and molecular phylogenies of kelp (*Laminariales, Phaeophyceae*), the world's largest protists, are generally at odds. This chapter discusses the current state of knowledge for relationships of the *Laminariales* relative to other orders of brown algae (*Phaeophyceae*) and among the families of kelp, in particular, the *Alariaceae, Laminariaceae* and *Lessoniaceae*. Phylogenetic relationships among fourteen kelp species, representing these three families were inferred from newly obtained and previously published (3′18S – ITS1 – 5.8S) rDNA sequences.

Kelp, large brown algae belonging to the order *Laminariales*, are perhaps the most unlikely members of the *Protista*. Contrary to other protists, kelp sporophytes may achieve great size [up to 45.7 m for *Macrocystis pyrifera* (L.) C. A. Agardh] (Abbott & Hollenberg 1976), they possess localized meristems which produce distinct structures (holdfasts, stipes, blades and sporophylls) and they have specialized cells for the active translocation of photosynthate from areas of high production to areas of high demand (Lobban 1978, Schmitz & Srivastava 1979).

Inclusion in the *Protista* is justified because kelp spores and gametes have heterokont flagellation which aligns them with group of single-celled and filamentous eukaryotes, generally considered protists. The value of heterokont flagellation in defining a very diverse subgroup of *Protista* is substantiated by molecular evidence. Phylogenetic analysis of actin genic regions and small-subunit ribosomal DNA tightly cluster autotrophic kelp, diatoms and chrysophytes (which have chlorophylls a and c) with the heterotrophic *Achlya bisexualis* Coker & Couch, a "lower fungus" (Bhattacharya & Druehl 1988, Bhattacharya & al. 1991)

In this chapter we will define the morphological and molecular relationships among selected kelp taxa and between kelp and closely related brown algal orders. We will, however, emphasize relationships among the three major kelp families (*Alariaceae, Laminariaceae* and *Lessoniaceae*) for which we present a phylogenetic analysis based on a combination of published and unpublished sequence data from a variety of taxa. Table 1 outlines the overall taxonomic hierarchy of kelp and indicates the specific genera discussed here.

Morphological relationships

The *Laminariales* (kelp) belong to the *Phaeophyceae*, a class of brown pigmented algae, distinguished from other similarly pigmented classes dominated by unicellular and simple filamentous representatives (e.g., *Bacillariophyceae, Chrysophyceae, Xanthophyceae*), by size, tissue differentiation and by having no known unicellular representatives.

Traditionally, the orders of *Phaeophyceae* are distinguished on the basis of life cycle type (heteromorphic or isomorphic alternation generations or an animal-like, gametic life cycle), reproduction (isogamous, anisogamous, oogamous), tissue organization (filamentous, pseudoparenchymatous, parenchymatous) and growth pattern (diffuse, apical, intercalary). Based on subsets of these criteria the kelp have been variously aligned with other brown algal orders (Kylin 1933, Papenfuss 1955, Wynne & Loiseaux 1976). For example, Wynne & Loiseaux (1976) placed the *Laminariales* in the subclass *Phaeophycidae* which contain those brown algae with an alternation of generations (in contrast to the *Cyclosporidae* which have gametic life cycles) (Fig. 1). The *Laminariales*, along with the *Scytosiphonales* and *Dictyosiphonales*, were separated from the remaining orders of the *Phaeophycidae* on the basis of haplopolystichous construction (where one generation is parenchymatous and the other is pseudoparenchymatous). The *Laminariales* are distinguished from the *Scytosiphonales* and *Dictyosiphonales*, by their oogamous reproduction and intercalary meristem.

Generally, the *Laminariales* are characterized by a large sporophytic generation that alternates with a microscopic gametophytic generation. The gametophytes, which are filamentous, are usually dioecious, and produce distinctive eggs and sperms. Usually, the parenchymatous sporophytes grow by means of an intercalary meristem and consist of holdfasts, stipes and blades. Meiospores are produced in clustered unilocular sporangia on the sporophytic thallus.

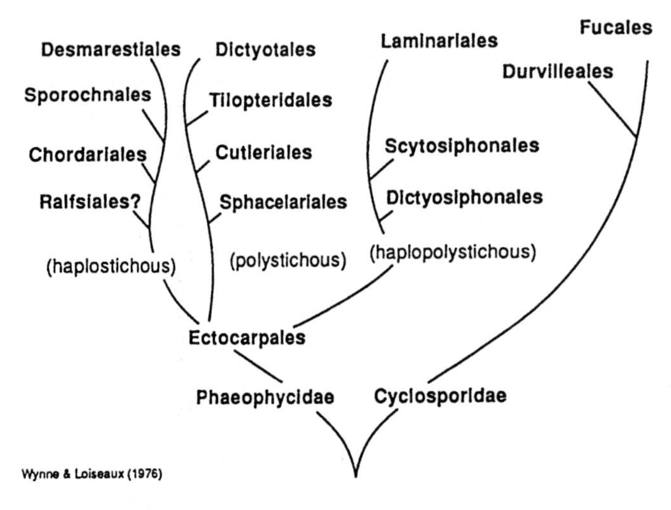

Fig. 1. A proposed phylogeny for the *Phaeophyceae* based on morphological and reproductive features (after Wynne & Loiseaux 1976)

The *Laminariales* are divided into six families. The families *Pseudochordaceae, Chordaceae* and *Phyllariaceae* (PCP) are distinguished by having an eyespot in their meiospores, a feature not shared with the remaining families: *Alariaceae, Laminariaceae* and *Lessoniceae* (ALL) (Kawai & Kurogi 1985, Henry & South 1987, Kawai 1992). Further, the PCP lack the more sophisticated translocation elements (trumpet hyphae) found in the ALL (Emerson & al. 1982) and the two family sets have different sexual pheromones (ALL have lamoxirene while *Chordaceae* and *Phyllariaceae* have multifidene and ectocarpene, respectively; Maier & al. 1984).

Among the PCP, the *Pseudochordaceae* have plurilocular antheridia and multicellular soral paraphyses whereas the *Chordaceae* and *Phyllariaceae* have single celled antheridia and unicellular soral paraphyses (Kawai & Kurogi 1985). The *Chordaceae* and *Phyllariaceae* may be distinguished by the structure of their translocating cells (the *Chordaceae* have elongated hyphae and the *Phyllariaceae* have solenocysts and allelocysts, Emerson & al. 1982) and by the different sexual pheromonal systems noted above (Maier & al. 1984).

The ALL families, all species of which lack eyespots in their meiospores, are traditionally defined on the basis of the sporophyte's gross morphology. Members of the *Alariaceae* have sporophylls, special blades bearing the clusters of unilocular sporangia, projecting from a stipe or from the blade base. The family consists of seven genera, two of which are monotypic, and 33 species. *Alariaceae* representatives employed in the following molecular assessment are: *Alaria marginata* has a moderate-sized stipe bearing two ranks of sporophylls and a blade with a conspicuous midrib (Fig. 2); *Pterygophora californica* has a dominant stipe bearing two sparse ranks of sporophylls and a diminutive blade with a poorly-defined, inconspicuous midrib; *Eisenia arborea* has, at maturity, a stout, bifurcating stipe with each bifurcation terminated by a dense scroll of sporophylls (the blade is lost early in development, Setchell 1905; but for an alternative interpretation see Saunders & Druehl 1993); *Egregia menziesii* has a profusely-branched stipe bearing two ranks of sporophylls and vegetative proliferations; at the end of each stipe is a blade lacking a midrib but bearing numerous lateral proliferations.

Members of the *Laminariaceae* have a simple, unbranched blade and stipe and produce sori (clusters of unilocular sporangia) on the vegetative blade. This family consists of ten genera, of which five monotypic, and approximately 47 species. The genera are morphologically distinct. The four genera used in the following molecular assessment are *Laminaria saccharina* which has a simple thallus consisting of a plain blade and distinct stipe and holdfast; *Hedophyllum sessile* has a plain blade which is frequently torn and, at maturity, the stipe is essentially non-existent; *Pleurophycus gardneri* has a distinct stipe and an untorn blade with one conspicuous midrib; *Costaria costata* has a distinct stipe and an untorn blade with five distinct midribs.

The *Lessoniaceae* consists of eight genera, of which five are monotypic, and 14 species. Species of the *Lessoniaceae* are characterized by a branched thallus which arises from an ontogenetic splitting at the meristematic transition region between the blade and stipe. Six representatives of the family were used in the following molecular study: *Postelsia palmaeformis* consists of a hollow stipe terminated with

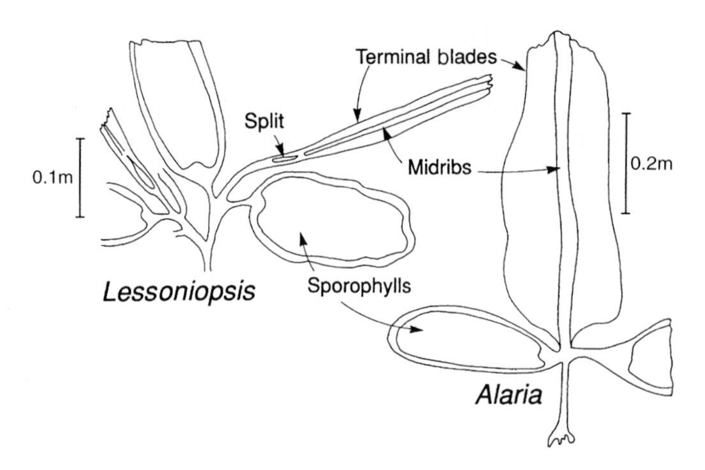

Fig. 2. Comparative gross morphology of *Lessoniopsis littoralis* (*Lessoniaceae*) and *Alaria marginata* (*Alariaceae*). Note that *Lessoniopsis* has the diagnostic features of both the *Alariaceae* (sporophylls) and the *Lessoniaceae* (splitting at the transition zone). See discussion for further details. (After DRUEHL & SAUNDERS 1992)

numerous, longitudinally furrowed blades which arise in dense-branched clusters; *Nereocystis leutkeana* consists of a long stipe which is inflated and hollow at its distal end and terminated with numerous smooth blades which arise in dense, branched clusters; *Dictyoneurum californicum* consists of highly reticulated, dichotomously branched blades which essentially arise directly from the holdfast; *Macrocystis integrifolia* is a large plant consisting of numerous fronds (stipe/blade complexes) arising in a more or less dichotomous fashion from a strap-like holdfast, sporophylls are the first blades encountered above the holdfast, reticulated blades bearing pneumatocysts are produced through splitting of an apical scimitar; *Lessoniopsis littoralis* (Fig. 2) consists of a stout, repeatedly branched system of stipes, each terminated by a long narrow, midribbed blade which has a pair of sporophylls near its base, whereas the morphologically similar kelp, *Lessonia nigrescens* has a stout, dichotomously branched system of stipes, each terminated by a single blade.

Molecular relationships

Kelp and other brown algae. Based on an analysis of partial 18S rDNA sequences, the brown algae, including kelp, are grouped into two main clusters (TAN 1995, Fig. 3) contrary to traditional views (Fig. 1 for example). Representatives of one cluster (Fig. 3 – left hand side) have normal pyrenoids, and they have been considered by some authors to belong to the *Ectocarpales* (FRITSCH 1954, GABRIELSON & al. 1989, TAN & DRUEHL 1994). The second cluster included brown algae which either lack pyrenoids or possess rudimentary ones, including the *Laminariales* (Fig. 3 – right hand side). Analysis of complete 18S rDNA sequences for selected taxa support the molecular distinctiveness of the two brown algal clusters (TAN & DRUEHL 1996) (Fig. 4). This same analysis consistently grouped representatives of the *Desmarestiales, Sporochnales* and *Laminariales*.

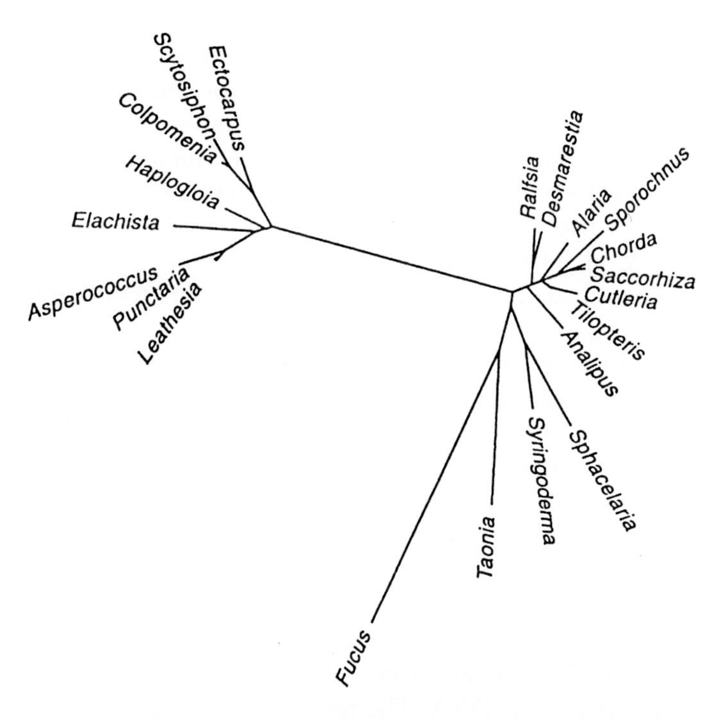

Fig. 3. Unrooted neighbor-joining tree based on partial (870 bp) 18S rDNA sequences (after TAN 1995). Contrary to traditional phylogenetic schemes, the studied taxa were separated by molecular comparisons into two clusters. The cluster on the left is composed to taxa which have pyrenoids and are now considered to belong to the *Ectocarpales* (sensu GABRIELSON & al. 1989). The cluster on the right includes taxa which either lack pyrenoids or have rudimentary ones, including the *Laminariales* (represented in this analysis by *Alaria marginata, Chorda tomentosa* and *Saccorhiza polyschides*).

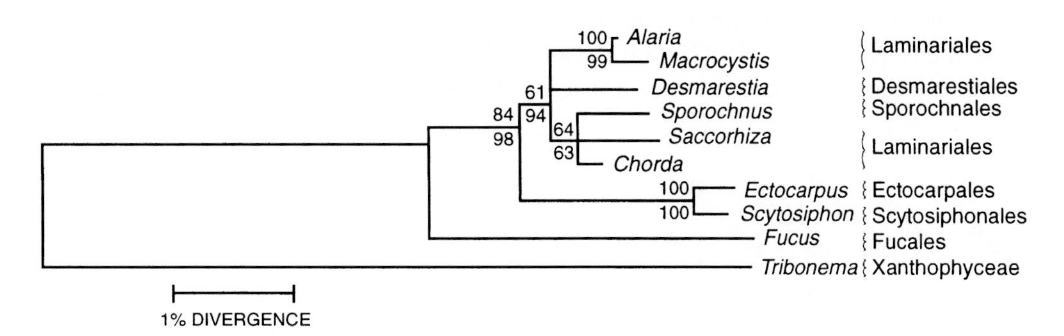

Fig. 4. A 50% majority-rule consensus tree inferred from the entire 18S rDNA sequence alignment by distance analysis with neighbor-joining and with parsimony methods. Top numbers are bootstrap values (100 replicates) from parsimony analysis and bottom numbers are bootstrap values (500 replicates) from the distance analysis. The scale bar = 1% sequence divergence (distance analysis). (After TAN & DRUEHL 1996)

These three orders share many features in common, but perhaps most striking is the observation that they all have sperms with a shorter anterior and a longer posterior flagellum. Sperms of other brown algae have a longer anterior and a shorter posterior flagellum (HENRY & COLE 1982, KAWAI 1992). The importance of the molecular linking of the *Laminariales, Sporochnales* and *Desmarestiales* is strengthened by CLAYTON's (1984) survey of evidence for a "common ancestor linking (these three orders)". It is also interesting to note that the recent phylogeny proposed by VAN DEN HOEK & al. (1995), based on their interpretation of the comparative morphology of extant taxa also grouped these three orders together in an assemblage.

Within the *Laminariales*, analysis of entire 18S rDNA sequences clearly distinguished *Alaria marginata* and *Macrocystis integrifolia* (representatives of the ALL complex) from *Saccorhiza polyschides* and *Chorda tomentosa* (PCP complex), thus supporting the notion that the 'advanced' ALL complex of species are phylogenetically isolated from the 'primitive' PCP families (MÜLLER & al. 1985, cf. TAN & DRUEHL 1996) (see Fig. 4). In fact, 18S data weakly allied the PCP complex with the order *Sporochnales* in lieu of the ALL families (TAN & DRUEHL 1996).

The ALL complex. The first molecular investigation of species included in the ALL complex was undertaken by FAIN & al. (1988) using restriction fragment length (RFLP) analysis of chloroplast (cp) DNA. The resulting phylogeny (Fig. 5) contradicts traditional perspectives on kelp evolution (Table 1) and presents a polyphyletic *Lessoniaceae* with *Lessoniopsis* and *Macrocystis* closely allied to *Alaria* whereas *Nereocystis* grouped with *Laminaria*.

SAUNDERS & DRUEHL (1991, 1992, 1993) provide plausible explanations, other than phylogenetic signal, to account for the discrepancies between the traditional morphological and cpDNA perspectives on relationships within the ALL complex. They also proposed methods to test these various hypotheses. SAUNDERS & DRUEHL (1992) attempted nucleotide sequence comparisons of 18S rDNA sequences for eight species of the *Alariaceae* and *Lessoniaceae*. From this study, as well as in consideration of previously published 18S data for *Costaria costata, (Laminariaceae* BHATTACHARYA & DRUEHL 1988), it was concluded that the 18S rRNA was too conservative for phylogenetic investigations within the ALL complex (SAUNDERS & DRUEHL 1992). SAUNDERS & DRUEHL (1993) then investigated the relatively more variable internal transcribed spacer (ITS) of the ribosomal cistron

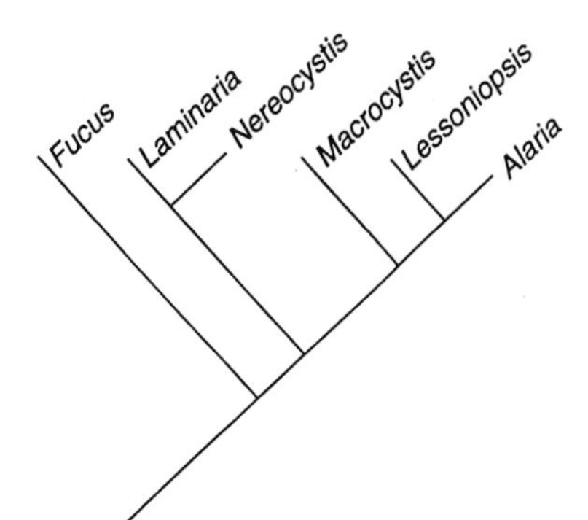

Fig. 5. A phylogeny of five kelp species and *Fucus gardneri* SILVA based on 53 restriction fragment length variants. The tree was generated by Wagner parsimony analysis (PHYLIP) using the "branch and bound" function. Branch lengths are arbitrary (after FAIN & al. 1988)

Table 1. Classical taxonomic hierarchy of kelp (*Laminariales*) and the specific representatives discussed in this chapter. *Published *rDNA* sequence data from these taxa used in the molecular phylogenetic study of Saunders & Druehl (1993) were incorporated in the present study. **New taxa included in the present analyses

Kingdom: *Protista*
- Division: *Heterokontophyta* sensu Van Den Hoek & al. (1995)
  - Class: *Phaeophyceae* Kjellman
    - Order: *Laminariales* Oltmanns
      - Families: *Pseudochordaceae* Kawai & Kurogi
        - *Pseudochorda nagaii* (Tokida) Inagaki
      - *Chordaceae* Dumortier
        - *Chorda tomentosa* Lyngbye**
      - *Phyllariaceae* Tilden
        - *Saccorhiza polyschides* (Lightfoot) Batters.
      - *Alariaceae* Setchell & Gardner
        - *Alaria marginata* Postels & Ruprecht*
        - *Eisenia arborea* Areschoug*
        - *Egregia menziesii* (Turner) Areschoug*
        - *Pterygophora californica* Ruprecht*
      - *Laminariaceae* Reichenbach
        - *Costaria costata* (Turner) Saunders*
        - *Hedophyllum sessile* (J. Agardh) Setchell**
        - *Laminaria saccharina* (L.) Lamouroux**
        - *Pleurophycus gardneri* Setchell & Saunders**
      - *Lessoniaceae* Setchell & Gardner
        - *Dictyoneurum californicum* Ruprecht*
        - *Lessonia nigrescens* Bory*
        - *Lessoniopsis littoralis* (Farlow & Setchell) Reinke*
        - *Macrocystis integrifolia* Bory*
        - *Nereocystis leutkeana* (Mertens) Postels & Ruprecht*
        - *Postelsia palmaeformis* Ruprecht*

(cf. Druehl & Saunders 1992) and recovered a phylogenetic tree strongly at odds with traditional concepts of taxonomic division within the ALL complex (Fig. 6). Aspects of the earlier cpDNA phylogeny (Fain & al. 1988) were observed in the ITS result, although the latter, which had a greater taxon sampling (ten versus five species), indicated that neither the *Alariaceae* or *Lessoniaceae* are monophyletic. Saunders & Druehl (1993) concluded that the *Alariaceae* (Group 1, Fig. 6) should be emended to include only those species of the ALL complex characterized by the combination of lamina with midribs and Group 1-type sporophylls (developing on the primary stipe and not derivatives of the lamina – exclusive to, and characteristic of all of the species included in the *Alariaceae* sensu Saunders & Druehl (1993).

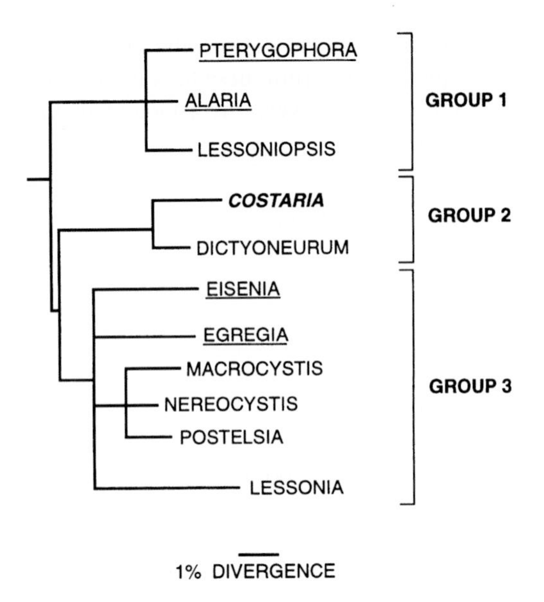

Fig. 6. Rooted distance tree derived with neighbor-joining and inferred from rDNA sequence data (including 3′18S – ITS1 – 5.8S gene regions) representing phylogenetic relationships among 11 kelp genera. Underlined generic epithets were traditionally in the *Alariaceae*. The single italicized generic epithet was originally in *Laminariaceae* and plain generic epithets were in *Lessoniaceae*. (After Saunders & Druehl 1993). Scale = 1% sequence divergence

The *Alariaceae* was thus restricted to the genera *Alaria, Lessoniopsis,* and *Pterygophora* (see Fig. 2). Group 2 sensu Saunders & Druehl (1993) (Fig. 6) contained the only member of the *Laminariaceae* included in the ITS study, *Costaria costata* as well as the monotypic genera, *Dictyoneurum* and *Dictyoneuropsis* of the *Lessoniaceae* (these latter two had identical ITS sequences), whereas Group 3 sensu Saunders & Druehl (1993) (Fig. 6) contained a variety of (ex) *Alariaceae* and *Lessoniaceae*. Although Group 2 contains a species with a vegetative midrib, *Dictyoneuropsis reticulata* (Saunders) G. M. Smith, this taxon lacks sporophylls of any type. Those members of Group 3 characterized by sporophylls were considered by these authors to have lamina-derived analogues, rather than homologues, of the Group 1-type sporophyll (cf. Saunders & Druehl 1993). Taxonomic conclusions, i.e. Group 2 consists of an expanded *Laminariaceae* to include *Dictyoneurum* and *Dictyoneuropsis*, whereas Group 3 contains in *Lessoniaceae* plus those traditional *Alariaceae* lacking Group 1-type sporophylls (Fig. 6). Further conclusions regarding group 3 were deferred by Saunders & Druehl (1993) because the *Laminariaceae* was represented by a single species which was not from the type genus *Laminaria*.

To build upon the molecular phylogeny of Saunders & Druehl (1993), the current analyses incorporated rDNA sequences (including the ITS1) from additional taxa of the *Laminariaceae* (*Laminaria saccharina, Hedophyllum sessile* and *Pleurophycus gardneri*), in order to enhance our understanding of phylogenetic relationships among the families of the ALL complex. The rDNA sequence from

Chorda tomentosa (*Chordaceae*) was also sequenced and designated as the outgroup taxon.

Materials and methods

Kelp DNA was extracted and purified from sporophytes of *Laminaria saccharina* (collected from Whyte Cliff Park, Vancouver, B. C. Canada), *Hedophyllum sessile* and *Pleurophycus gardneri* (Whiffen Spit, Vancouver Island, B. C., Canada) and *Chorda tomentosa* (Magdalen Islands, Que., Canada) according to protocol of Mayes & al. (1992) Double-stranded PCR products were amplified and sequenced using the primers and conditions as outlined in Saunders & Druehl (1993). Resulting sequences (spanning the 3′ 18S – ITS1 – 5.8S rDNA) were aligned with those previously published for a number of taxa representing the *Alariaceae, Laminariaceae* and *Lessoniaceae* by Saunders & Druehl (1993 see Table 1), using multiple manual alignment (SeqPup, Gilbert 1995). Regions of ambiguous alignment were removed and the data were subjected to three different phylogenetic analyses: Distance analysis: calculated distance between taxa were corrected with the Kimura two parameter model (Kimura 1980). One-thousand bootstrap replicates of the data matrix were generated and subjected to Neighbour-joining (Saitou & Nei 1987, algorithms found in PHYLIP vers. 3.5 c, Felsenstein 1993). Parsimony analysis: generated using PAUP (Swofford 1990), Maximum likelihood: generated with fastDNAml (Olsen & al. 1994).

Results

The resulting sequence alignment contained 468 unambiguously aligned nucleotide sites for 15 kelp taxa. rDNA sequences included 100 base pairs of the 3′ end of the 18S gene, complete ITS1 and 5.8S gene sequences from each species. As expected, the majority of sequence variation occurred in the ITS1 region.

All three methods of phylogenetic analysis produced trees with similar topologies (Fig. 7 a–c). Each tree displayed three groups of identical composition. Group 1 consisted of two lineages in the parsimony tree (Fig. 7: a); an *Alaria-Pterygophora-Lessoniopsis* polytomy and *Pleurophycus*. In the distance and likelihood trees (Fig. 7: b & c), Group 1 consisted of three lineages: *Alaria-Pterygophora* (both traditionally classified in the *Alariaceae*), *Lessoniopsis* (*Lessoniaceae*) and *Pleurophycus* (*Laminariaceae*). Group 1 was highly supported by bootstrap values in both the parsimony and distance derived consensus trees (94% & 97%, respectively), (Fig. 7: a & b).

Group 2, consisted of *Costaria costata* (*Laminariaceae*) and *Dictyoneurum californicum* (*Lessoniaceae*) (Fig. 7 a, b & c) and was highly supported with bootstrap values of 99% each (Fig. 7 a & b).

The remaining taxa formed a third variously supported and largely unresolved group, identical in composition to the third clade of Saunders & Druehl (1993 see Fig. 6), except for the addition of *Laminaria saccharina* and *Hedophyllum sessile* (two laminariacean taxa). In the parsimony tree, Group 3, supported by a bootstrap value of 74%, consists of *Lessonia* and a polytomy of *Egregia*, *Eisenia, Postelsia-Macrocystis* with *Nereocystis*, *Laminaria* and *Hedophyllum* (Fig. 7 a). Distance and maximum likelihood trees present 6 and 5 lineages of Group 3 taxa respectively (Fig. 7 b & c).

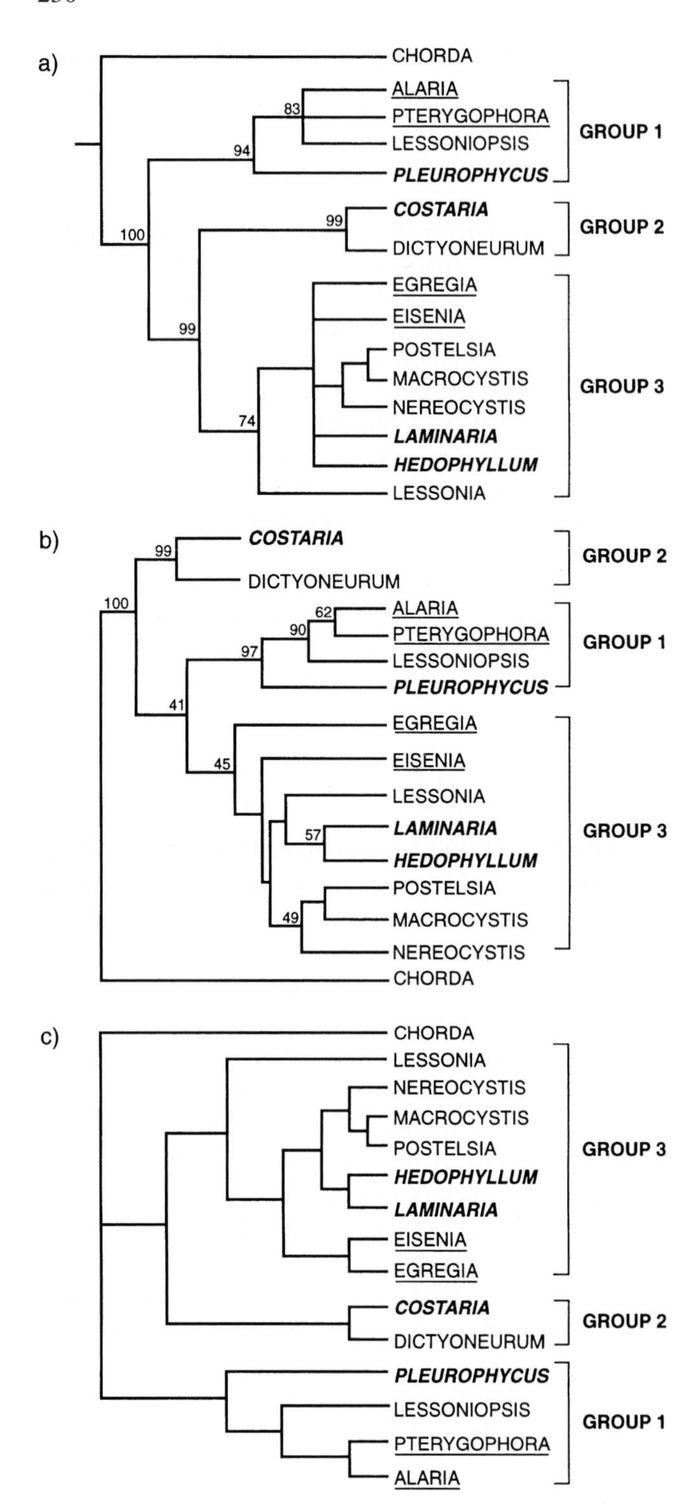
a)
CHORDA
ALARIA
PTERYGOPHORA
LESSONIOPSIS
PLEUROPHYCUS
GROUP 1
83
94
COSTARIA
DICTYONEURUM
GROUP 2
99
100
EGREGIA
EISENIA
POSTELSIA
MACROCYSTIS
NEREOCYSTIS
LAMINARIA
HEDOPHYLLUM
LESSONIA
GROUP 3
99
74
b)
COSTARIA
DICTYONEURUM
GROUP 2
99
100
ALARIA
PTERYGOPHORA
LESSONIOPSIS
PLEUROPHYCUS
GROUP 1
62
90
97
41
EGREGIA
EISENIA
LESSONIA
LAMINARIA
HEDOPHYLLUM
POSTELSIA
MACROCYSTIS
NEREOCYSTIS
GROUP 3
45
57
49
CHORDA
c)
CHORDA
LESSONIA
NEREOCYSTIS
MACROCYSTIS
POSTELSIA
HEDOPHYLLUM
LAMINARIA
EISENIA
EGREGIA
GROUP 3
COSTARIA
DICTYONEURUM
GROUP 2
PLEUROPHYCUS
LESSONIOPSIS
PTERYGOPHORA
ALARIA
GROUP 1

Discussion

Consistent with the findings of SAUNDERS & DRUEHL (1993) each of our current analyses resolved three groups of species for the ALL complex (Fig. 7). In addition to the earlier recognition of the *Alariaceae* (SAUNDERS & DRUEHL 1993) and *Lessoniaceae* (FAIN & al. 1988, SAUNDERS & DRUEHL 1993) as polyphyletic, we clearly establish that the *Laminariaceae* are similarly not a natural group (compare Table 1 to Fig. 7). Separation of the 'advanced' kelp genera into families based on gross morphology, viz. presence of sporophylls in *Alariaceae* ontogenetic splitting in the *Lessoniaceae*, and the absence of these features for the *Laminariaceae*, does not yield a phylogenetic system of classification. SETCHELL & GARDNER (1925: 631) noted inconsistencies in the familial arrangement of the kelp from its inception – most notably for *Lessoniopsis:* "...*Lessoniopseae* might perhaps be placed with equal propriety either under *Lessoniaceae* or under *Alariaceae*...". Despite this caveat, the taxonomic system established for these genera has endured and is now entrenched in kelp literature. In an earlier review, DRUEHL & SAUNDERS (1992, also see SAUNDERS & DRUEHL 1993) listed a number of other inconsistencies between the characteristics of extant kelp and the taxonomic system in which they were placed. For example, splitting is not exclusive to the *Lessoniaceae*, but is also seen in other kelp representatives; *Arthrothamnus bifidus* (GMEL.) RUPRECHT and *Streptophyllopsis kurioshioensis* (SEGAWA) KAJIMURA, classified in the *Laminariaceae*, are characterized by splitting at the transition zone to yield two blades. Where does the inconsistency between the molecular data and traditional interpretations leave the kelp systematist? Can anything be salvaged of the traditional system, or should the slate be cleaned allowing for a new start for kelp taxonomy? Only a careful consideration of the three groups will elucidate these questions:

Group 1. SAUNDERS & DRUEHL (1993, Fig. 6) recognized an assemblage which included *Alaria, Lessoniopsis* and *Pterygophora*. They reasoned that this was a natural assemblage that could be defined by morphological criteria, viz. lamina with a midrib and Group 1-type sporophylls (derived from the primary stipe). Although *Dictyoneuropsis*, Group 2, (represented by *Dictyoneurum* in Fig. 6 because of an identical sequence) has lamina with a midrib, the Group 1-type sporophylls are absent. According to the interpretation of these authors, Group 1 sporophylls are exclusive to, and shared by all species included in Group 1. A

Fig. 7. Phylogenetic trees inferred from rDNA sequence data (including 3′18S – ITS1 – 5.8S gene regions) representing relationships among 14 kelp genera. *a* Parsimony tree generated with PAUP (50 random replications). Strict consensus of six trees (length = 338, Cl = .707, Rl = .639). 1000 bootstrap replicates (jumble = 5) were completed with PHYLIP. No scale. *b* Distance tree derived using the Kimura 2 parameter correction with neighbor-joining. 1000 bootstrap replicates (PHYLIP). Scale = 1% sequence divergence. The branch supporting Group 2 has been slightly lengthened for the sake of clarity. *c* Maximum-likelihood tree constructed with fastDNAml (OLSEN & al. 1994); derived from ten random additions with this tree recovered three or four times. Scale = 0.02. *Chorda tomentosa* (*Chordaceae*) is the designated outgroup taxon. Underlined generic epithets were traditionally in the *Alariaceae*, bold-italicized generic epithets (excluding *Chorda*) were in Lessoniaceae

number of Group 3 species have sporophylls (*Eisenia*, *Egregia, Macrocystis*); sporophylls of *Eisenia* and *Macrocystis* were considered analogous to the Group 1 sporophylls being derived from the lamina and not from the stipe (SAUNDERS & DRUEHL 1993). *Egregia*'s sporophylls were presumed to be derived from the stipe (DRUEHL, pers. comm.). These same species lack midribs on their vegetative lamina. In the current report we add *Pleurophycus*, characterized by a broad midrib and no sporophylls to Group 1. Although this grouping is strongly supported by all of the molecular analyses (Fig. 7), the distinct morphological designation accredited to Group 1 is now compromised. Perhaps Group 1 sporophylls evolved with Group 1 subsequent to the divergence of *Pleurophycus*. The only other genus of kelp potentially characterized by Group 1-type sporophylls is *Undaria* (HARVEY) SURINGAR. Preliminary molecular results position *Undaria* as the earliest divergence in Group 1 (SAUNDERS, unpubl.), and it is more parsimonious to conclude that there has been a secondary loss of this feature in the immediate ancestor to *Pleurophycus*.

Group 2. This group has not changed its constitution since the work of SAUNDERS & DRUEHL (1993) and the reader is referred to that study for a brief discussion of this group. The three members of the *Laminariaceae* investigated herein failed to ally with Group 2 indicating that the caution expressed by SAUNDERS & DRUEHL (1993) in framing taxonomic conclusions for Group 2 and Group 3 was justified.

Group 3. This association contains a myriad of genera from all three traditional families of the ALL complex including representatives from the type genera of the *Laminariaceae* and *Lessoniaceae*. Few relationships are resolved by the molecular data among the genera included in this group to date (Fig. 7). The current study contributes little to advance understanding of this complex group over the discussions presented by SAUNDERS & DRUEHL (1992) and SAUNDERS & DRUEHL (1993). In fact, the opposite is true; this group has become even more complicated in its composition with the addition of *Laminaria* and *Hedophyllum* of the *Laminariaceae*. SAUNDERS & DRUEHL (1993) did speculate on the historical biogeography of this group, arguments in part based on an inferred alliance between *Eisenia* and the Asiatic kelp, *Ecklonia radiata* (C. AG.) J. AGARDH (the inferrence of lamina derived sporophylls for *Eisenia* was also a result of this putative alliance to *Ecklonia*). We now have ITS data (SAUNDERS, unpubl.) for *E. radiata* which supports this hypothesis.

Conclusions

Problems with the segregation of the genera of the ALL complex into the three families *Alariaceae, Laminariaceae* and *Lessoniaceae* have been noted since the inception of this taxonomic system (SETCHELL & GARDNER 1925). Accumulating molecular data ranging from early chloroplast investigations (FAIN 1986, FAIN & al. 1988), 18S RFLP and sequence studies (SAUNDERS & DRUEHL 1991, 1992), to the more detailed ITS investigations (SAUNDERS & DRUEHL 1993) embellished herein only serve to aggravate this taxonomic conundrum. Group 1 and Group 2 are solidly supported by the molecular data, and although they lack obvious morphological synapomorphies, features should be forthcoming with renewed nonmolecular investigation in light of our findings. The absence of clearly defined morphological features with which to assign taxa to superordinant groups does not rest well with human need to fit everything unambiguously into its place. Even for Group 1, the best circumscribed on the basis of a group-specific sporophyll type in the study of Saunders & Druehl (1993), requires an ad hoc loss of this conservative feature for its newest member, *Pleurophycus*.

Add to the problems highlighted in the previous paragraph the recognition that the kelp, on the basis of nonmolecular (Estes & Steinberg 1988, Lüning & tom Dieck 1990) and molecular (Saunders & Druehl 1991, 1992, 1993; Saunders & Druehl 1992) data, are a very recently evolved group whose multitude of monotypic genera is possibly a manifestation of over-classification (Saunders & al. 1992) and the situation for resolving kelp taxonomy appears bleak. To the contrary, we look to the resolution of this conundrum with excitement. There is a great deal of work still to be completed and there are many questions still to be answered – a feast for the scientific mind! Firstly, more taxa must be included in the ITS phylogeny. A recent analysis of rDNA sequences (including the ITS1 and rDNA regions) from ten genera of the *Laminariaceae* and 14 species of *Laminaria* indicates considerable overlap in levels of sequence variation among these putative species and genera (Mayes & Druehl unpubl.).

Future investigations should include exploration of other nuclear coding regions so that phylogenies can be constructed based on a comparison of a greater number of nucleotide sites. Ultimately, comparisons of all available data (e.g. molecular, morphological, anatomical, physiological, life history, biogeographic) should be conducted in order to arrive at the most robust phylogenetic hypothesis. According to Theriot (1989: 409), "It is unlikely that several independent lines of evidence should support the same hypothesis through chance alone".

This research and summation were supported by NSERC Canada grants to L. D. D. and G. W. S.

References

Abbott, I. A., Hollenberg, G. J., 1976: Marine algae of California. – Stanford, Ca. Stanford University Press.

Bhattacharya, D., Druehl, L. D., 1988: Phylogenetic comparison 8 the small-subunit ribosomal DNA sequence of *Costaria costata* (*Phaeophyta*) with those of other algae, vascular plants and oomycetes. – J. Phycol. **24**: 539–543.

– Stickel, S. K., Sogin, M. L., 1991: Molecular phylogenetic analysis of genic regions from *Achlya bisexualis* (*Oomycota*) and *Costaria costata* (*Chromophyta*). – J. Molec. Evol. **33**: 525–536.

Clayton, M. N., 1984: Evolution of the *Phaeophyta* with particular reference to the *Fucales*. – Prog. Phycol. Res. **3**: 11–46.

Druehl, L. D., Saunders, G. W., 1992: Molecular explorations in kelp evolution. – Prog. Phycol. Res. **8**: 47–83.

Emerson, C. J., Buggeln, R. G., Bell, A. K., 1982: Translocation in *Saccorhiza dermatodea* (*Laminariales, Phaeophyceae*): anatomy and Physiology. – Canad. J. Bot. **60**: 2164–2184.

Estes, J. A., Steinberg, P. D., 1988: Predation, herbivory and kelp evolution. – Paleobiology **14**:19–36.

Fain, S. R., 1986: Molecular evolution in the *Laminariales*: restriction analysis of chloroplast DNA. – Ph. D. thesis. Simon Fraser University, Canada.
– Druehl, L. D., Baillie, D. L., 1988: Repeat and single copy sequences are differently conserved in the evolution of kelp chloroplast DNA. – J. Phycol. **24**: 292–302.
Felsenstein, J., 1993: PHYLIP (phylogeny inference package) version 3.5 c. – Distributed by the author. Department of Genetics, University of Washington, Seattle, USA.
Fritsch, F. E., 1945: The structure and reproduction of the algae. **2**. – Cambridge: Cambridge University Press.
Gabrielson, P. W., Scagel, F. F., Widdowson, T. B., 1989: Keys to the benthic marine algae and seagrasses of British Columbia, Southeast Alaska, Washington and Oregon. – Phycol. Contrib. **4**: 45–57.
Gilbert, D. G., 1995: SeqPup, a biological sequence editor and analysis program for Macintosh computers. – Published electronically on the Internet, available via anonymous ftp to ftp. bio. indiana. edu.
Henry, E. C., Cole, K., 1982: Ultrastructure of swarmers in the *Laminariales* (*Phaeophyceae*). II. Sperm. – J. Phycol. **18S**: 570–579.
– South, G. R., 1987: *Phyllariopsis* gen. nov. and a reappraisal of the *Phyllariaceae* Tilden 1935 (*Laminariales, Phaeophyceae*). – Phycologia **26**: 9–16.
Van Den Hoek, C., Mann, D. G., Jahns, H. M., 1995: Algae, an introduction to phycology. – Cambridge: Cambridge University Press.
Kawai, H., 1992: A summary of the morphology of chloroplasts and flagellated cells in the *Phaeophyceae*. – Korean J. Phycol. **7**: 33–43.
– Kurogi, M., 1985: On the life history of *Pseudochorda nagaii* (*Pseudochordaceae* fam. nov.) and its transfer from the *Chordariales* to the *Laminariales* (*Phaeophyta*). – Phycologia **24**: 289–296.
Kimura, M., 1980: A simple method for estimating evolutionary rates of base substitutions through comparative studies of nucleotide sequences. – J. Molec. Evol. **16**: 111–120.
Kylin, H., 1933: Über die Entwicklungsgeschichte der Phaeophyceen. – Lund Univ. Årsskr. N. F. **29** 1.
Lüning, K., Tom Dieck, I., 1990: The distribution and evolution of the *Laminariales*: North Pacific – Atlantic relationships. – In Garbary, D. J., South, G. R., (Eds): Evolutionary biogeography of the marine algae of the north Atlantic, pp. 187–204. – Berlin: Springer.
Lobban, C. S., 1978: Translocation of ^{14}C in *Macrocystis pyrifera* (giant kelp). – Pl. Physiol. **61**: 585–589.
Maier, I., Müller, D. G., Gassmann, G., Boland, B., Marner, F.-J., Janice, L., 1984: Pheromone-triggered gamete release in *Chorda tomentosa*. – Naturwissenschaften **71**: 48–49.
Mayes, C., Saunders, G. W. S., Tan, I., Druehl, L. D., 1992: DNA extraction methods for kelp (*Laminariales*) tissue. – J. Phycol. **17**: 712–716.
Müller, D. G., Clayton, M. N., Germann, G., 1985: Sexual reproduction and life history of *Perithalia caudata* (*Sporochnales, Phaeophyta*). – Phycologia **24**: 467–473.
Olsen, G. J., Matsuda, H., Hagstrom, R., Overbeek, R., 1994: FastDNAml: a tool for construction of phylogenetic trees of DNA sequences using maximum likelihood. – CABIOS **10**: 41–48.
Papenfuss, G. F., 1955: Classification of the algae. – In: A century of progress in the natural sciences 1853–1953, pp. 115–224. – San Francisco: Academy of Science.
Saitou, N., Nei, M., 1987: The neighbor-joining method: a new method for reconstructing phylogenetic trees. – Molec. Biol. Evol. **4**: 406–425.
Saunders, G. W., Druehl, L. D., 1991: Restriction enzyme mapping of the nuclear ribosomal cistron in selected *Laminariales* (*Phaeophyta*): a phylogenetic assessment. – Canad. J. Bot. **69**: 2647–2654.

– – 1992: Nucleotide sequences of the small-subunit ribosomal RNA genes from selected *Laminariales* (*Phaeophyta*): implications for kelp evolution. – J. Phycol. **28:** 544–549.

– – 1993: Revision of the kelp family *Alariaceae* and the taxonomic affinities of *Lessoniopsis* Reinke (*Laminariales, Phaeophyta*). – Hydrobiologia **260/261**: 689–697.

– Kraft, G., Tan, I. H., Druehl, L. D., 1992: When is a family not a family? – BioSystems **28**: 109–116.

Schmitz, K., Srivastava, L. M., 1979: Long distance transport in *Macrocystis integrifolia.* II. Tracer experiments with ^{14}C and ^{32}P. – Pl. Physiol. **63**: 1003–1009.

Setchell, W. A., 1905: Post-embryonal stages of the *Laminariaceae.* – Univ. Calif. Publ. Bot. **2**: 115–138.

– Gardner, N. L., 1925: The marine algae of the Pacific coast of North America. III. *Melanophyceae.* – Univ. Calif. Publ. Bot. **8**: 383–739.

Swofford, D. L., 1990: PAUP: Phylogenetic analysis using parsimony, vers. 3.0. – Illinois Natural History Survey, 607 E Peabody Dr., Champaign, Illinois, USA 61820.

Tan, I. H. A., 1995: Ribosomal DNA phylogeny of the plant division *Phaeophyta.* – Simon Fraser University, British Columbia, Canada, Ph. D. thesis.

– Druehl, L. D., 1994: A molecular analysis of *Analipus* and *Ralfsia* (*Phaeophyceae*) suggests the order *Ectocarpales* is polyphyletic. – J. Phycol. **30**: 721–729.

– – 1996: A ribosomal DNA phylogeny supports the close evolutionary relationships among the *Sporochnales, Desmarestiales* and *Laminariales* (*Phaeophyceae*). – J. Phycol. **32**: 112–118.

Theriot, E., 1989: Phylogenetic systematics for phycology. – J. Phycol. **25**: 407–411.

Wynne, M. J., Loiseaux, S., 1976: Phycological reviews 5. Recent advances in life history studies of the *Phaeophyta.* – Phycologia **15**: 435–452.

Small-subunit ribosomal RNA sequences from selected dinoflagellates: testing classical evolutionary hypotheses with molecular systematic methods

G. W. Saunders, D. R. A. Hill, J. P. Sexton, and R. A. Andersen

Key words: *Dinoflagellata, Dinophyceae, Gonyaulacales, Gymnodiniales, Noctilucales, Peridiniales, Prorocentrales,* – Phylogeny, small-subunit rRNA, systematics, taxonomy.

Abstract: The dinoflagellates are a large, richly diverse group of protists with marine and freshwater representatives, photosynthetic and heterotrophic nutritional modes, toxic and non-toxic isolates and some species that form resting cysts that can be found in the fossil record dating back to the Triassic or perhaps earlier. Traditional classification and phylogeny of the dinoflagellates has been based largely on the structure of their cell wall – or amphiesma. More recently, however, a number of molecular phylogenies have emerged that challenge the more traditional perspectives. A review of these molecular results is presented with comparative reference to the long-standing traditional views.

In dealing with dinoflagellate phylogeny the systemataist is confronted with a fascinating study in conflicts; one in which the occurrence of anomalies is commonplace and their explanation is challenging. Dinoflagellates for the most part are unicellular and have two distinct flagella. These protists are abundant in marine and freshwater habitats and are diverse in structure. Their importance is manifest in two major ways. Firstly, they produce potent toxins that may result in human sickness or death due to shellfish poisoning. Secondly, several endosymbiotic species (usually of the genus *Symbiodinium*) are responsible for the photosynthetic activity necessary for growth and survival of the reef-building corals. Whereas about half the dinoflagellates are photosynthetic, there are many voracious predators (phagotrophs) and parasites. Normal dinoflagellate pigmentation includes chlorophylls a and c_2, with peridinin as the main carotenoid. There are, however, many exceptions which, in combination with the predatory nature of dinoflagellates, indicates that there have been numerous endosymbioses occurring between dinoflagellates and other eukaryotes (Lucas & Vesk 1990, Schnepf 1992). Dinoflagellates also have a complex cell wall structure – termed the amphiesma – which incorporates cortical alveoli and in its different forms is largely responsible for the myriad variety of cell morphologies.

Dinoflagellates have an external mitotic spindle and all but the nonphotosynthetic, marine, parasitic syndineans (Ris & Kubai 1974, Corliss 1984, Fensome & al. 1993) possess an unusual nucleus (dinokaryon) characterized by permanently

"Predinoflagellate"	***Oxyrrhis***
DIVISION Dinoflagellata	
Subdivision Syndinea	
Class Syndiniophyceae	
Order Syndiniales -	***Syndinium***
Subdivision Dinokaryota	
Class Blastodiniphyceae	
Order Blastodiniales -	***Blastodinium***
Class Noctiluciphyceae	
Order Noctilucales	
Family Noctilucaceae -	***Noctiluca***
Class Dinophyceae	
Subclass Gymnodiniphycidae	
Order Gymnodiniales	
Family Gymnodiniaceae -	***Gymnodinium***
	Amphidinium
	Gyrodinium
	Lepidodinium
Family Polykrikaceae -	***Polykrikos***
Order Suessiales	
Family Symbiodiniaceae -	***Symbiodinium***
Subclass Peridiniphycidae	
Order Gonyaulacales	
Family Gonyaulacaceae -	***Gonyaulax***
Family Ceratiaceae -	***Ceratium***
Family Ceratocoryaceae -	***Ceratocorys***
Family Crythecodiniaceae -	***Crypthecodinium***
Family Goniodomaceae -	***Alexandrium***
Family Pyrocystaceae -	***Pyrocystis***
Order Peridiniales	
Family Peridiniaceae -	***Peridinium***
	Pentapharsodinium
Family Heterocapsaceae -	***Heterocapsa***
Subclass Dinophysiphycidae	
Order Dinophysiales	
Family Dinophysiaceae -	***Dinophysis***
Subclass Prorocentrophycidae	
Order Prorocentrales	
Family Prorocentraceae -	***Prorocentrum***
Subclass Uncertain	
Order Phytodiniales	
Family Phytodiniaceae -	***Gloeodinium***
	(as *Hemidinium*)
Order Thoracosphaerales	
Family Thoracosphaeraceae -	***Thoracosphaera***

Fig. 1. The practical taxonomic scheme used in this report (from FENSOME & al. 1993)

condensed chromosomes and a lack of histone proteins in at least some stage of their life history. These hypothetically primitive features led to the theory that the dinoflagellates were the first of the extant protist groups to evolve from the eukaryotic line following its divergence from the prokaryotes and to the coining of the term "Mesokaryote" (DODGE 1965, 1966; LOEBLICH 1976). In order to maintain consistency throughout the text, all of the phylogenies that we will consider will be translated to reflect the taxonomic system of FENSOME & al. (1993). A schematic of this system including those taxa discussed in this chapter is presented (Fig. 1).

Molecular systematics and the affinities of the dinoflagellates

Despite the existence of the many interesting characteristics discussed in the previous section, the *Dinophyceae* are relatively unstudied by the modern techniques of molecular systematics that have been applied extensively, and with exciting results, to other major algal lineages (cf. Chapters 3, 5 and 6 in this volume). It is difficult to understand why these powerful techniques have not been brought to bear on dinophycean taxonomy with the same force as for other protistan lineages and one of our objectives is to remedy this inequity.

Early molecular studies of dinoflagellates were largely concerned with the origin (s) of this group relative to other protistan lines, rather than the relationships among the various dinoflagellate lineages. This is to be expected because a key proposal concerning the phylogenetic affiliation of the dinoflagellates is the "Mesokaryote" hypothesis (Dodge 1965, 1966). Dodge considered carefully the unusual chromosomes and mitosis of the dinoflagellates and proposed that this group of protists are intermediate between the prokaryotes and eukaryotes, hence the designation Mesokaryote. Other opinions on the affinities of dinoflagellates included proposed alliances to the chromophyte algae (Chadefaud 1950, Cavalier-Smith 1975, Taylor 1976, Ragan & Chapman 1978, Roberts & Roberts 1991) and to the ciliates (Corliss 1975; Taylor 1976, 1978). Although these last two hypotheses have gained contemporary favor (ciliates & apicomplexans close relatives; chapter 8 in this volume, cf. Cavalier-Smith 1993; with the heterokont chromophytes more distant (Van de Peer & al. 1993, Cavalier-Smith & al. 1995, Van der Auwera & al. 1995), they lack the allurement of the Mesokaryote hypothesis. It is not surprising, therefore, that most of the early molecular investigations considering dinoflagellates had the expressed objective of testing this hypothesis.

Hinnebusch & al. (1981) presented the first substantial molecular test of the Mesokaryote hypothesis with their 5S investigation of *Crypthecodinium*. They concluded that the dinoflagellates are only distantly related to the prokaryotes, having strongest affiliation with "higher" eukaryote lineages, and not positioned at the base of the eukaryote line as predicted by the Mesokaryote hypothesis. Their data supported a view that the dinokaryotic nucleus was derived by reduction from the typical eukaryote state and was not an ancestral condition intermediate between that of prokaryotes and eukaryotes. Investigations of the primary and secondary structure of dinoflagellate U5 small nuclear RNA were also, albeit indirectly, interpreted as evidence against the Mesokaryote hypothesis (Liu & al. 1984). Two years later, Herzog & Maroteaux (1986: 8644) published a more detailed discussion of the available 5S data and concluded that the dinoflagellate *Crypthecodinium* is most closely related to the ciliate lineage. Ironically Herzog & Maroteaux (1986) also provided the first molecular data interpreted in support of the Mesokaryote theory.

Herzog & Maroteaux (1986) determined the nuclear small-subunit (SSU) ribosomal RNA (rRNA) sequence for the dinoflagellate *Prorocentrum micans*. In that study comparisons of SSU rRNA secondary structure, as well as relationships inferred from a distance-derived phylogenetic tree, were interpreted in support of an early divergence for the dinoflagellates relative to the other eukaryote lineages supporting the Mesokaryote hypothesis. In the years that followed, however, phylogenetic trees inferred from partial SSU (CEDERGREN & al. 1988, JOHNSON & al. 1988, JOHNSON & BAVERSTOCK 1989), partial LSU (large subunit rRNA) (BAROIN & al. 1988; QU & al. 1988; LENAERS & al. 1989, 1991) and reanalyses of complete SSU data (GUNDERSON & al. 1987, SOGIN 1989) all pointed to a monophyletic group including the apicomplexans, ciliates and dinoflagellates, this combined assemblage evolving from the crown rather than base of the eukaryote line. Subsequent to the 1986 study of HERZOG & MAROTEAUX, complete dinoflagellate SSU sequences were published for *Crypthecodinium cohnii* and *Symbiodinium pilosum* (GAJADHAR & al. 1991, SADLER & al. 1992). Robust methods of phylogenetic analysis, in combination with additional apicomplexan and ciliate SSU rRNA sequences to complement the new dinoflagellate data, indicated that the SSU phylogenies strongly support an alliance of the dinoflagellates with the ciliates and apicomplexans. This combined assemblage, termed the Alveolates (cf. CAVALIER-SMITH 1993), branches from within the eukaryote crown and is not remotely related to early eukaryote lineages. This data strongly rejects the Mesokaryote hypothesis of DODGE (1966) which, nonetheless, was a plausible proposal in its time and was important in thrusting an algal lineage into the center stage of macroevolutionary debate.

Relationships among the dinoflagellate lineages

The next SSU sequence published for a dinoflagellate was that of *Alexandrium tamarense* (DESTOMBE & al. 1992). This sequence was determined as a "first step" towards establishing a molecular system for understanding relationships within this taxonomically complex genus (DESTOMBE & al. 1992). DESTOMBE & al. were followed by a series of investigations dedicated to meeting this aim (SCHOLIN & al. 1993, 1994, 1995; SCHOLIN & ANDERSON 1994). These studies are paradigm examples of the versatility of ribosomal RNA sequences in providing phylogenetic information down to the level of species and strains. The presence of an SSU rRNA pseudogene was reported in *Alexandrium fundyense* (SCHOLIN & al. 1993) and subsequently used as a genetic marker via a restriction fragment length polymorphism (RFLP) assay (SCHOLIN & ANDERSON 1994). The pseudogene was found in three species of *Alexandrium* and the RFLP assay revealed groups that interestingly clustered by geographic origin rather than by morphospecies. A similar result was obtained using variable domain LSU rDNA sequences (SCHOLIN & al. 1994) and the work was then extended to an elegant study of the molecular evolution, population structure and dispersal of three toxic *Alexandrium* species (known as the "*tamarensis* species complex") in the North American and West Pacific regions (SCHOLIN & al. 1995). The reader is also directed to a recent publication by ADACHI & al. (1996).

The first molecular investigation with the explicit objective of resolving relationships among the various lineages of dinoflagellates was that of LENAERS & al. (1991). The LSU divergent domains D1 and D8 were sequenced for 13 dinoflagellate species. This study suffered from two shortcomings: (1) the unavoidable curse of being first – poor taxon sampling; and, (2) the short region of sequence (only 401 bp) included in the phylogenetic analyses. Nevertheless, this study is important because it introduces some unorthodox concepts concerning the relationships among the dinoflagellate lineages, relationships that are largely congruent with the more robust phylogenies that followed (discussed below). Important findings include the early divergence of *Oxyrrhis*, *Crypthecodinium* and *Noctiluca* relative to the other dinoflagellate lineages, support for the "*Gonyaulacales*" (including *Alexandrium* and *Pyrocystis*, but not for *Gonyaulax polyedra*) as distinct from the *Peridiniales* (TAYLOR 1980), and a recently evolved complex including the *Gymnodiniales, Peridiniales* and *Prorocentrales* (hereafter the GPP complex). The molecular results of LENAERS & al. (1991) failed to support any of the hypotheses on dinoflagellate phylogeny based on interpretations of morphological and ultrastructural observations (Fig. 2). These included: (1) a plate increase model which would place the *Prorocentrales* as ancestral to the other lineages (LOEBLICH 1976, TAYLOR 1980); (2) a plate reduction model which considered the *Gymnodiniales* as the ancestral stock with the *Prorocentrales* as being derived (DÖRHÖFER & DAVIES 1980, EATON 1980, cf. LOEBLICH 1984); or, (3) the plate fragmentation model which again considered the *Prorocentrales* ancestral (BUJAK & WILLAMS 1981).

More recently, ZARDOYA & al. (1995) investigated the LSU variable domains D1, D2, D9 and D10 for ten species of dinoflagellates. ZARDOYA & al. (1995) concluded that the dinoflagellates could be divided into three groups viz. the *Prorocentrales, Gymnodiniales* and the *Gonyaulacales* (as *Peridiniales*). This conclusion, however, is based on multiple species of only three genera, one genus from each of the lineages they recognized, and their results are thus not surprising. Their best supported tree (ZARDOYA & al. 1995: fig. 3B) places the *Gonyaulacales* as sister to the *Gymnodiniales/Prorocentrales* group consistent with other molecular results.

At the same time that the then LSU investigations were considering the "big picture" of dinoflagellate phylogeny, SSU investigations were being developed to address specific aspects of dinoflagellate evolution. In addition to the discussion of *Alexandrium* above, emphasis was placed on the zooxanthellae – "unicellular algae that occur as endosymbionts in many....marine invertebrate species" (ROWAN & POWERS 1991) – most of which are dinoflagellates referred to the single genus *Symbiodinium.* Investigations using RFLP analysis of partial SSU sequences, and subsequently sequence data for partial SSU rRNAs (ROWAN & POWERS 1991, 1992; cf. ROWAN 1991) generated evidence for relatively substantial divergence at the molecular level between isolates of *Symbiodinium* reinforcing the assertion that this genus consisted of a number of distinct species. Although not an immediate objective of ROWAN & POWERS (1992), a preliminary phylogeny based on partial SSU sequences was generated and is insightful for discussing relationships among the major dinoflagellate lineages. Their SSU phylogeny is similar to those discussed above for the LSU data. A distinct *Gonyaulacales* (*Ceratium*) lineage was resolved, as well as a diverse *Symbiodinium* complex that was sister to the GPP complex sensu stricto (excluding *Suessiales*, Fig. 1). For the first time *Peridinium foliaceum* was included in a molecular study and it branches early relative to most dinoflagellates, not from a more recent lineage including the other "peridinioid" genera (e.g., *Heterocapsa*).

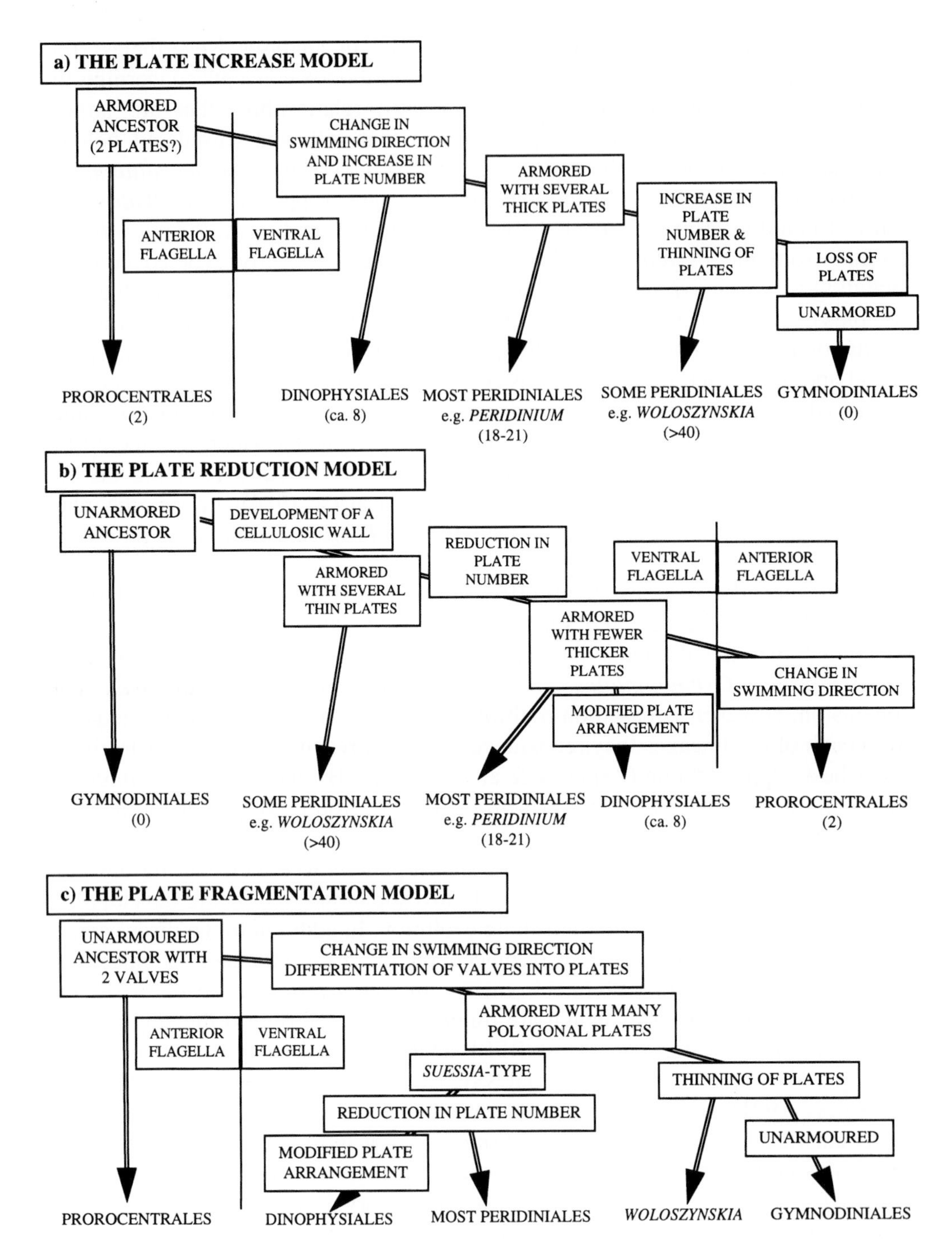

Fig. 2. Schematic representation of the *a* PIM *b* PRM and, *c* PFM hypotheses of dinoflagellate evolution (modified from BUJAK & WILLIAMS 1981)

McNally & al. (1994) presented four complete SSU sequences, one species each for the genera *Amphidinium* and *Gloeodinium* and two speices of *Symbiodinium*, adding to the short list of complete SSU sequences available at that time (ten species, with the *Amphidinium carterae* sequence unavailable for subsequent investigations, representing only six genera). Phylogenetic analysis of this data again provided support for the Alveolata as a monophyletic assemblage. Within the *Dinophyceae*, early lineages comprised the genera *Alexandrium*, *Amphidinium*, and *Crypthecodinium* which were variously (and always with low bootstrap support) related depending on the method of phylogenetic inference and the assigned outgroup. Next, a monophyletic and strongly supported *Symbiodinium* complex of species was resolved as sister group to the genera *Prorocentrum* and *Gloeodinium* (*Phytodiniales*), but relationships among these three taxa were only weakly supported and varied with the method of phylogenetic inference. To explore dinoflagellate relationships further, McNally & al. (1994) combined the data that they had generated with the partial sequences reported previously by Rowan & Powers (1992). The alignment contained only 480 sites and there was virtually no support for the topology generated from this data. Exceptions included resolution for relationships among most species in the strongly supported *Symbiodinium* complex, and recognition of monophyletic genera for the included species of both *Amphidinium* and *Heterocapsa*. There was moderate support for an alliance between *Alexandrium* and *Crypthecodinium*. In summary, *Amphidinium* diverged early and associated with the *Gonyaulacales*, this combined assemblage sister to the GPP complex. This topology is largely consistent with all of the molecular phylogenies discussed previously in this review.

Despite the general agreement among the various published SSU and LSU phylogenies, the support that the various data sets present for the inferred phylogenies is weak to nonexistent. This is due in a large part to the partial sequence data on which many of the previous studies are based (as few a 199 aligned characters, Lenaers & al. 1991) and poor taxon sampling (maximum of 20 species representing 11 genera in the partial SSU tree of McNally & al. 1994). McNally & al. (1994) recognized that one remedy to this problem was the acquisition of complete SSU sequences for a diversity of dinoflagellates. We could only recover 14 complete SSU sequences for dinoflagellates from GenBank, 11 discussed previously in this section and three unpublished *Alexandrium* sequences (Table 1). It was the paucity of complete dinoflagellate SSU sequences that prompted us to undertake an extended molecular survey of the dinoflagellates. It was our intention to generate sufficient dinoflagellate sequence data to entice a few of our peers into continuing this work, but admittedly, curiosity about the phylogenetic relationships among the major lineages of dinoflagellates, as well as the distribution of biological features, e.g., dinokaryotic nucleus, bioluminescence, were strong motivations. We have determined complete SSU sequence data for 17 species representing 13 genera of dinoflagellates, as well as partial SSU data for one additional species of yet another genus (Table 1). This brings the total in our alignment to 31 complete dinoflagellate SSU sequences representing 18 genera. We also analyze an alignment of partial dinoflagellate sequences with 41 representatives of 20 genera.

Table 1. Sources of the SSU sequences for species included in our multiple alignment; [1]*R* see reference for details, [2]Melbourne University Culture Collection, [3]Provasoli-Guillard National Center for the Culture of Marine Phytoplankton, [4]partial SSU sequence only

Species	Algal source	GenBank	Reference
Sarcocystis muris (Blanchard) Alexieff	R[1]	M34846	Gajadhar & al. (1991)
Perkinsus marinus (Mackin, Owen & Collier) Leuine	R	X75762	Fong & al. (1993)
P. sp.	R	L07375	Goggin & Barker (1993)
Alexandrium fundyense Balech	R	U09048	Scholin & al. (1993)
A. margalefi Balech	R	U27498	Bergquist & Reeves, unpubl.
A. minutum Halim	R	U27499	Bergquist & Reeves, unpubl.
A. ostenfeldii (Paulsen) Balech & Tangen	R	U27500	Bergquist &Reeves, unpubl.
A. tamarense (Lebour) Balech	R	X54946	Destombe & al. (1992)
A. tamarense?	MUCC99[2]	AF022191	Current study
Amphidinium belauense Trench	R	L13719	McNally & al. (1994)
Ceratium fusus (Ehrenberg) Dujardin	R	M88510[4]	Rowan & Powers (1992)
C. fusus	CCMP154[3]	AF022153	Current study
C. tenue Ostenfeld & Schmidt	MUCC248	AF022192	Current study
Ceratocorys horrida Stein	CCMP157	AF022154	Current study
Crypthecodinium cohnii Seligo	R	M34847	Gajadhar & al. (1991)
Dinophysis acuminata Claparéde & Lachmann	field isolate	–[4]	Current study
Gloeodinium viscum Banaszak, Iglesias-Prieto & Trench	R	L13716	McNally & al. (1994)
Gonyaulax spinifera Dangeard	CCMP409	AF022155	Current study
Gymnodinium catenatum Graham	MUCC273	AF022193	Current study
G. fuscum Ehrenberg	MUCC282D	AF022194	Current study
G. galatheanum Braarud	R	M88511[4]	Rowan & Powers (1992)
G. mikimotoi Miyake & Kominami	MUCC098	AF022195	Current study
G. simplex Lohmann	R	M88512[4]	Rowan & Powers (1992)
G. varians Maskall	R	M88513[4]	Rowan & Powers(1992)
G. sp.	MUCC284	AF022196	Current study
Gyrodinium impudicum Fraga & Bravo	MUCC276D	AF022197	Current study
Heterocapsa ildefina (Loeblich) Morrill & Loeblich III	R	M88516[4]	Rowan & Rowers (1992)
H. niei (Loeblich) Morrill & Loeblich	R	M88515[4]	Rowan & Powers (1992)
H. triquetra (Ehrenberg) Balech	MUCC285	AF022198	Current study
Lepidodinium viride Watanabe, Suda, Inouye, Sawaguchi & Chihara	MUCC247D	AF022199	Current study
Noctiluca scintillans (Macartier) Kofoid & Swezy	field isolate	AF022200	Current study
Pentapharsodinium tyrrhenicum (Balech) Montresor, Zingone & Marino	MUCC097	AF022201	Current study
Peridinium foliaceum (Stein) Biechler	R	M88517[4]	Rowan & Powers (1992)
P. sp.	field isolate	AF022202	Current study
Polykrikos schwartzii Büschli	field isolate	–	Current study
Prorocentrum micans Ehrenberg	R	M14649	Herzog & Maroteaux (1986)
P. minimum (Pavillard) Schiller	R	M88520[4]	Rowan & Powers (1992)
Pyrocystis noctiluca Murray & Schütt	CCMP732	AF022156	Current study
Symbiodinium corculorum Trench	R	L13717	McNally & al. (1994)
S. meandrinae Trench	R	L13718	McNally & al. (1994)
S. microadriaticum Freudenthal	R	M88521	Rowan & Powers (1992)
S. pilosum Trench & Blank	R	X62650	Sadler & al. (1992)
S. pulchrorum Trench	R	M88509	Rowan & Powers (1992); cf. McNally & al. (1994)
Thoracosphaera heimii (Lohmann) Kamptier	R	M88514[4]	Rowan & Powers (1992)

Material and methods

The sources of the dinoflagellates investigated in this study are provided in Table 1. Total genomic DNA was isolated by the phenol: chloroform procedure of SAUNDERS (1993) and the SSU rDNA polymerase-chain-reaction amplified as two to four overlapping fragments (SAUNDERS & KRAFT 1994, 1996; SAUNDERS & al. 1995, 1997). Primer information has been published previously (SAUNDERS & KRAFT 1994, SAUNDERS & al. 1997). Amplification reactions (Gene-Amp kit following manufacturer's recommendations, Perkin Elmer Cetus, Norwalk, CT) were performed in an automated thermal cycler as outlined in SAUNDERS & al. (1997).

PCR products were agarose-gel purified through low melting temperature agarose with DNA retrieved using the Wizard™ PCR Preps DNA Purification System (Promega, Madison, Wisconsin). Purified DNA was sequenced with the Taq DyeDeoxy™ Terminator Cycle Sequencing Kit (Applied Biosystems, ABI, division of Perkin Elmer Cetus). Sequence reactions were completed with both of the PCR primers for each fragment as well as appropriate internal primers (SAUNDERS & KRAFT 1994, 1996; SAUNDERS & al. 1997). Automated DNA sequencing used the Applied Biosystems model 373A DNA sequencer. Alternatively, the SSU data for *Ceratium fusus, Ceratocorys horrida, Gonyaulax spinifera* and *Pyrocystis noctiluca* were determined as described previously (POTTER & al. 1997).

The partial and complete SSU sequences included in our multiple alignment (Table 1) were manually added using the SeqPup computer program (GILBERT 1995). The multiple alignment of complete SSU sequences contained 34 species and 1787 bp excluding the 5′ and 3′ primer regions (SAUNDERS & KRAFT 1994) as well as ambiguously aligned regions. This data set was converted to a distance matrix using the Kimura two-parameter correction (KIMURA 1980) of the DNADIST computer program of PHYLIP (FELSENSTEIN 1995). A phylogenetic tree was constructed from the distance matrix using the neighbor-joining algorithm (SAITOU & NEI 1987) of PHYLIP. The multiple alignment was also subjected to parsimony analysis with the PAUP computer package (version 3.1.1 for the Macintosh, Swofford (1993)). All variable nucleotide characters were equally weighted and alignment gaps were treated as a fifth base. Most parsimonious cladograms were sought by random (50 replicates) sequential addition of taxa with the heuristic search option. Unrooted trees were calculated with the dinoflagellates subsequently rooted with reference to the outgroup taxa (*Sarcocystis, Perkinsus* spp.). The SEQBOOT option of PHYLIP, in combination with DNADIST/NEIGHBOR and DNAPARS was empolyed to complete 1000 replicates of bootstrap resampling (FELSENSTEIN 1985) each for the distance and parsimony analyses respectively. Maximum likelihood analysis was applied to the multiple alignment using the fastDNAml algorithm for the Power Macintosh (OLSEN & al. 1994).

A second alignment was constructed including all of the available partial SSU sequences and the homologous regions from the complete sequences. This alignment included 44 species of dinoflagellates and 497 nucleotide positions. This alignment was analyzed only by the distance procedures outlined above because resolution was too low among the included taxa to justify further analyses.

Results

The methods used in the current study successfully generated complete SSU sequence data for the 17 dinoflagellate isolates indicated in Table 1. In one case, *Polykrikos*, the sequence was difficult to determine resulting in a number of ambiguously assigned sites. Although this sequence is included in the following phylogenetic analyses (removing *Polykrikos* did not alter relationships) it is in need of confirmation. We attempted to obtain sequence data from other dinoflagellate isolates during the course of this investigation, but the data were considered excessively "noisy" and unusable. All of the samples which failed to yield clean sequence data were cells of heterotrophic species isolated from the wild for DNA extraction.

All methods of phylogenetic analyses applied to the complete-SSU sequence alignment resolved *Noctiluca* as the earliest lineage (Figs. 3–5) with weak (54% bootstrap replicates) to moderate (83 replicates) support in the distance (Fig. 3) and parsimony (Fig. 4) phylogenies respectively. In the distance tree there is weak support for *Amphidinium* as sister to a gonyaulacalean lineage including *Alexandrium, Ceratium, Crypthecodinium, Gonyaulax,* and *Pyrocystis. Ceratocorys*

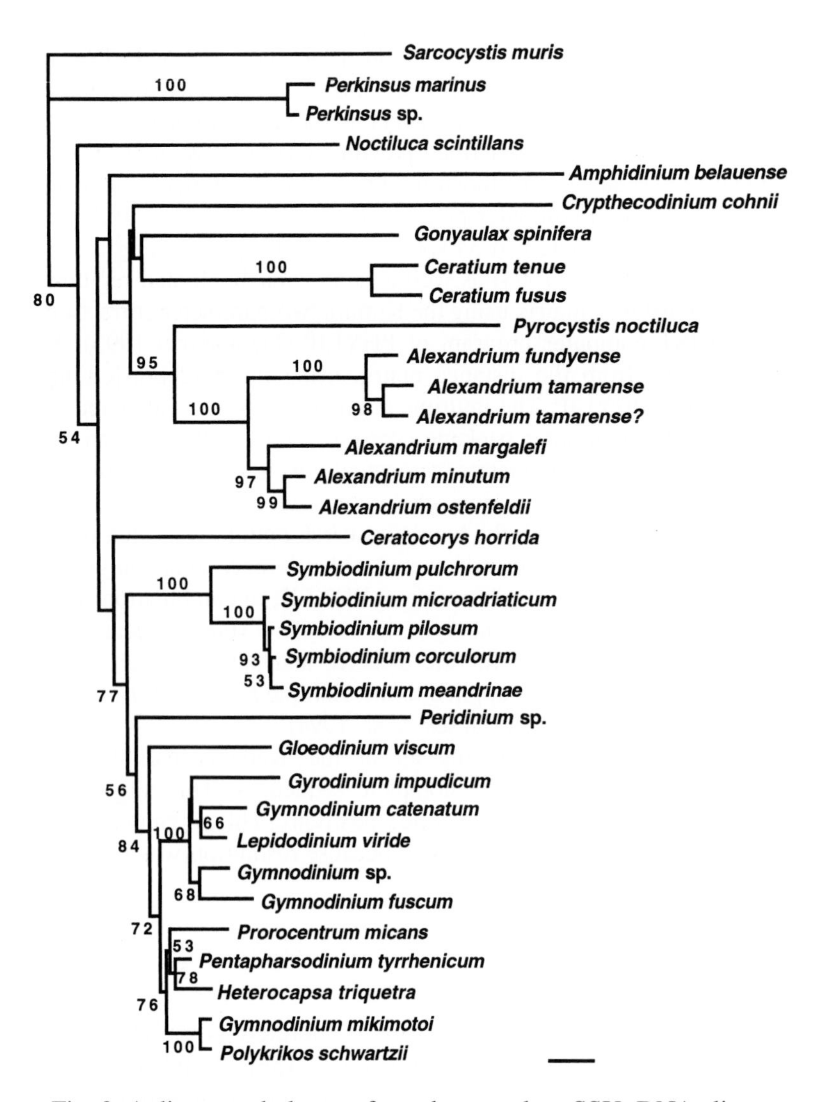

Fig. 3. A distance phylogeny from the complete-SSU rDNA alignment. Numbers at internal branches indicate percentage of bootstrap support (1000 replicates). Values <50% have not been included. Bar: 1% divergence

did not join this group in the distance tree, but was weakly allied to the GPP complex, however, it did join the gonyaulacalean clade in the distance bootstrap tree (not shown). For the GPP complex an assemblage for the species of *Symbiodinium* was on a separate lineage to all of the remaining species. The GPP complex sensu stricto includes a diversity of taxa such as *Peridinium, Gloeodinium*, a strongly supported (100 replicates) lineage for a number of *Gymnodinium* spp., including the type *G. fuscum* as well as *Lepidodinium*, and a final group of peridinialean (*Heterocapsa, Pentapharsodinium*), gymnodinialean (*Gymnodinium, Polykrikos*), and prorocentralean genera (Fig. 3).

Parsimony analysis yielded a single most parsimonious solution (length = 2415 steps, consistency index = 0.537, retention index = 0.606). This phylogeny (Fig. 4), although similar in many respects to the distance result (Fig. 3), presented some differences in relationships that were only weakly supported in both trees. Most notably *Ceratocorys* joined the other *Gonyaulacales, Amphidinium* grouped with

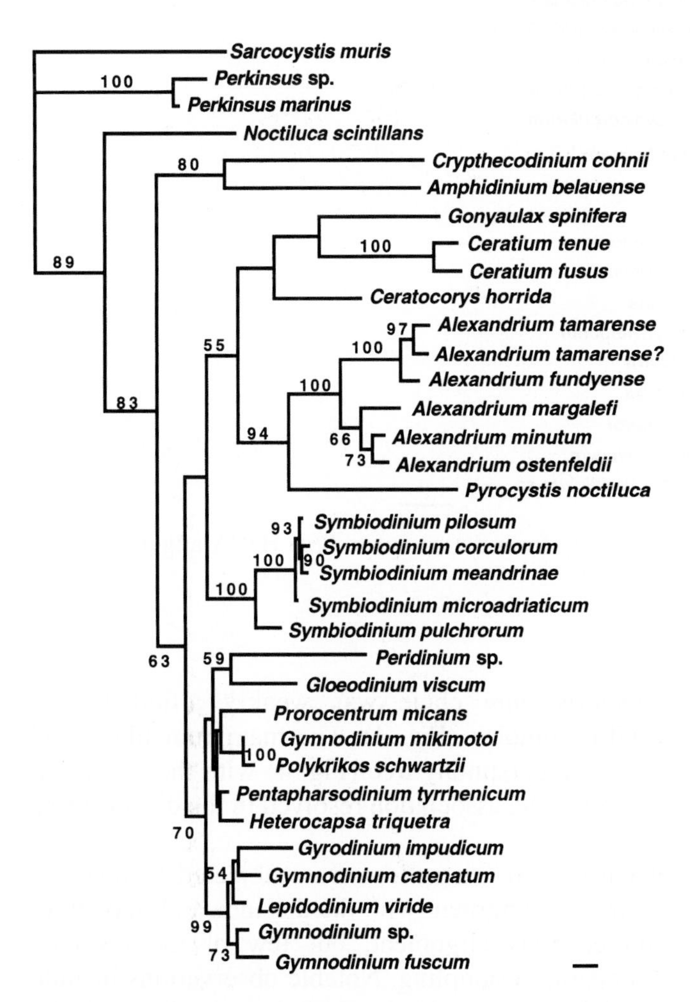

Fig. 4. A parsimony phylogeny from the complete-SSU rDNA alignment. Numbers at internal branches indicate percentage of bootstrap support (1000 replicates). Values <50% have not been included. Bar: 20 steps

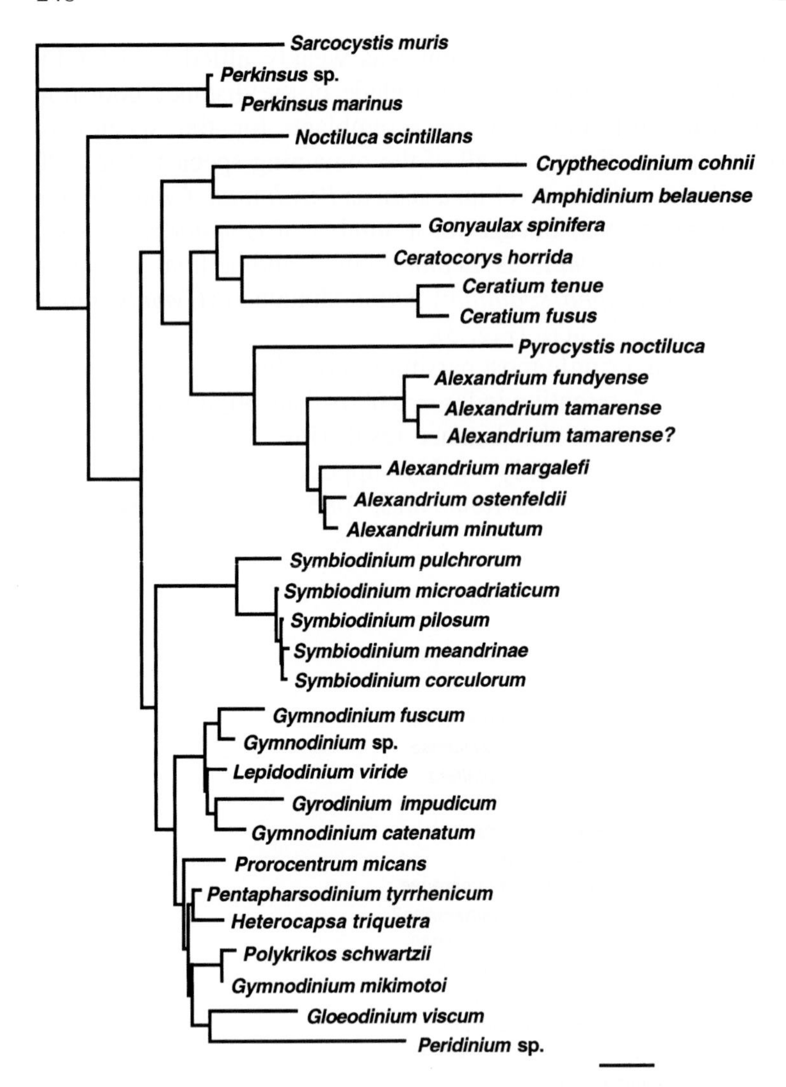

Fig. 5. A maximum likelihood phylogeny from the complete-SSU rDNA alignment. Bar: 0.02 units

Crypthecodinium, and the *Symbiodinium* clade was weakly allied to the *Gonyaulacales* rather than the GPP complex (Fig. 4). The maximum likelihood result (Fig. 5) was most similar to the parsimony tree (Fig. 4) with the exception that the *Symbiodinium* clade reverted to the association resolved in the distance tree (Fig. 3).

The multiple alignment of partial-SSU sequences was subjected to distance analysis (Fig. 6). The calculated tree is congruent with the distance and maximum likelihood results for the complete-SSU alignment, but few of the resolved relationships are supported by bootstrap resampling. Notable observations include the association of *Peridinium foliaceum* with *Amphidinium* in lieu of the *Peridinium* sp. determined herein, and the *Symbiodinium* clade is still supported but now contains species included in *Gymnodinium*. Few relationships are resolved

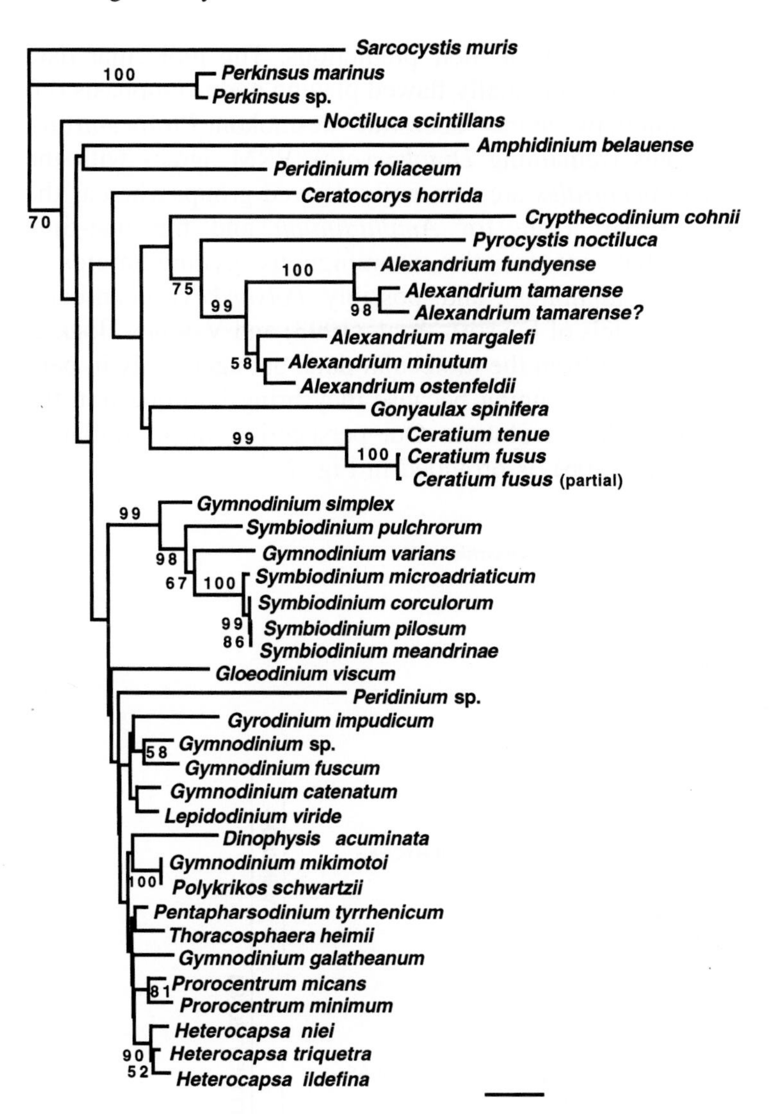

Fig. 6. A distance tree inferred from the partial-SSU rDNA alignment. Numbers at internal branches indicate percentage of bootstrap support (1000 replicates). Values <50% have not been included. Bar: 3% divergence

in the remainder of the GPP complex, but monophyly of *Heterocapsa* (including *Cachonina*, cf. Morrill & Loeblich (1981)) and *Prorocentrum* is recognized (Fig. 6).

Discussion

Three traditional perspectives on dinoflagellate phylogeny, viz. the plate increase model (PIM) (Loeblich 1976, Taylor 1980), the plate reduction model (PRM) (Dörhöfer & Davies 1980, Eaton 1980, cf. Loeblich 1984), and the plate fragmentation model (PFM) (Bujak & Williams 1981), are summarized in Fig. 2. Aspects of all of these models are supported by the current molecular analyses, however, all have conflicts with many of their predictions. The molecular data indicates that PIM and PFM are fundamentally flawed placing undue emphasis (cf. Dodge & Bibby 1973) on a split between the "ancestral" desmokont *Prorocentrales* and the "advanced" dinokonts (remaining *Dinokaryota*). PRM agrees with the molecular result that the *Prorocentrales* are a recently derived group, whereas the *Gymnodiniales,* at least in part (true for *Amphidinium* and the one-time gymnodinialean *Noctiluca*, but not for the unambiguous gymnodinioids – *Gymnodinium*, *Gyrodinium*, *Lepidodinium*, and, possibly, *Polykrikos*), is an early divergence. The more recent models of Fensome & al. (1993) and Van den Hoek & al. (1995) harness stronger support from the molecular data, but again only in part. In short, the molecular data are challenging because they bring into question the established models of dinoflagellate evolution. Our perspective, most similar to that of Fensome & al. (1993: 33, 206), is provided in Fig. 7.

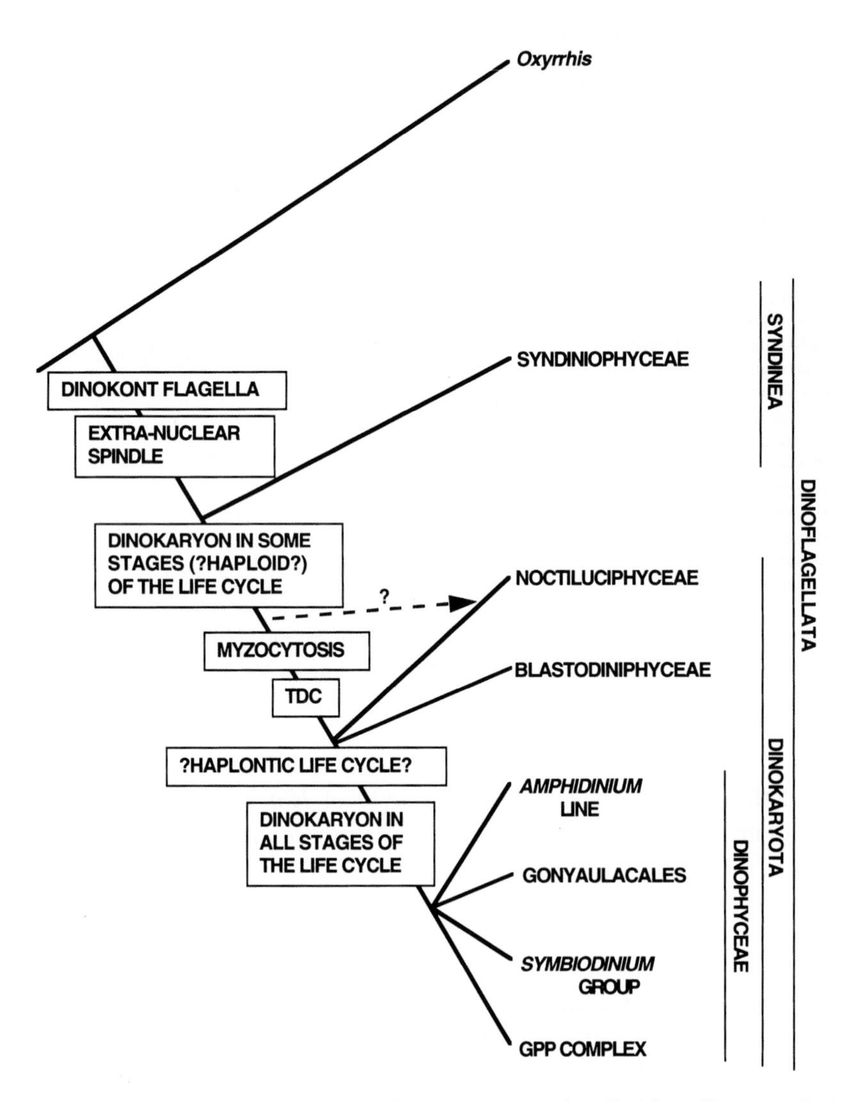

Fig. 7. A speculative hypothesis showing early dinoflagellate evolution. See text for abbreviations

***Oxyrrhis, Syndinea* and *Blastodiniphyceae (Dinokaryota)*.** All of the recent evolutionary models that consider *Oxyrrhis*, position it as an early divergence (Figs. 1, 7), sister to the *Dinoflagellata* (FENSOME & al. 1993: 32–33). This is supported by observations of more typical eukaryotic mitosis and chromosome-type (hereafter TEMC) for *Oxyrrhis* relative to the division *Dinoflagellata* (TRIEMER 1982, cf. LOEBLICH 1984), lack of a typical dinoflagellate transverse flagellum (FENSOME & al. 1993), cladistic analyses of non-molecular data (ROBERTS 1991, ROBERTS & ROBERTS 1991), and by the LSU molecular phylogeny (LENAERS & al. 1991). Unfortunately, no SSU data are available to date for this taxon and it is not included in the current study. This is an aspect of the alveolate SSU data set in need of further investigation. Similar arguments can be made regarding the *Syndinea* which are considered to be an early divergence from the dinoflagellate line and sister to the *Dinokaryota* (Figs. 1, 7) owing to TEMC in all stages of the life history and the absence of a "typical" dinoflagellate chloroplast (TDC) (i.e., peridinin and chlorophyll c_2-containing with three membranes), but with a typical dinoflagellate transverse flagellum in the motile stage (RIS & KUBAI 1974, FENSOME & al. 1993).

Considering the wide distribution of the TDC among the *Dinokaryota* (Fig. 1), it is parsimonious to conclude that the TDC was acquired once and at an early point in dinoflagellate evolution (TAYLOR 1980, VAN DEN HOEK & al. 1995), perhaps by a phagotrophic *Oxyrrhis*-type (LOEBLICH 19984) or *Noctiluca*-type ancestor (Fig. 7). If the endosymbiont was a prokaryote (primary endosymbiont event), then the outer TDC membrane would be homologous to the host food-vacuole membrane, the middle membrane would be derived from the outer membrane of a gram negative bacterium, and the inner membrane would represent the plasma membrane of the prokaryotic endosymbiont (GIBBS 1990). In other acknowledged primary endosymbiont event (s), however, chloroplasts with only two membranes – the outer is homologous to the food vacuole and the inner to the plasma membrane of the symbiotic prokaryote (GIBBS 1990) – have evolved. Alternatively the ancestral phagotrophic dinoflagellate could have acquired the TDC from a phototrophic eukaryote with one of the four expected membranes subsequently lost (GIBBS 1978, SCHNEPF 1992). This model, however, is not parsimonious and requires an ad hoc membrane loss. A limitation to both models was discussed by SCHNEPF (1992): there are no currently recognized prokaryote or eukaryote lineages to provide the putative endosymbiont in support of one or the other of these hypotheses. There are some similarities between the TDC and prasinophyte chloroplasts (starch for storage, chlorophyll c, VAN DEN HOEK & al. (1995), and similar plastid DNA configuration, COLEMAN (1985)). Perhaps an early (possibly extinct) prasinophyte-type eukaryote was the endosymbiont supporting the phagotrophic eukaryote/ phototrophic eukaryote model. It is likely that the prerhodophyte plastid had similar features (see chapter 6).

Another scenario consistent with the eukaryote/phototrophic eukaryote model, discussed by SCHNEPF (1992; also see SCHNEPF & ELBRÄCHTER (1988), could yield a TDC without ad hoc membrane loss. Myzocytosis is widespread in the *Dinokaryota* (Fig. 7) and must have evolved early in the group (ELBRÄCHTER 1991). It occurs in the parasitic *Blastodiniales* (CACHON & CACHON 1987, ELBRÄCHTER 1991) which are considered an early divergence owing to TEMC in stages of the life cycle (SOYER 1971, LOEBLICH 1976, CACHON & CACHON 1977). Although blastodinialean species have yet to be included in molecular investigations, LSU and SSU phylogenies indicate that other dinoflagellates characterized by TEMCs, viz. *Oxyrrhis* and *Noctiluca*, are early divergences and this pattern will probably hold true for the *Blastodiniales*. The *Blastodiniales* include species characterized by TDCs (FENSOME & al. 1993) raising the distinct possibility of an ancestor to the *Dinokaryota* (Fig. 1) characterized by myzocytosis and TDCs (Fig. 7). If a myzocytotic dinoflagellate ingested a phototrophic eukaryote (early prasinophyte or rhodophyte?) from which the TDC has evolved, three membranes would be predicted: outer derived from the host food-vacuole membrane, middle and inner from the chloroplast membranes of the phototrophic eukaryote. This scenario yields the TDC without ad hoc hypotheses of membrane loss.

LARSEN (1988) provides convincing arguments that in fact kleptochloroplasts and possibly a few non TDC chloroplasts have been acquired myzocytotically by dinoflagellates, notably *Amphidinium*, from cryptophytes. Interestingly, cryptophytes share starch deposition, plastid DNA configuration and chlorophyll c (specifically chlorophyll c_2 with dinoflagellates) with prasinophytes and dinoflagellates (COLEMAN 1985, VAN DEN HOEK & al. 1995) and have clearly acquired their plastid from an early rhodophyte lineage (see chapter 6). *Amphidinium* branches early in the molecular tree and speculations of a myzocytotic *Amphidinium*-type ancestor ingesting a cryptophyte to yield the TDC come to mind. *Amphidinium*, however, lacks life history stages characterized by TEMC and probably diverged subsequent to the TDC-containing *Blastodiniales* (Fig. 7), and myzocytotic ingestion of a cryptophyte should yield a five-membrane organelle (cf. LARSEN 1988). Ad hoc hypotheses of membrane loss could be invoked as in the *Amphidinium*/ "cryptophyte" endosymbiont event reported by WILCOX & WEDEMAYER (1985) in which a chloroplast with three membranes is considered a derivative of a cryptomonad, but it is more parsimonious to avoid such ad hoc hypotheses. Nevertheless, circumstantial evidence indicates that membrane loss may be a common event in the evolution of non TDC chloroplasts in dinoflagellates (e.g., WILCOX & WEDEMAYER 1985, SCHNEPF & ELBRÄCHTER 1988, see LUCAS & VESK 1990: 355). In summary, there are wondrous tales to be framed on the evolution of the TDC and other dinoflagellate chloroplasts. What is urgently wanting is data. Molecular data, in particular gene sequence data, must be determined from both TDC and other dinoflagellate chloroplasts to elucidate the origins of these organelles.

Noctiluciphyceae. *Noctiluca* is the earliest divergence among the dinoflagellates included in our SSU phylogeny (Figs. 3–6), a result consistent with the LSU tree (LENAERS & al. 1991). Although considered by many to be a member of the *Gymnodiniales*, FENSOME & al. (1993) recognized a fundamental divergence between *Noctiluca* and the *Dinophyceae* and erected the *Noctiluciphyceae*. *Noctiluca* is different from other dinoflagellates in its retention of putatively ancestral features (Fig. 7) including: (1) an apparently diplontic rather than haplontic life cycle (ZINGMARK 1970, PFIESTER & ANDERSON 1987); and, (2) by the presence of TEMC in the "diploid" vegetative stage of the life history with the derived dinokaryotic nucleus confined to the "haploid gametes" (cf. SOYER 1972). It is interesting to speculate that the decreased significance of the haploid stage in an ancestral "predominantly-diploid dinoflagellate" resulted in the reductive evolution of the gamete nucleus, which with the change to haplontic life cycles in the *Dinophyceae* has placed this reduced nuclear state into the vegetative stage and into prominence in macroevolutionary studies (Fig. 7).

Dinophyceae. The remainder of the relationships resolved by molecular data among species of the *Dinophyceae* are not in close agreement with any of the proposed models of dinoflagellate evolution (cf. FENSOME & al. 1993: appendix B, for a detailed summary). Phylogenetically, four lineages can be recognized in the *Dinophyceae* for the species included in the molecular investigations: (1) *Amphidinium* line; (2) *Gonyaulacales* (including *Crypthecodinium*); (3) *Symbiodinium* group; and, (4) the GPP complex sensu stricto (Fig. 7).

Amphidinium line. The symbiotic 'gymnodinialean' *Amphidinium belauense* emerges early (Figs. 3–6) but does not combine with other symbiotic species or, more importantly, with the other members of the *Gymnodiniales* (Fig. 1). As MCNALLY & al. (1994) suggested, and they had a second species confirming its position (*A. carterae*), *Amphidinium* is too distant from the other gymnodinialeans for inclusion in this order – the taxonomy of *Amphidinium* needs to be reevaluated. Such an early divergence for *Amphidinium* may come as a surprise on first glance, but reports by OAKLEY & DODGE (1974) of distinct kinetochores associated with chromosomes of *Amphidinium carterae* strongly urge further study of the genus *Amphidinium* to elucidate its phylogenetic affinities.

Gonyaulacales. The gonyaulacalean species *Crypthecodinium cohnii* also emerges early and the relationships among this species, *Amphidinium*, and the remaining *Gonyaulacales* vary with the analysis (Figs. 3-6). Based on chromosome features in *Amphidinium* it is probable that this genus diverged prior to the *Gonyaulacales*. *Crypthecodinium* usually joins the *Gonyaulacales* in the molecular analyses consistent with nuclear (KUBAI & RIS 1969), as well as morphological observations (FENSOME & al. 1993). *Ceratocorys* inexplicably groups with the GPP complex in the full-gene distance analysis (Fig. 3), whereas it joins the other gonyaulacalean taxa in all other analyses. The latter result is consistent with the morphology of *Ceratocorys* (FENSOME & al. 1993). Despite these few anomalies, there seems to be support in the molecular data for the *Gonyaulacales* (TAYLOR 1980) as constituted in FENSOME & al. (1993), but they are a sister to the GPP complex sensu lato and should not be considered an order within the *Peridiniphycidae* (Table 1, Fig. 7).

Symbiodinium group. The strongest feature of the full-gene phylogeny is the resolution of a *Symbiodinium* complex, which gives good support to their placement in the separate order *Suessiales* (FENSOME & al. 1993). The molecular data do not necessitate that this order be included in the subclass *Gymnodiniphycidae* (Fig. 7), but rather this is a major lineage of the *Dinophyceae* sister to the GPP complex sensu stricto and worthy of subclass status in the system of FENSOME & al. (1993).

GPP complex. The GPP complex sensu stricto is in greatest need of further investigation. In all of the models proposed for dinoflagellate evolution, the GPP complex includes those organisms considered to represent both the most ancestral and the most derived species (compare Fig. 2 to Figs. 3–6). As noted previously, the opinion that the *Prorocentrales* is some how ancestral and sister to the bulk of the other dinoflagellates is not supported by any of the available molecular data. The *Prorocentrales* is a recently derived lineage associated with a diversity of gymnodinioid and peridinioid species, as well as other taxa of uncertain affinity (compare Fig. 1 to Figs. 3–6) within the *Dinophyceae*. Although the PRM considers *Prorocentrum* derived (Fig. 2), this lineage did not evolve from other *Dinophyceae* as depicted in this model. DODGE & BIBBY (1973) concluded that prorocentralean species were similar in many respects to the dinokonts, and recent ultrastructural investigations have highlighted the derived nature of the *Prorocentrales* relative to other dinoflagellates (ROBERTS & al. 1995 and references therein) buttressing the molecular results.

In the present understanding of the genus *Gymnodinium*, there are photosynthetic species and heterotrophic species, marine and freshwater species, toxic and non-toxic species and variously pigmented species. *Lepidodinium viride* has grass green chloroplasts due to the presence of chlorophyll b (WATANABE & al. 1990) and its emergence on a branch with *Gymnodinium catenatum* in the SSU tree (Figs. 3, 6) is a case in point. These two species both possess classic gymnodinioid amphiesmal structure to the extent of sharing an almost identical apical groove structure. *Gymnodinium catenatum* is, however, a chainforming, toxin-producing, cyst-forming alga with TDCs, whereas *Lepidodinium viride* is solitary, non-toxic, not known to produce cysts, has chlorophyll b and possesses superficial organic scales (WATANABE & al. 1990). This somewhat anomalous possession of green chloroplasts might be explained by a relationship with a prasinophyte endosymbiont as indicated by a more detailed pigment analysis (WATANABE & al. 1991). It is interesting to speculate that the scales were also acquired during this endosymbiotic event.

On a practical taxonomic level, how should systematists deal with chimeric entities? *Lepidodinium* clearly falls within a cluster of *Gymnodinium* species (including the type species, *G. fuscum*) in our molecular results as well as on the basis of amphiesmal structure. Should the genera be merged recognizing *L. viride* as an anomaly, or separate artificial genera maintained for practical taxonomic purposes? These are questions that will continue to arise, especially as the artificial gymnodinialean genera, e.g., *Gymnodinium, Gyrodinium* and *Katodinium*, are examined in more detail.

The *Peridiniales* are still poorly represented – by just *Peridinium* (= *Kryptoperidinium* = *Glenodinium*) *foliaceum*, an undescribed species of *Peridinium*, *Pentapharsodinium* and three species of *Heterocapsa*. *P. foliaceum* emerges early in the partial SSU phylogeny (Fig. 6), far removed from all of the other peridinialean species which are contained in the GPP complex (Figs. 3–6). *P. foliaceum* needs a careful re-evaluation by both non-molecular and molecular techniques. Certainly, many more members of the *Peridiniales* need to be examined by molecular systematists before any further taxonomic proposals are framed for species of the GPP complex.

The relatively isolated position of "*Gloeodinium viscum*" is reassuring. It is the only representative of the *Phytodiniales* investigated at the molecular level and with its unusual life history deserves an isolated spot. *Thoracosphaera* also has an unusual life history including a stage enclosed in a case composed of calcium carbonate (FENSOME & al. 1993). It is contained in its own order *Thoracosphaerales* and emerges near the top of the tree along with *Pentapharsodinium* (produces calcium carbonate cysts) although the support for this relationship is not great. There are other dinoflagellates that have cysts composed of calcium carbonate, e.g., *Scrippsiella*, and although they are arguably of peridinialean origin, it would be of some interest to include more of these in the molecular tree.

Conclusions

The dinoflagellates are a marvelously diverse group of protists. Numerous aspects of their evolution remain to be explored and we hope that this chapter, if nothing else, has highlighted this point. Of particular interest to us is the variety observed in chloroplast type and the still uncertain origin of the TDC. For the dinoflagellate hosts themselves, representatives of the parasitic *Syndinea* and *Blastodiniphyceae* should be investigated to determine their affinities relative to the *Noctiluciphyceae* and *Dinophyceae*. The GPP complex is probably a relatively recent lineage, however, in includes an immensely morphologically diverse assemblage of dinoflagellates. More GPP species need to be investigated to elucidate relationships among the component taxa and to unravel the evolution of the myriad of morphotypes in this group. Lastly, the relatively late divergence of such a taxonomically diverse assemblage as the GPP complex of species will undoubtedly have an effect on how results of palaeontological investigations are interpreted (Fensome & al. 1993, 1996).

We thank Dr R. Fensome for helpful comments on this manuscript and Isabelle Strachan, Dr Dan Potter and Kate Ross for technical assistance. During preparation of this review GWS was supported initially by an Australian Research Council QEII Fellowship and subsequently by funds from the Natural Sciences and Engineering Research Council of Canada. RAA was supported by an Office of Naval Research Grant N00014-92-J-1717.

References

Adachi, M., Sako, Y., Ishida, Y., 1996: Analysis of *Alexandrium* (*Dinophyceae*) species using sequences of the 5.8S ribosomal DNA and internal transcribed spacer regions. – J. Phycol. **32**: 424–432.

Baroin, A., Perasso, R., Qu, L.-H., Brugerolle, G., Bachellerie, J.-P., Adoutte, A., 1988: Partial phylogeny of the unicellular eukaryotes based on rapid sequencing of a portion of 28S ribosomal RNA. – Proc. Natl. Acad. Sci. USA **85:** 3474–3478.

Bujak, J. P., Williams, G. L., 1981: The evolution of dinoflagellates. – Canad. J. Bot. **59**: 2077–2087.

Cachon, J., Cachon, M., 1977: Observations on the mitosis and on the chromosome evolution during the lifecycle of *Oodinium*, a parasitic dinoflagellate. – Chromosoma **60**: 237–251.

– 1987: Parasitic dinoflagellates. – In Taylor, F. J. R., (Ed.): The biology of dinoflagellates. Botanical monographs **21**, pp. 571–610. – Oxford: Blackwell Scientific.

Cavalier-Smith, T., 1975: Review article. The origin of nuclei and of eukaryote cells. – Nature **256**: 463–468.

– 1993: Kingdom *Protozoa* and its 18 phyla. – Microbiol. Rev. **57**: 953–994.

– Chao, E. E., Allsopp, M. T. E. P., 1995: Ribosomal RNA evidence for chloroplast loss within *Heterokonta*: Pedinellid relationships and a revised classification of Ochristan algae. – Archiv Protistenk. **145**: 209–220.

Cedergren, R., Gray, M. W., Abel, Y., Sankoff, D., 1988: The evolutionary relationships among known life forms. – J. Molec. Evol. **28**: 98–112.

Chadefaud, M., 1950: Les cellules nageuses des algues dans l'embranchement des Chromophycées. Compt. Rend. Acad. Sci. Paris **231**: 788–790.

Coleman, A. W., 1985: Diversity of plastid DNA configuration among classes of eukaryote algae. – J. Phycol. **21**: 1–16.

Corliss, J. O., 1975: Nuclear characteristics and phylogeny in the protistan phylum *Ciliophora*. – Biosystems **7**: 338–349.

– 1984: The kingdom *Protista* and its 45 phyla. – Biosystems **17**: 87–126.

Destombe, C., Cembella, A. D., Murphy, C. A., Ragan, M. A., 1992: Nucleotide sequence of the 18S ribosomal RNA genes from the marine dinoflagellate. *Alexandrium tamarense* (*Gonyaulacales, Dinophyta*). – Phycologia **31**: 121–124.

Dodge, J. D., 1965: Chromosome structure in the dinoflagellates and the problem of the mesokaryotic cell. – Excerpta Med. Int. Congr. Ser. **91**: 339–341.

– 1966: The *Dinophyceae*. – In Godward, M. B. E., (Ed.): The chromosomes of the algae, pp. 96–115. – London: Arnold.

– Bibby, B. J., 1973: The *Prorocentrales* (*Dinophyceae*) I. A comparative account of fine structure in the genera *Prorocentrum* and *Exuviella*. – Bot. J. Linn. Soc. **67**: 175–187.

Dörhöfer, G., Davies, E. H., 1980: Evolution of archeopyle and tabulation in rhaetogonyaulacinean dinoflagellate cysts. pp. 1–91. – Toronto: Royal Ontario Museum Life Sci. Misc. Publ.

Eaton, G. L., 1980: Nomenclature and homology in peridinialean dinoflagellate plate patterns. – Palaeontology **23**: 667–688.

Elbrächter, M., 1991: Food uptake mechanisms in phagotrophic dinoflagellates and classification. – In Patterson, D. J., Larsen, J., (Eds): The biology of free-living heterotrophic flagellates, pp. 303–312. – Oxford: Clarendon Press.

Felsenstein, J., 1985: Confidence limits on phylogenies: an approach using the bootstrap. – Evolution **39**: 783–791.

– 1995: Phylip-Phylogeny inference package (version 3.57c). Distributed by the author, Department of Genetics, University of Washington, Seattle.

Fensome, R. A., Taylor, F. J. R., Norris, G., Sarjeant, W. A. S., Wharton, D. I., Williams, G. L., 1993: A classification of living and fossil dinoflagellates. – Micropaleontol. Spec. Publ. **7**.

– Macrae, R. A., Moldowan, J. M., Taylor, F. J. R., Williams, G. L., 1996: The early Mesozoic radiation of dinoflagellates. – Paleobiology **22**: 329–338.

Fong, D., Rodriguez, R., Koo, K., Sun, J., Sogin, M. L., Bushek, D., Littlewood, D. T., Ford, S. E., 1993: Small subunit ribosomal RNA gene sequence of the oyster parasite *Perkinsus marinus*. – Molec. Mar. Biol. Biotechnol. **2**: 346–350.

Gajadhar, A. A., Marquardt, W. C., Hall, R., Gunderson, J., Ariztia-Carmona, E. V., Sogin, M. L., 1991: Ribosomal RNA sequences of *Sarcocystis muris*, *Theileria annulata* and *Crypthecodinium cohnii* reveal evolutionary relationships among apicomplexans, dinoflagellates and ciliates. – Molec. Biochem. Parasit. **45**: 147–154.

Gibbs, S. P., 1978: The chloroplasts of *Euglena* may have evolved from symbiotic green algae. – Canad. J. Bot. **56**: 2883–2889.

– 1990: The evolution of algal chloroplasts. – In Wiessner, W., Robinson, D. G., Starr, R. C., (Eds): Exp. Phycol. **1**. Cell walls and surfaces, reproduction, photosynthesis, pp. 145–157. – Berlin: Springer.

GILBERT, D. G., 1995: SeqPup. a biological sequence editor and analysis program for Macintosh computers. – Published electronically on the Internet, available via anonymous ftp to ftp. bio. indiana. edu.

GOGGIN, C. L., BARKER, S. C., 1993: Phylogenetic position of the genus *Perkinsus* (*Protista, Apicomplexa*) based on small subunit ribosomal RNA. – Molec. Biochem. Parasit. **60**: 65–70.

GUNDERSON, J. H., ELWOOD, H., INGOLD, A., KINDLE, K., SOGIN, M. L., 1987: Phylogenetic relationships between chlorophytes, chrysophytes, and oomycetes. – Proc. Natl. Acad. Sci. USA **84:** 5823–5827.

HERZOG, M., MAROTEAUX, L., 1986: Dinoflagellate 17S rRNA sequence inferred from the gene sequence: evolutionary implications. – Proc. Natl. Acad. Sci. USA **83**: 8644–8648.

HINNEBUSCH, A. G., KLOTZ, L. C., BLANKEN, R. L., LOEBLICH, A. R. III, 1981: An evaluation of the phylogenetic position of the dinoflagellate *Crypthecodinium cohnii* based on 5S rRNA characterization. – J. Molec. Evol. **17**: 334–347.

JOHNSON, A. M., BAVERSTOCK, P. R., 1989: Rapid ribosomal RNA sequencing and the phylogenetic analysis of protists. – Parasitology Today **5**: 102–105.

– ILLANA, S., HAKENDORF, P., BAVERSTOCK, P. R., 1988: Phylogenetic relationships of the apicomplexan protist *Sarcocystis* as determined by small subunit ribosomal RNA comparison. – J. Parasit. **74**: 847–860.

KIMURA, M., 1980: A simple method for estimating evolutionary rates of base substitutions through comparative studies of nucleotide sequences. – J. Molec. Evol. **16**: 111–120.

KUBAI, D. F., RIS, H., 1969: Division in the dinoflagellate *Gyrodinium cohnii* (SCHILLER). A new type of nuclear reproduction. – J. Cell Biol. **40**: 508–528.

LARSEN, J., 1988: An ultrastructural study of *Amphidinium poecilochroum* (*Dinophyceae*), a phagotrophic dinoflagellate feeding on small species of cryptophytes. – Phycologia **27**: 366–377.

LENAERS, G., MAROTEAUX, L., MICHOT, B., HERZOG, M., 1989: Dinoflagellates in evolution. A molecular phylogenetic analysis of large-subunit ribosomal RNA. – J. Molec. Evol. **29**: 40–51.

– SCHOLIN, C., BHAUD, Y., SAINT-HILAIRE, D., HERZOG, M., 1991: A molecular phylogeny of dinoflagellate protists (*Pyrrophyta*) inferred from the sequence of 24S rRNA divergent domains D1 and D8. – J. Molec. Evol. **32**: 53-63.

LIU, M. H., REDDY, R., HENNING, D., SPECTOR, D., BUSCH, H., 1984: Primary and secondary structure of dinoflagellate U5 small nuclear RNA. – Nucl. Acids Res. **12**: 1529–1542.

LOEBLICH, A. R., III, 1976: Dinoflagellate evolution: speculation and evidence. – J. Protozool **23**: 13–28.

– 1984: Dinoflagellate evolution. – In SPECTOR, D. L., (Ed.): Dinoflagellates, pp. 481–522. – Orlando: Academic Press.

LUCAS, I. A. N., VESK, M., 1990: The fine structure of two photosynthetic species of *Dinophysis* (*Dinophysiales, Dinophyceae*). – J. Phycol. **26**: 345–357.

MCNALLY, K. L., GOVIND, N. S., THOMÉ, P. E., TRENCH, R. K., 1994: Small-subunit ribosomal DNA sequence analyses and a reconstruction of the inferred phylogeny among symbiotic dinoflagellates (*Pyrrophyta*). – J. Phycol. **30**: 316–329.

MORRILL, L.C., LOEBLICH, A. R., III, 1981: A survey for body scales in dinoflagellates and a revision of *Cachonina* and *Heterocapsa* (*Pyrrophyta*). – J. Plankton. Res. **3**: 53–65.

OAKLEY, B. R., DODGE, J. D., 1974: Kinetochores associated with the nuclear envelope in the mitosis of a dinoflagellate. – J. Cell Biol. **63**: 322–325.

OLSEN, G. J., MATSUDA, H., HAGSTROM, R., OVERBEEK, R., 1994: FastDNAml: a tool for construction of phylogenetic trees of DNA sequences using maximum likelihood. – CABIOS **10**: 41–48.

PFIESTER, L. A., ANDERSON, D. M., 1987: Dinoflagellate reproduction. – In TAYLOR, F. J. R., (Ed.): The biology of dinoflagellates. Botanical monographs **21**, pp. 611–648. – Oxford: Blackwell Scientific.

POTTER, D., LAJEUNESSE, T. C., SAUNDERS, G. W., ANDERSEN, R. A., 1997: Convergent evolution masks extensive biodiversity among marine coccoid picoplankton. – Biodiversity & Conservation. **6**: 99–107.

QU, L.-H., PERASSO, R., BAROIN, A., BRUGEROLLE, G., BACHELLERIE, J-P., ADOUTTE, A., 1988: Molecular evolution of the 5′-terminal domain of large-subunit rRNA from lower eukaryotes. A broad phylogeny covering photosynthetic and non-photosynthetic protists. – BioSystems **21:** 203–208.

RAGAN, M. A., CHAPMAN, D. J., 1978: A biochemical phylogeny of the protists. – New York: Academic Press.

RIS, H., KUBAI, D. F., 1974: An unusual mitotic mechanism in the parasitic protozoan *Syndinium* sp. – J. Cell. Biol. **60**: 702–720.

ROBERTS, K. R., 1991: The flagellar apparatus and cytoskeleton of diro flagellates: organization and use in systematics. – In PATTERSON, D. J., LARSEN, J., (Eds.): The biology of free-living heterotrophic flagellates, pp. 285–302. – Oxford: Clarendon Press.

– ROBERTS, J. E., 1991: The flagellar apparatus and cytoskeleton of the dinoflagellates. A comparative overview. – Protoplasma **164**: 105–122.

– HEIMANN, K., WETHERBEE, R., 1995: The flagellar apparatus and canal structure in *Prorocentrum micans* (*Dinophyceae*). – Phycologia **34**: 313–322.

ROWAN, R., 1991: Minireview. Molecular systematics of symbiotic algae. – J. Phycol. **27**: 661–666.

– POWERS, D. A., 1991: A molecular genetic classification of zooxanthellae and the evolution of animal-algal symbioses. – Science **251**: 1348–1351.

– 1992: Ribosomal RNA sequences and the diversity of symbiotic dinoflagellates (zooxanthellae). – Proc. Natl. Acad. Sci. USA **89**: 3639–3643.

SADLER, L. A., MCNALLY, K. L., GOVIND, N. S., BRUNK, C. F., TRENCH, R. K., 1992: The nucleotide sequence of the small subunit ribosomal RNA gene from *Symbiodinium pilosum*, a symbiotic dinoflagellate. – Curr. Genet. **21**: 409–416.

SAITOU, N., NEI, M., 1987: The neighbor-joining method: a new method for reconstructing phylogenetic trees. – Molec. Biol. Evol. **4:** 406–425.

SAUNDERS, G. W., 1993: Gel purification of red algal genomic DNA: an inexpensive and rapid method for the isolation of polymerase chain reaction-friendly DNA. – J. Phycol. **29**: 251–254.

– KRAFT, G. T., 1994: Small-subunit rRNA gene sequences from representatives of selected families of the *Gigartinales* and *Rhodymeniales* (*Rhodophyta*). I. Evidence for the Plocamiales ord. nov. – Canad. J. Bot. **72:** 1250–1263.

– 1996: Small-subunit rRNA gene sequences from representatives of selected families of the *Gigartinales* and *Rhodymeniales* (*Rhodophyta*). 2. Recognition of the *Halymeniales* ord. nov. – Canad J. Bot. **74**: 694–707.

– POTTER, D., PASKIND, M. P., ANDERSEN, R. A., 1995: Cladistic analyses of combined traditional and molecular data sets reveal an algal lineage. – Proc. Natl. Acad. Sci. USA. **92**: 244–248.

– HILL, D. R. A., TYLER, P. A., 1997: Phylogenetic affinities of *Chrysonephele palustris* (*Chrysophyceae*) based on the nuclear small-subunit rRNA gene. – J. Phycol. **33**: 132–134.

SCHNEPF, E., 1992: From prey via endosymbiont to plastid: comparative studies in dinoflagellates. – In LEWIN, R. A., (Ed.): Origins of plastids, pp. 53–76. – New York, London: Chapman & Hall.

– Elbrächter, M., 1988: Cryptophycean-like double membrane-bound chloroplast in the dinoflagellate, *Dinophysis* Ehrenb.: evolutionary, phylogenetic and toxicological implications. – Bot. Acta **101**: 196–203.

Scholin, C. A., Anderson, D. M., 1994: Identification of group- and strain-specific genetic markers for globally distributed *Alexandrium* (*Dinophyceae*). I. RFLP analysis of SSU rRNA genes. – J. Phycol. **30**: 744–754.

– – Sogin, M. L., 1993: Two distinct small-subunit ribosomal RNA genes in the North American toxic dinoflagellate *Alexandrium fundyense* (*Dinophyceae*). – J. Phycol. **29**: 209–216.

– Herzog, M., Sogin, M., Anderson, D. M., 1994: Identification of group- and strain-specific genetic markers for globally distributed *Alexandrium* (*Dinophyceae*). II. Sequence analysis of a fragment of the LSU rRNA gene. – J. Phycol. **30**: 999–1011.

– Hallegraeff, G. M., Anderson, D. M., 1995: Molecular evolution of the *Alexandrium tamarense* 'species complex' (*Dinophyceae*): dispersal in the North American and West Pacific regions. – Phycologia **34**: 472–485.

Sogin, M. L., 1989: Evolution of eukaryotic microorganisms and their small subunit ribosomal RNAs. – Amer. Zool. **29**: 487–499.

Soyer, M.-O., 1971: Structure du noyau des *Blastodinium* (Dinoflagellés parasites). Division et condensation chromatique. – Chromosoma **33**: 70–114.

– 1972: Les ultrastructures nucléaire de la Noctiluque (Dinoflagellés libre) au cours de la sporogenése. – Chromosoma **39**: 419–441.

Swofford, D. L., 1993: PAUP-phylogenetic analysis using parsimony, version 3.1.1. – Champaign; Illinois Natural History Survey, University of Illinois.

Taylor, F. J. R., 1976: Flagellate phylogeny: a study of conflicts. – J. Protozool. **23**: 28–40.

– 1978: Problems in the development of an explicit phylogeny of the lower eukaryotes. – BioSystems **10**: 67–89.

– 1980: On dinoflagellate evolution. – BioSystems **13**: 65–108.

Triemer, R. E., 1982: A unique mitotic variation in the marine dinoflagellate *Oxyrrhis marina* (*Pyrrophyta*). – J. Phycol. **18**: 399–411.

Van den Hoek, C., Mann, D. G., Jahns, H. M., 1995: Algae: an introduction to phycology. – Cambridge: Cambridge University Press.

Van de Peer, Y., Neefs, J.-M., de Rijk, P., De Wachter, R., 1993: Evolution of eukaryotes as deduced from small ribosomal subunit RNA sequences. – Biochem. Syst. Ecol. **21**: 43–55.

Van der Auwera, G., De Baere, R., Van de Peer, Y., De Rijk, P., Van den Broeck, I., De Wachter, R., 1995: The phylogeny of the *Hyphochytriomycota* as deduced from ribosomal RNA sequences of *Hyphochytrium catenoides*. – Molec. Biol. Evol. **12**: 671–678.

Watanabe, M. M., Suda, S., Inouye, I., Sawaguchi, T., Chihara, M., 1990: *Lepidodinium viride* gen. et sp. nov. (*Gymnodiniales, Dinophyta*), a green dinoflagellate with a chlorophyll a- and b-containing endosymbiont. – J. Phycol. **26**: 741–751.

– Sasa, T., Suda, S., Inouye, I., Takaichi, S., 1991: Major carotenoid composition of an endosymbiont in a green dinoflagellate, *Lepidodinium viride*. – J. Phycol. **27**[Suppl.]: 75.

Wilcox, L. W., Wedemayer, G. J., 1985: Dinoflagellate with blue-green chloroplasts derived from an endosymbiotic eukaryote. – Science **227**: 192–194.

Zardoya, R., Costas, E., Løpez-Rodas, V., Garrido-Pertierra, A., Bautista, J. M., 1995: Revised dinoflagellate phylogeny inferred from molecular analysis of large-subunit ribosomal RNA gene sequences. – J. Molec. Evol. **41**: 637–645.

Zingmark, R. G., 1970: Sexual reproduction in the dinoflagellate *Noctiluca miliaris* Suriray. – J. Phycol. **6**: 122–126.

Plastids in apicomplexan parasites

GEOFFREY I. MCFADDEN, ROSS F. WALLER, MICHAEL E. REITH,
and NAOMI LANG-UNNASCH

Key words: *Apicomplexa, Plasmodium, Toxoplasma.* – Endosymbiosis, plastids, malaria, toxoplasmosis.

Abstract: The discovery in malarial and toxoplasmodial parasites of genes normally occurring in the photosynthetic organelle of plants and algae has prompted speculation that these so-called protozoans might harbour a vestigial plastid. The plastid-like parasite genes occur on an extrachromosomal, maternally inherited, 35 kb DNA circle with an architecture reminiscent of plastid genomes. The 35kb genome is distinct from the 6–7 kb linear mitochondrial genome. Localization of the 35kb genome within the parasite cells by high resolution in situ hybridization has identified a multi membrane-bound organelle in which the plastid-like genome resides. Since phylogenetic trees incorporating genes from the 35 kb genome group them within the plastid radiation, we believe that the parasite organelle is a reduced plastid, that is probably no longer photosynthetic. Combined molecular and ultrastructural evidence indicate plastids to be widespread among apicomplexan parasites. The origin and role of the plastid in obligate intracellular parasites is completely unknown. The potential utility of the plastid as a parasite-specific target for therapeutic agents is examined.

The phylum *Apicomplexa* (previously included in the phylum *Sporozoa*) is a group of unicellular endoparasites. All members possess a specialised apical structure (the apical complex) involved in penetration of the host cell. The phylum includes approximately 4600 described species, but this is probably only a small fraction of the total radiation of the apicomplexan parasites (LEVINE 1988). The *Apicomplexa* are of enormous veterinary and medical significance being responsible for a wide variety of diseases including coccidiosis, cryptosporidiosis, babesiosis (Texas cattle fever), theileriosis (East Coast Fever), toxoplasmosis, and, most damaging of all, malaria. Until recently, there would have been no place for these so-called 'protozoans' in a book on plastids. Recent results, however, reveal that apicomplexans do in fact contain a plastid. This chapter reviews the molecular genetic data that first alerted biologists to a plastid in apicomplexans. In reviewing the literature more broadly, we also show that there exists abundant evidence of the plastid being a character common to most apicomplexan groups. Several excellent reviews (JEFFRIES & JOHNSON 1996; FEAGIN 1994, 1995; WILSON & al. 1991, 1994; PALMER 1992a) served as our springboard in assembling this chapter.

Apicomplexan phylogeny

The evolutionary relationship of apicomplexans to other protozoans was initially rather obscure. Electron microscopists had noted that the apparently triple-layered plasma membrane of apicomplexans, created by the subtending sheath of ER, resembles the alveolar sacs of ciliates and dinoflagellates (Vivier & Desportes 1990). Little was made of this similarity, however, until molecular phylogenies included apicomplexan taxa. Evolutionary trees inferred from nuclear-encoded 18S rRNA genes confirmed the alliance of ciliates, dinoflagellates, and apicomplexans, consistently grouping them as a monophyletic lineage (Gajadhar & al. 1991, Wolters 1991, Barta & al. 1991, Sadler & al. 1992, Gagnon & al. 1993, Nelissen & al. 1995, Bhattacharya & al. 1995, McFadden & al. 1994, Cavalier-Smith & al. 1996). This triumvirate has been provisionally named the "Alveolates" (a name referring to the subplasmalemmal membranous sacs producing the tri-layered membrane first noted by electron microscopists). Phylogenetic trees inferred from protein sequence data also group apicomplexans with ciliates (e.g., Baldauf & Palmer 1993, Keeling & Doolittle 1996), but none of these trees yet include dinoflagellates. Recent 18S rRNA trees consistently position the apicomplexans as sister taxon to the dinoflagellates (Nelissen & al. 1995, Bhattacharya & al. 1995, Cavalier-Smith 1995a, Cavalier-Smith & al. 1996, Van de Peer & al. 1996, Siddal & al. 1995, Pawlovski & al. 1996, Silberman & al. 1996, Kumar & Rzhetsky 1996), a group in which many members contain plastids (Taylor 1990).

Plasmodium cells contain two extrachromosomal DNAs

The discovery of extrachromosomal DNAs in *Plasmodium* initiated a trail of investigation that eventually led to the identification of a plastid in apicomplexans. Hind-sight allows us to record Kilejian (1975) as the first to observe the extrachromosomal DNAs of apicomplexans. Using electron microscopy, Kilejian (1975) observed circular DNA molecules, approximately 27 kb in length, in *Plasmodium lophurae* Coggeshall (a causative agent of malaria in birds). *Plasmodium* organelle research being in its infancy, Kilejian (1975) naturally interpreted these molecules as part of the mitochondrial genome, and was therefore unable to realise their significance. However, extrachromosomal DNA of *Plasmodium* became the subject of more intense scrutiny when *Iain Wilson's* group began molecular genetic studies on two, AT-rich DNA elements recovered from isopycnic density gradient fractionation of total *Plasmodium knowlesi* Sinton & Mulligan DNA (Williamson & al. 1985). One of the high-AT fractions was revealed to be a linear molecule of tandemly-repeated 6 kb elements (Aldritt & al. 1989, Vaidya & al. 1989, Feagin 1992). The other fraction is a circular DNA of length 35 kb (Gardner & al. 1988) equivalent to the cruciform circles observed by Kilejian (1975). Both fractions appeared to represent organellar genomes since they contain rRNAs with eubacterial-like sequences (Gardner & al. 1988; Vaidya & al. 1989; Wilson & al. 1991; Feagin & al. 1991, 1992; Palmer 1992a). Again, it was assumed that both fractions derive from the mitochondrion (Gardner & al. 1988, Aldritt & al. 1989, Vaidya & al. 1989, Bzik 1991). Bipartite mitochondrial genomes are rare but have been described in several plants (Rush & Misra 1985,

SMALL & al. 1989). Such a case in *Plasmodium*, however, failed to gain support since only the linear 6 kb element cofractionated with the mitochondria (WILSON & al. 1992). Mitochondrial origin of the 6 kb element is further supported by the presence of three genes (*coxI*, *coxIII*, and *cob*) for respiratory chain polypeptides (ALDRITT & al. 1989, VAIDAY & al. 1989, FEAGIN 1992), which are typically encoded by mitochondrial genomes in organisms examined thus far (PALMER 1992b, GRAY 1992).

Provenance of the 35kb circle therefore remained mysterious. Two possibilities were evident: either the 35 kb circle was a highly unusual component of the mitochondrion, or it was the genome of an hitherto unrecognised organelle in *Plasmodium*.

Is the 35 kb circle a plastid DNA?

As the characterisation of the 35 kb DNA circle of *Plasmodium falciparum* WELCH progressed, several pieces of evidence spawned a new hypothesis of its provenance – that the 35 kb circle represents a vestigial plastid genome (WILSON & al. 1991, PALMER 1992a).

I. The architecture of the 35 kb circle (Fig. 1), which has an inverted repeat containing the large and small subunit rRNA genes, is reminiscent of plastid DNAs (GARDNER & al. 1991a, 1993). (Organisation of the rRNAs within the repeat is unusual, however, in that the small and large subunits are transcribed divergently; GARDNER & al. 1991a, 1993).

II. The predicted folding of the small subunit rRNA shares a secondary structure feature unique to plastids (GARDNER & al. 1991a).

III. Several tRNAs are contranscribed with the rRNA genes in a precursor transcript, a trait seen also in plastids (GARDNER & al. 1991a, 1994a).

IV. The 35 kb circle carriers an operon of *rpo* genes (Fig. 1) encoding subunits of a bacterial-like DNA-dependent RNA polymerases (GARDNER & al. 1991b, 1994b). The rpoC gene is split (*rpo*C1 and *rpo*C2 in Fig. 1) as it occurs in cyanobacteria and plastids. An *rpo* operon is a universal feature of plastid genomes but has not yet been observed in mitochondrial genomes (PALMER 1992b). The latter use a nuclear-encoded, T3/T7 phage-like, RNA polymerase for transcription (CERMAKIAN & al. 1996).

V. The 35 kb genome includes an open reading frame (ORF 470) with high similarity to a gene from the plastid DNA of red algae, diatoms, and *Cyanophora*, and the cyanobacterium *Synechocystis* 6803 (WILLIAMSON & al. 1994, WILSON 1993). A homologue of ORF 470 also occurs in *Mycobacterium leprae* (PIETROKOVSKI 1994), and as a plastid-targeted nuclear gene in *Arabidopsis* (GenBank accession no. F14083).

VI. The 35 kb genome displays uniparental inheritance (CREASEY & al. 1994) as is typical of plastids and mitochondria.

VII. Phylogenetic analysis of gene sequences from the 35 kb circle group these genes with those of plastids (HOWE 1992; GARDNER & al. 1993, 1994b; DELWICHE & PALMER, pers. comm.).

VII. The 35 kb circle gene map (PREISER & al. 1995, WILSON & al. 1996) suggests the presence of modified S10, str, spc, and alpha operons. This "super

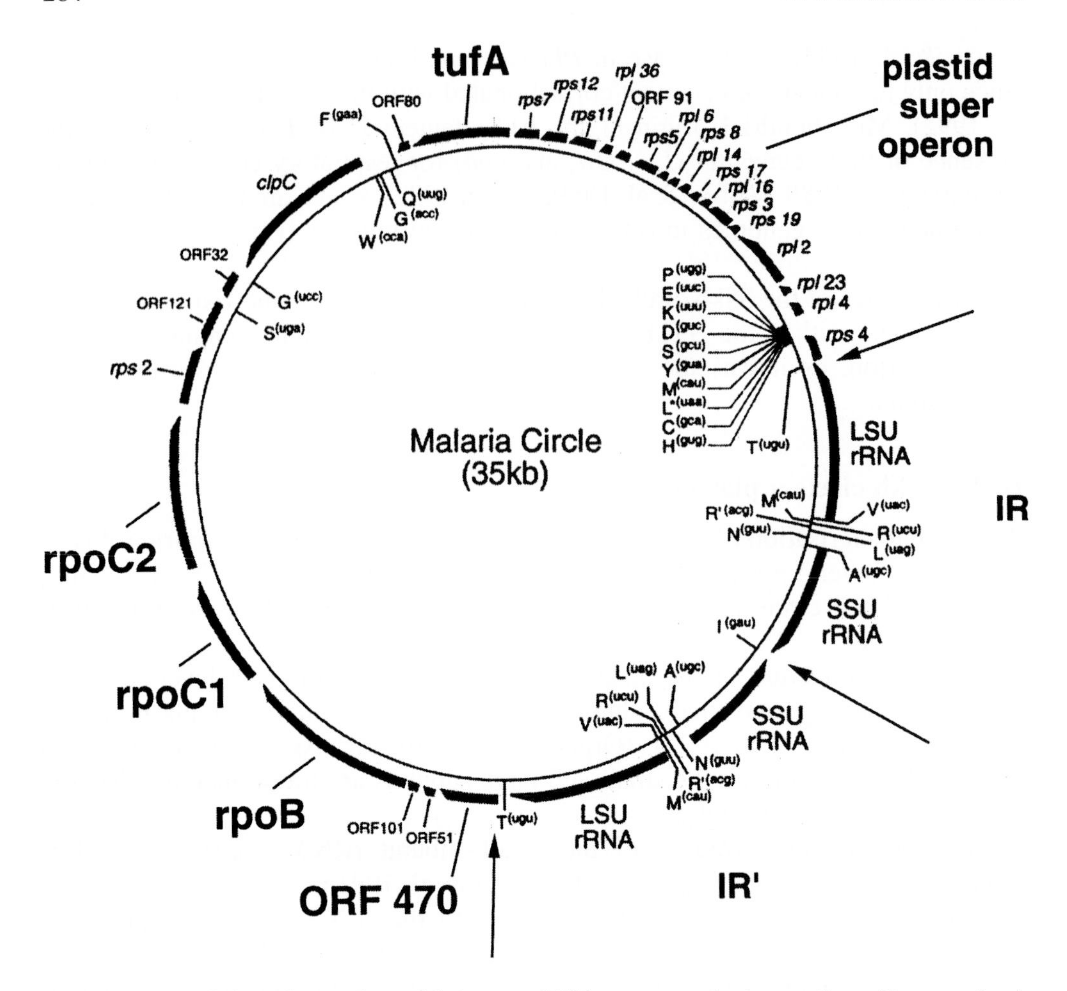

Fig. 1. Map of the *Plasmodium falciparum* 35 kb genome (redrawn from Preiser & al. 1995, Wilson & al. 1996) showing features characteristic of plastid genomes in bold type. The inverted repeats are indicated by arrows and labelled IR and IR′. The small single copy region between the two inverted repeats contains the unique *trnI*. The original map (Preiser & al. 1995) showed *rpo*D (the sigma factor) downstream of rpoC but this gene is more similar to rpoC2, the bipartite β' subunit of DNA-dependent RNA polymerase (Wilson & al. 1996)

operon", comprised of several bacterial operons with numerous genes excised and rearranged (Fig. 2), is recognised as a feature of certain plastids (Reith & Mulholland 1993, Douglas 1994, Harris & al. 1994). Its apparent presence in *Plasmodium* (Fig. 2) is strongly indicative of a common origin with other plastids.

In concert, these data emphatically suggest that *Plasmodium* contains a plastid genome, the inescapable corollary being that it resides in an organelle homologous to plastids.

Evidence for a plastid-like genome in other apicomplexans

The plastid-like genome of *P. falciparum* is the most comprehensively studied, but it does not represent a novelty amongst the apicomplexan parasites. In several other

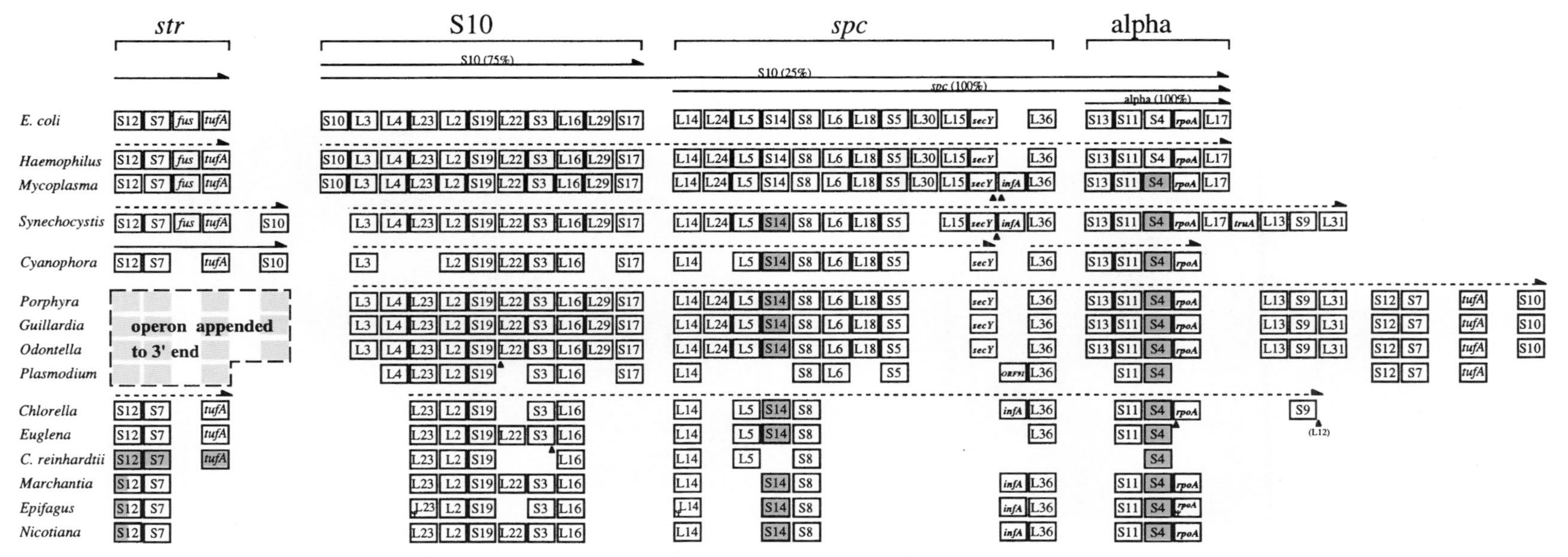

Fig. 2. Conservation of a plastid "super-operon" of ribosomal protein and translation-related genes derived from the *E. coli*-type operons *str*, S10, *spc*, and alpha. Solid arrows indicate known operons (read-through transcription is indicated for the S10 and *spc* operons of *E. coli*), whereas dashed arrows indicate putative plastid operons inferred from conserved arrays of genes on the bacterial or plastid genomes. The plastid "super-operon" resembles the bacterial condition with several modifications. These derivations reflect different plastid alliances and can be listed as follows. (1) Several gene deletions/translocations from the plastid operon have occurred, with the non-green plastid operons being the least derived, *Plasmodium* being intermediate, and the green plastid operons the most depleted. (2) Common to cyanobacteria, cyanelles and plastids is the appendance of the S10 gene to the *str* operon (this state is also seen amongst the archaebacteria and hence is likely to represent the ancestral state). (3) In non-green plastids and *Plasmodium*, the *str* operon is appended to the 3′ end of the alpha operon, whereas in green plastids, and cyanelles it remains unlinked or distal to the main gene cluster. Since, the ancestral arrangement is likely to have been *str*/S10/*spc* (Keeling & al. 1994), the appendance of *str* is likely to represent a synapomorph for red algal, brown algal, and apicomplexan plastids. Northern blot analysis of malaria RNA using *tufA* as a probe identifies extremely long transcripts (Wilson & al. 1996) suggesting that the entire 'super operon' could be transcribed polycistronically in *Plasmodium*. (4) The selective conservation of *secY* in the non-greens, and *infA* in the green plastids indicates a dichotomy between these groups. At this site *Plasmodium* contains an ORF with sequence not obviously similar to either gene. However, ORF91 is conspicuous for its similar size to other *infA* genes. In summary, the *Plasmodium* "super-operon" resembles those of plastids, sharing characters present in both green and non-green plastids.

Shaded boxes indicate genes (or parts of genes) that have been lost from the corresponding operons but are present elsewhere in a given plastid genome. Idiosyncratic gene insertions into the operons are indicated with a triangle (s). Pseudogenes are indicated by the symbol psi (ψ). Gene maps constructed for *Escherichia coli, Cyanophora, Porphyra, Euglena, Chlomydomouas, reinhardii, Marchantia, Epifagus,* and *Nicotiana* from *Harris* & al. (1994) and references therein; *Haemophilus* (Fleischmann al. 1995); *Mycoplasma* (Fraser & al. 1995); *Synechocystis* (Cyanobase 1996); *Odontella* (Kowallik & al. 1995); and *Plasmodium* (Preiser & al. 1995, Wilson & al. 1996)

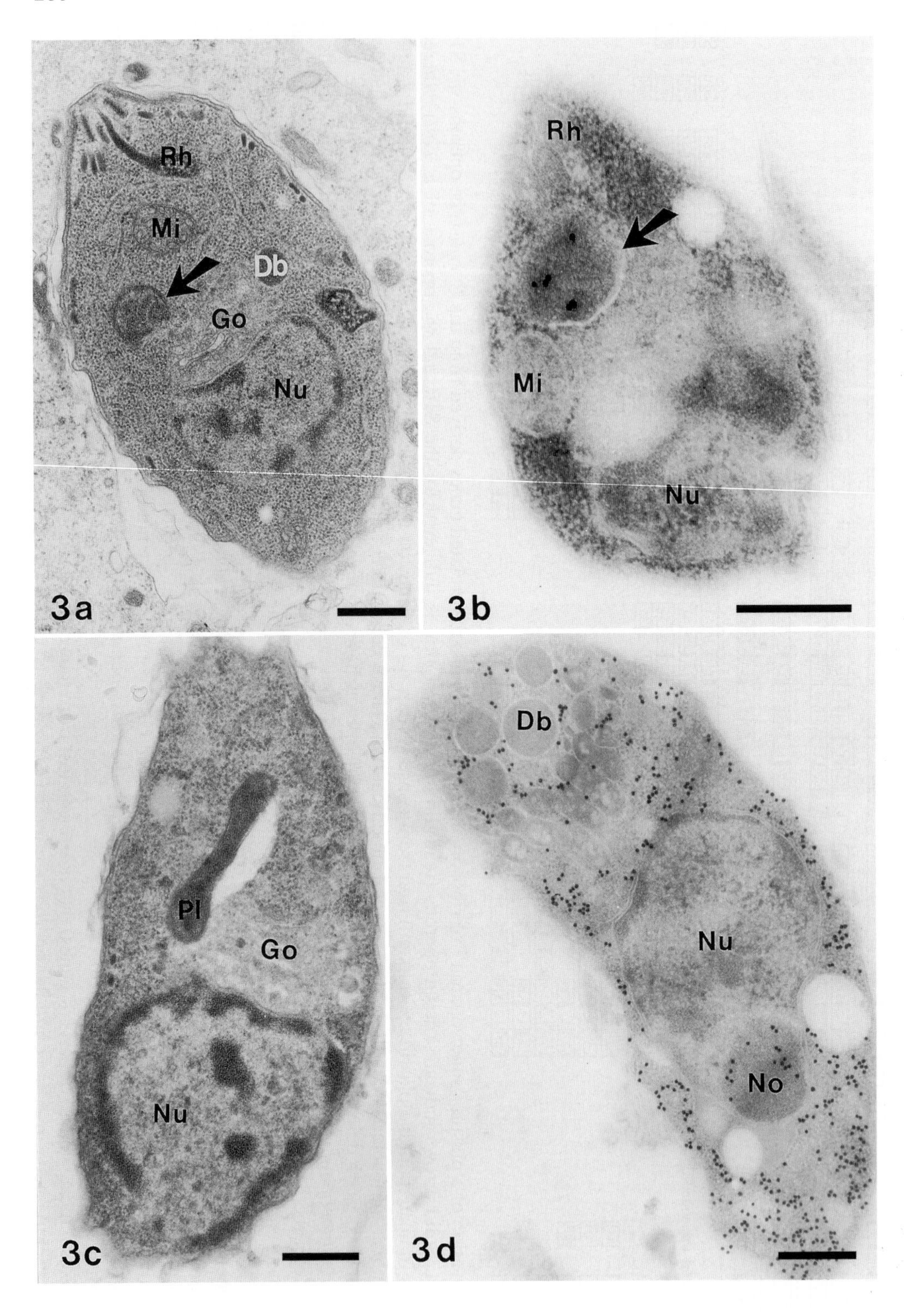
Rh
Mi
Db
Go
Nu
3a
Rh
Mi
Nu
3b
Pl
Go
Nu
3c
Db
Nu
No
3d

Plasmodium species similar circular DNAs are evident by electron microscopy (Kilejian 1975; Dore & al. 1983; Joseph & al. 1989; Williamson & al. 1985, 1996). The causative agent of toxoplasmosis, *Toxoplasma gondii* Nicolle & Manceaux, also contains a circular DNA of size 35 kb that has been partially characterised revealing an inverted repeat containing rRNA genes (Borst & al. 1984, Egea & Lang-Unnasch 1995, Beckers & al. 1995). The coccidian parasite *Eimeria tenella* Raillet & Lucet also contains a circular DNA of about 35 kb with evidence of an inverted repeat structure (Wilson & al. 1993). In addition, the piroplasm *Babesia bovis* (the parasite responsible for babesiosis in cattle) contains a small subunit rRNA gene with high sequence identity to the genes of *Plasmodium* and *Toxoplasma* encoded in the 35 kb circle (Gozar & Bagnara 1993; 1995). From the taxa studied to date, three major groups of the *Apicomplexa* are represented; the haemosporins (*Plasmodium* species), the coccidia (*Eimeria tenella* and *Toxoplasma gondii*), and the piroplasms (*Babesia bovis*) Starcovie. The presence of a plastid-like genome in representatives of each of these groups suggests that it is a likely feature of many other members of the Apicomplexa.

Identification of a plastid in *Toxoplasma*

While the molecular genetic evidence pointed clearly to the presence of a plastid, formal proof of its existence required proof that the 35 kb genome was harboured within a plastid-like structure. We decided to do this using in situ hybridization to localise transcripts from the 35 kb genome (McFadden & al. 1996). *Toxoplasma gondii* was selected as the model cell, and the small subunit rRNA gene transcripts as the target to be localised. The rRNA gene was cloned by PCR (Egea & Lang-Unnasch 1995), and an antisense RNA probe labelled with biotin as previously described (McFadden 1991). The probe was hybridised to ultrathin sections of tachyzoites from *Toxoplasma gondii* RH grown in fibroblasts. The probe labels a small, irregular to ovoid organelle located anterior to the nucleus in the mid-region of the cell (Figs. 3–5). The labelling experiments clearly indicate this organelle as the repository for the 35 kb genome. This result does not eliminate the possibility, though highly unlikely, that the transcripts from the 35 kb genome are produced in another location and are then transported into the organelle.

The plastid is bounded by at least two membranes (Figs. 4a, b, d, 5b, c). Endoplasmic reticulum often associates with the plastid and creates an impression

Fig. 3. Longitudinal sections of tachyzoites of *Toxoplasma gondii* RH strain. *a* Standard electron micrograph of parasite cell inside a fibroblast showing nucleus (Nu), Golgi apparatus (Go), mitochondrion (Mi), rhoptries (Rh), dense body (Db), and the single plastid anterior and eccentric to the nucleus (arrow). *b* Similar section to *a* labelled for transcripts of plastid-like rRNA from the 35 kb genome. Gold label is restricted to an ovoid structure corresponding to the plastid (arrow). *c* Cell about to enter endodyogeny. The plastid (Pl) has elongated and constricted in the middle to assume a dumbbell shape. The dilation of the plastid-bounding membranes is a fixation artefact. *d* Cell labelled with nuclear-encoded, cytoplasmic, 18S, rRNA gene. Transcripts occur throughout the cytoplasm and in the nucleolus (No). No transcripts are observed in the nucleoplasm (Nu) or dense bodies (Db). Bars: 400 nm

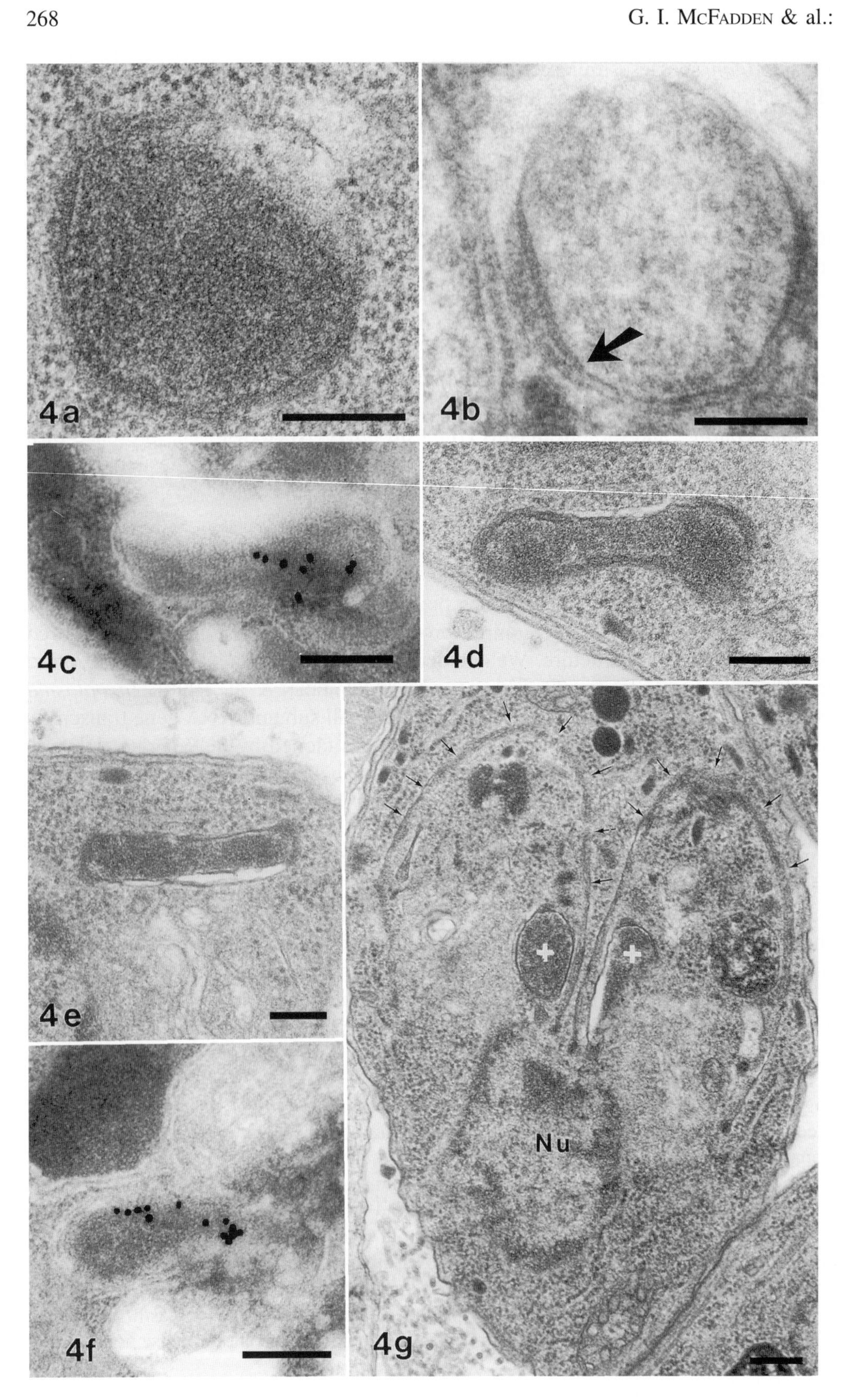
4a
4b
4c
4d
4e
4f
4g
Nu

of four surrounding membranes in some regions (Fig. 5c). The plastid is distinct from the tubulocristate mitochondria (Fig. 3a). The plastid contents are largely homogeneous, the only discernible structures within the organelle are particles (Fig. 4a, d) of comparable size to the 70S ribosomes of plastids, mitochondria and bacteria. The occurrence of ribosomes is consistent with the plastid genome encoding components of ribosomes (Preiser & al. 1995) and adds to the evidence of the plastid containing the machinery to express its information content.

During the cell cycle of *Toxoplasma*, the plastid undergoes conformational and positional changes (Fig. 6). In the interphase condition, each cell contains one plastid that lies in an eccentric position just anterior to the nucleus (Figs. 3a–c, 6; Ogino & Yoneda 1966, Scholtyseck & Piekarski 1965, Sheffield & Melton 1968). When in this position, the plastid is largest in diameter, though the outline is somewhat irregular. When the meront is about to enter endodyogeny (the process by which two daughter cells are produced within a mother cell), the plastid metamorphoses to a cylindrical shape (Figs. 3c, 4c-f). At this stage the contents become more condensed and the inner membrane (s) take on a darker, thicker appearance (Figs. 4e, 5b). This plastid phase was termed the Hohlzylinder (a German word meaning hollow cylinder) by Scholtyseck (Scholtyseck & Piekarski 1965). The Hohlzylinder-stage plastid moves to occupy a more central position directly anterior to the nucleus (Figs. 3c, 6), whereupon it begins to constrict in the middle and assumes a configuration like a dumbbell (Figs. 4c, d, 5b, c). We believe this stage to be a division phase as the next observable images show two plastids, now reverted to the more ovoid-shaped interphase condition, being partitioned into the nascent daughter merozoites by the forming apical complexes and alveoli (Fig. 4g).

Mystery organelles in other apicomplexans

The identification of a plastid in *Toxoplasma* (McFadden & al. 1996) demonstrates the presence of a previously unidentified third genetic compartment in an apicomplexan. As the *Toxoplasma* plastid is the repository for the 35 kb genome, we can anticipate that in *Plasmodium* a similar organelle, that harbours the 35 kb genome, will be identified. Electron microscopic studies of *Plasmodium* cells contain intriguing reports of unidentified organelles. Termed 'spherical bodies' (Aikawa 1966), 'double-walled vesicles' (Hackstein & al. 1995), or 'vacuoles' (Rudzinska & Vickerman 1968), these structures have two or more limiting

Fig. 4. Plastid morphology throughout the cell cycle. *a* Interphase plastid showing ovoid shape, granular contents, and two bounding membranes. *b* Interphase plastid showing two bounding membranes, the inner of which appears to bifurcate (arrow). *c* Dumbbell-shaped plastid labelled for plastid-like 16sRNA. *d* Similar dumbbell-shaped plastid to that shown in *c* showing two membranes and 18nm particles within. *e* Plastid at the 'Hohlzylinder' stage showing dense contents and thick inner membrane(s). *f* Plastid at the 'Hohlzylinder' stage labelled for plastid-like 16sRNA. *g* Cell undergoing endodyogeny (compare with Fig. 6). The nucleus (Nu) has not yet divided but the daughter apical complexes and alveolar layers are forming (arrows). The plastid has divided and a daughter plastid (+) occurs in each forming daughter cell. Bars: 200 nm

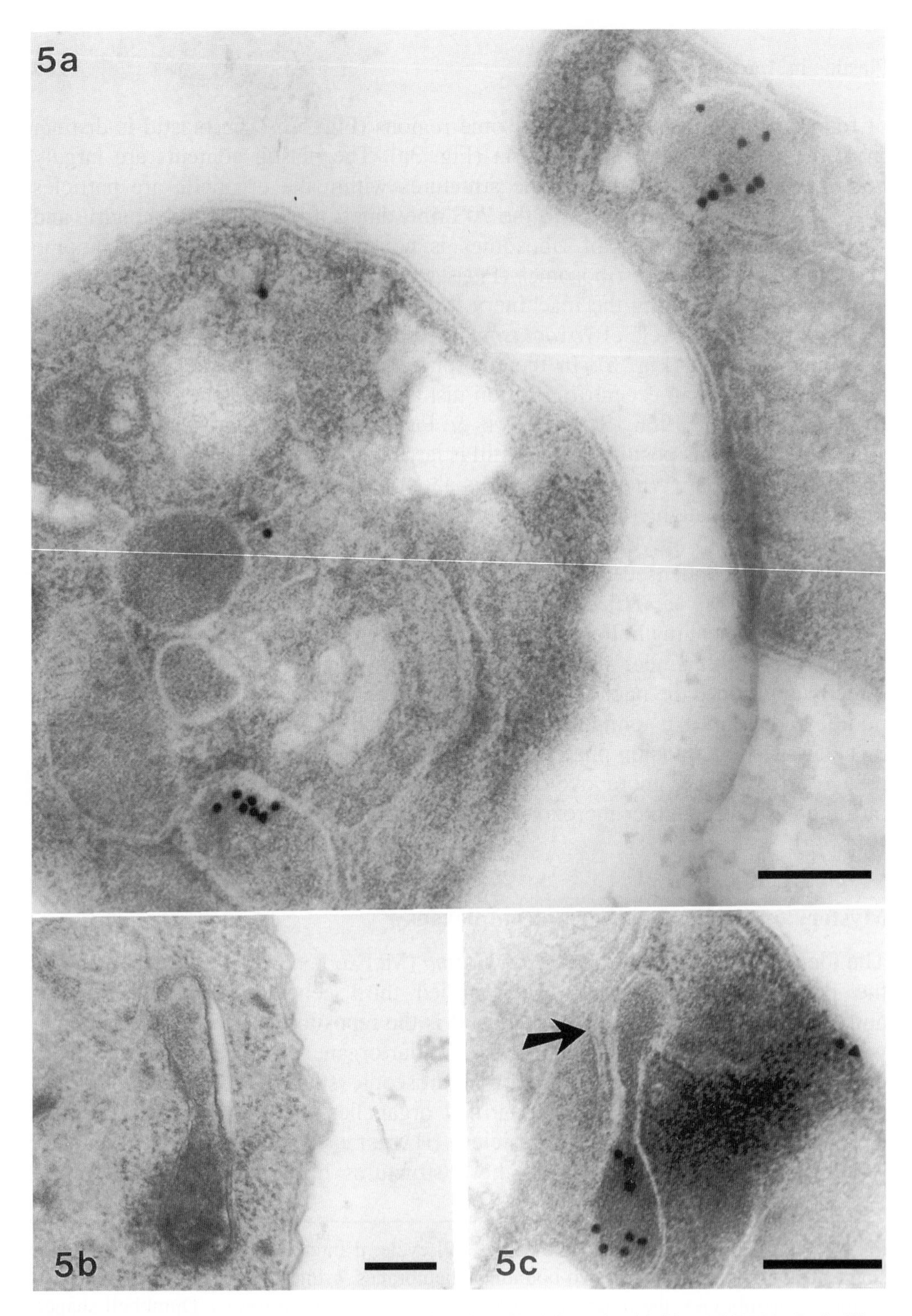

Fig. 5. *a* Tangential sections of two parasites labelled for plastid-like 16sRNA. Labelling is confined to the ovoid plastids in each cell. *b* Dumbbell-shaped plastid showing two bounding membranes. The outer membrane has artificially pulled away to on the right hand side to reveal the thick inner membrane. *c* Dumbbell-shaped plastid labelled for plastid-like 16sRNA. The labelling is restricted to one half of the dumbbell. Endoplasmic reticulum (arrow) is appressed to the left side of the upper half of the dumbbell giving an impression of four surrounding membranes but in the lower half of the dumbbell there are clearly only two membranes (represented as white lines in this non-osmicated preparation). Bars: 200 nm

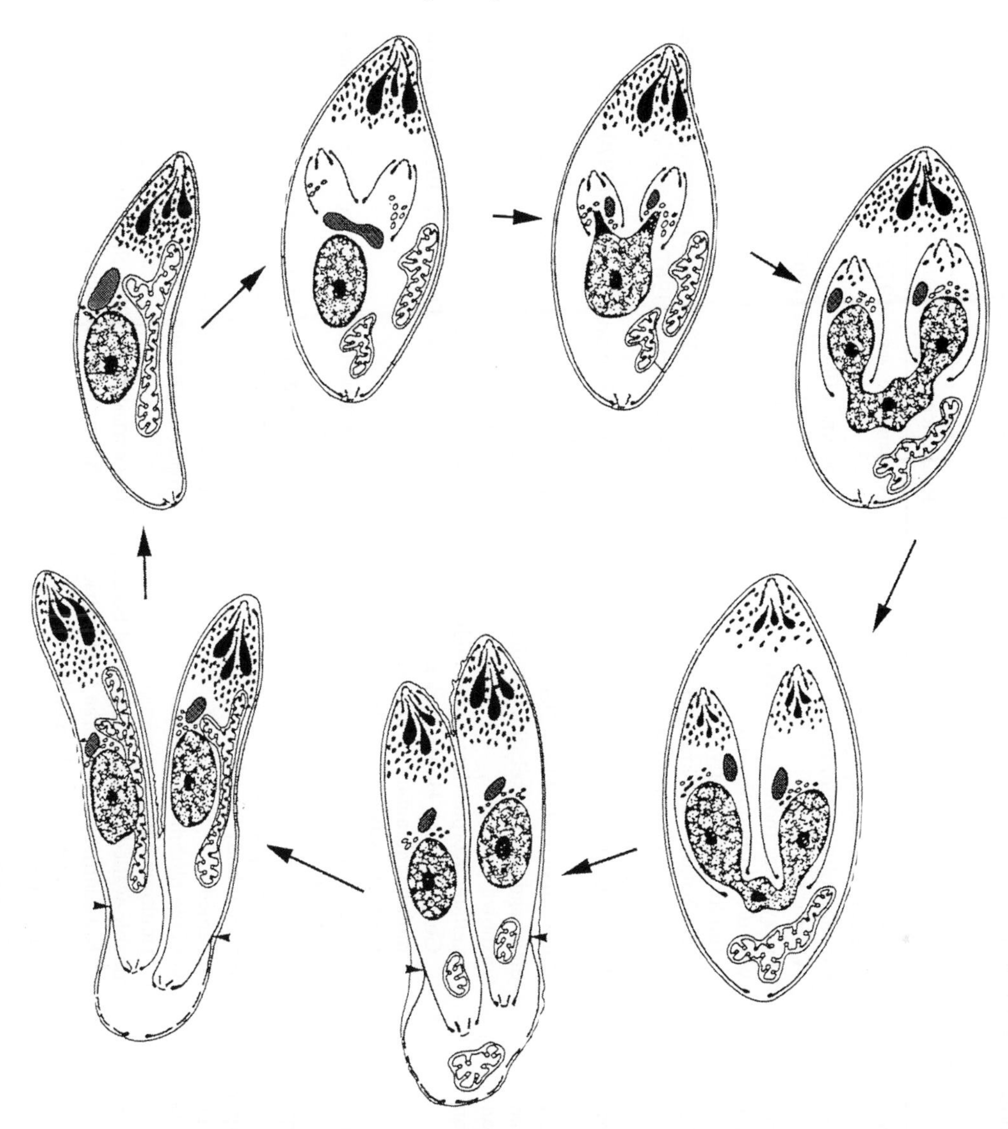

Fig. 6. Scheme showing stages of endodyogeny in *Toxoplasma* (redrawn from Scholtyseck 1979). At interphase, the single ovoid plastid lies anterior to the nucleus. At the onset of endodyogeny the plastid constricts in the middle to form a dumbbell shape. This constricted form is thought to undergo binary fission and the daughter plastids resume the ovoid shape and are partitioned semi-conservatively into the forming daughter cells

membranes and granular contents. The unidentified structures in avian-*Plasmodium* parasites are typically appressed to the mitochondrion (Aikawa 1966). The developmental origin of these 'vacuoles' or 'spherical bodies' is not known, but they are invariably present (Rudzinska & Vickerman 1968), and are thus strong candidates to contain the 35 kb circular genome. In some micrographs of *Plasmodium*, these structures are dumbbell-shaped (e.g., Rudzinska & Vickerman 1968: Figs. 2, 6), reminiscent of the apparently dividing plastids that we observe at the onset of endodyogeny in *Toxoplasma*.

Table 1. List of 'mystery organelles' in the *Apicomplexa* literature. electron microscopists report a multi (occasionally triple) membrane-bound compartment-often with granular (ribosome-like?) contents – as a consistent cell feature in representatives of each major apicomplexan category. The various 'mystery organelles' are of unknown significance but are potential homologues of the plastid-like structure of *Toxoplasma* and *Plasmodium*

Organism	Host(s)	Category	'Mystery organelle'	References
Dehornia	annelids	eugregarine	vésicule plurimembranaires (multimembraned vesicle)	Porchet-Hennere (1972)
Selenidium	annelids	eugregarine	vacuoles plurimembranaires	Schrevel (1971)
Coelotropha	annelids	eugregarine	V_3 (vacuole type 3)	Vivier & Hennere (1965)
Haemogregarina	vertebrates	haemogregarine	Hohizylinder (hollow cylinder)	Siddal (1992)
Hepatozoon	reptiles	haemogregarine	vésicule plurimembranaires (multimembraned vesicle)	Vivier & al. (1972)
Eimeria	various	coccidian	grosse Vakuole mit kräftiger Wandung (big vacuole with strong wall); paranuclear body	Roberts & al. (1970), Scholtyseck & Piekarski (1965), Colley (1967)
Aggregata	cephalopod/crustacean	coccidian	dark round bodies	Porchett-Hennere & Richard (1970), Bardele (1966)
Sarcocystis	vertebrates	coccidian	Golgi adjunct; double-walled organelle	Scholtyseck (1979), Hackstein & al. (1995)
Globidium	sheep	coccidian	dickwandiger Vesikel (thick-walled vesicle)	Scholteyseck (1979)
Besbijtia	vertebrates	coccidian	unknown structure	Sheffield (1966)
Toxoplasma	mammals/birds	coccidian	lamellärer Körper (lamellated body), Golgi adjunct, Hohizylinder, multimembranous organelle; rundlich-ovale Organelle	van der Zypen & Piekarski (1967), sheffield & melton (1968), Dubremetz 1995, Wildführ 1966
Frenkelia	mammals (vole)	haemosporin	vesicle with granular contents/Golgi adjunct	Scholtyseck (1979), Kepka & Scholtyseck (1970)
Plasmodium	lizards, birds, primates	haemosporin	double-walled organelle; vacuole; spherical bodies; 'plastid'	Hackstein & al. 1995, Rudzinska & Vickerman (1968), Aikawa (1966)
Haemoporoteus	vertebrates	haemosporin	spherical body	Sterling & De Giusti 1972
Leucocytozoon	birds	haemosporin	unnamed	Aikawa & al. (1970)
Anthemosoma	rodent	piroplasm	vésicule plurimembranaires (multimembraned vesicle)	Vivier & Petitprez (1969)
Babesia	cattle	piroplasm	double-walled organelle; spheroid bodies; Konzentrischer Membran (concentric membranes); double membraned structure	Hackstein & al. 1995, Scholtyseck 1979, Buttner (1968), Rudzinska & Trager (1962)
Theileria	vertebrates	piroplasm	Konzentrische Membran (concentric membranes); double walled organelle	Buttner (1967), Melhorn & Schein 1984

What of the other apicomplexans? The molecular genetic data has already indicated that a plastid occurs in three major apicomplexan groups, and examination of the ultrastructural literature reveals a plethora of tantalising evidence for a plastid in many apicomplexans. In fact, the apicomplexan plastid has apparently been staring us in the face for decades. For instance, the cover of SCHOLTYSECK's "Fine structure of parsitic *Protozoa*" (1979) depicts *Eimeria perforans* LEUCKART with what is almost certainly a plastid to the right of the nucleus. Furthermore, the ultrastructural literature is littered with references to 'mystery organelles' in a wide selection of apicomplexans (Table 1). These 'mystery organelles', which go by various names, were observed to be a consistent cellular feature (consult references in Table 1). The origin and function of these structures is not known, and it is not certain that they are even homologues. Nevertheless their conspicuous presence prompted PORCHET-HENNERE (1972), in reviewing the shreds of information concerning these 'mystery organelles' in apicomplexans, to comment that they are 'un organite d'un grand intérêt.' We predict that many of these 'mystery organelles' will eventually be identified as plastids.

It is encouraging to note that electron microscopists did not overlook the plastid. Indeed, they made careful and accurate observations, including cell cycle stages equivalent to those that we observed in *Toxoplasma* (e.g., OGINO & YONEDA 1966, SHEFFIELD & MELTON 1968, SHEFFIELD 1966, KEPKA & SCHOLTYSECK 1970, SCHOLTYSECK 1979). However, in the absence of molecular genetic evidence for a plastid, it never occurred to any of them that the structures they observed in these 'protozoans' could be a remnant photosynthetic organelle. Hence, they were forced to grope for descriptive terms to name what they observed (Table 1). SIDDALL (1992) advocated adoption of SCHOLTYSECK's term 'Hohlzylinder' for the apicomplexan plastid, but we prefer the recommendation of PALMER (1992a) to simply refer to it as a plastid.

It is noteworthy that two of the simplest, and arguably most primitive apicomplexans, have not yet been observed to contain a 'mystery organelle'. The bivalve parasite *Perkinsus*, and the hetorotrophic flagellate *Colpodella* (previously *Spiromonas*) have been examined in some detail electron microscopically, but are not reported to contain any organelle that might be compared to a plastid (PERKINS & MENZEL 1967; PERKINS 1969, 1976; MACDONALD & DARBYSHIRE 1977; BRUGEROLLE & MIGNOT 1979; FOISSNER & FOISSNER 1984; AZEVEDO 1989; MYL'NIKOV 1991; KLEINSCHUSTER & al. 1994; SIMPSON & PATTERSON 1996). Considering the putatively basal position of these two flagellates in the apicomplexan phylogenetic tree, the question of whether these two parasites contain a plastid-like organelle is one of great importance to our understanding of the evolution of the *Apicomplexa*. One can hypothesize three alternative explanations: (1) *Perkinsus* and *Colpodella* do contain a plastid but it has been overlooked, (2) *Perkinsus* and *Colpodella* represent lineages that diverged prior to the acquisition of a plastid by later-evolving apicomplexans, or (3) the original progenitor apicomplexan possessed a plastid, and *Perkinsus* and *Colpodella* have secondarily lost the organelle. Until our knowledge of the presence or absence of a plastid in the spectrum of apicomplexans is improved, it is not possible to favour any one of these scenarios.

Cyclospora – a green apicomplexan?

Cyclosporosis is a diarrhoeal disease caused by the human intestinal pathogen *Cyclospora*. Molecular phylogenetic study (Relman & al. 1996) has confirmed earlier classification based on sporulation data (Ortega & al. 1993) indicating that *Cyclospora* is a coccidian closely related to *Eimeria*. Several reports of patients diagnosed with cyclosporosis describe round, pigmented organisms in the diarrhoeic stools of travelers and AIDS patients (Long & al. 1990, 1991; Hart & al. 1990; Hoge & al. 1993). The organisms are green, autofluorescent, and contain an organelle with thylakoid-like structures claimed to be similar to plastids (Long & al. 1990). Unfortunately, the identification of these orgainsms as *Cyclospora* (Long & al. 1990, 1991; Hart & al. 1990; Hoge & al. 1993) is at best equivocal, the key diagnostic feature being the 'spore' diameter. Without either sporulation data or some molecular diagnosis we cannot be certain that these round, pigmented organisms associated with these diarrhoeral diseases are really *Cyclospora*, rather than the green alga *Prototheca* (Kaplan 1978) for instance which is also associated with diarrhoea. It would be useful to examine organisms unequivocally identified as *Cyclospora* (Relman & al. 1996, Ortega & al. 1993) for evidence of a plastid.

Function of a plastid in apicomplexans

To date, molecular genetic and ultrastructural data provide a congruence of evidence for a plastid in the three apicomplexan groups; coccidia, haemosporins and piroplasms. These three lineages are all obligate endocellular parasites and, apparently, non-photosynthetic. They are estimated to have diverged from their last common ancestor (which presumably possessed a plastid) between 300 million and 800 million years ago (Escalante & Ayala 1995). The apicomplexan plastid is therefore extremely ancient and has probably been non-photosynthetic for a considerable time. In addition to photosynthesis, plant plastids are responsible for a range of biosynthetic processes plus storage of carbohydrates, proteins, and pigments (Whatley 1977). The non-photosynthetic portions of plants always retain the plastids, usually in a modified form (Whately 1977).

Conversion from an autotrophic lifestyle to a parasitic one has occurred numerous times during plant and algal evolution (Kuijt 1969, Goff 1982). Interestingly, parasitic plants and algae retain their plastids (Walsh & al. 1980, Goff 1982), although they are much reduced (De Pamphilis & Palmer 1989, 1990; Wolfe & al. 1992). A similar situation is observed in several flagellates that have converted secondarily to heterotrophy. Again these organisms (e.g., *Astasia longa* Pringsheim and *Polytoma*) retain a plastid complete with a reduced plastid genome (Harris & al. 1994). In either case, it is apparent that the organelle has some indispensable, but as yet unidentified, cellular function other than photosynthesis (De Pamphilis & Palmer 1990, Goff 1982, Harris & al. 1994, Wolfe & al. 1992). Hence, it seems likely that the apicomplexan parasites retain a plastid as an evolutionary hangover from a former autotrophic lifestyle (Palmer 1992a). Some potential roles for the plastid in parasites are discussed below.

Haem synthesis? Parasitic plants are suggested to retain the plastid because in plants it is the only known site for haem synthesis (Howe & Smith 1991, Harris & al. 1994, Wolfe & al. 1992). Plants, most algae, and cyanobacteria are unusual in that they use a haem synthesis pathway originating from glutamate rather than the classic Shemin pathway originating from glycine (Kannangara & al. 1988). It is likely that this pathway was introduced into the plant/algal lineage by the endosymbiosis of a cyanobacterium, and that it replaced the classic Shemin pathway, which presumably pre-existed in the host cell prior to the establishment of the endosymbiosis. Since the apicomplexan plastid derives from the same endosymbiotic event, the glutamate-based pathway for haem synthesis would most probably also have been introduced into the apicomplexan lineage.

Despite being inundated with haemoglobin from the host blood cells, *Plasmodium* is apparently unable to use this haem in its metabolism (Raventos-Suarez & al. 1982). Indeed, the surfeit of decomplexed haem is one of the parasite's major problems and its neutralization is a key part of its metabolism (Slater & Cerami 1992, Olliaro & Goldberg 1995). It is therefore ironic that de novo haem synthesis is essential for parasite viability (Surolia & Padmanaban 1992). However, since *Plasmodium falciparum* is apparently able to synthesis the haem precursor δ-aminolevulinate via the Shemin pathway from glycine and succinyl CoA (Wilson & al. 1996, Surolia & Padmanaban 1992), which is different to the plastid and cyanobacterial pathway originating with glutamate (Kannangara & al. 1988), it seems unlikely that haem synthesis in the plastid is its raison d'etre. Nevertheless, haem synthesis could be occurring in the plastid and we note that the *Plasmodium* plastid genome encodes $tRNA^{Glu}$ (Preiser & al. 1995), a requisite cofactor for synthesis of tetrapyroles in plants, algae and select bacteria (Kannangara & al. 1988). The retention of *trnE* in parasitic plants (Wolfe & al. 1992), and albino plants with otherwise severely depleted plastid genomes (Harada & al. 1992), attest to its indispensability. The *Plasmodium* plastid $tRNA^{Glu}$, which shares an intriguing T-loop modification with the plant plastid $tRNA^{Glu}$ (Preiser & al. 1995), could be used for both translation and synthesis of δ-aminolevulinate from glutamate. Although the complete sequence of the *Plasmodium* plastid genome reveals no enzymes for haem synthesis (Wilson & al. 1996), it has been established in plants that most of the enzymes in the haem synthesis pathway are imported into plastids (e.g., Grimm 1990, Mock & al. 1995, Lim & al. 1994) so haem synthesis in the apicomplexan plastid cannot yet be discounted. Herbicides, such as acifluorfen, oxadiazon and diphenyl ethers, which inhibit the plastid enzyme protoporphyrinogen oxidase, can be exploited as tools to investigate haem synthesis in apicomplexans.

Fatty acid synthesis? Another function attributable to plant plastids is fatty acid synthesis (Dennis & Miernyck 1982, Dennis 1989, Douce & al. 1989). The first committed step in de novo fatty acid synthesis is the conversion of acetyl CoA to malonyl CoA by the enzyme acetyl CoA carboxylase, which in plants and certain algae is a chloroplast enzyme (Dennis & Miernyck 1982, Dennis 1989). The enzyme has four subunits. One subunit (accD) is encoded by the chloroplast genome of broad leaf plants and the other subunits are presumably nuclear-encoded. In certain algae three subunits of acetyl CoA carboxylase (*accA*, *accB*, *accD*) plus two other downstream components of fatty acid synthesis (*acp*P=acyl carrier protein, *fab*H = β-ketoacyl-acyl carrier protein synthase III) are chloroplast-encoded (Reardon & Price 1995). The *Plasmodium* plastid genome does not encode any ACC carboxylase subunits or other fatty acid synthesis machinery (Wilson & al. 1996), but an enzyme could be targeted into the plastid from the nucleocytoplasmic compartment, as apparently occurs in land plants (Konishi & Sasaki 1994). A wide selection of herbicides in the 'fops' and 'dims' categories inhibit plant chloroplast acetyl CoA carboxylase and could be tested with *Plasmodium* and *Toxoplasma*.

Amino acid synthesis? Plants synthesize essential amino acids in their chloroplasts (Reardon & al. 1989, Lea & al. 1989) so this metabolic process is another potential raison d'être for the vestigial plastid in apicomplexan parasites. Again, herbicides (e.g., glyphosate and sulfometuron methyl) that inhibit these functions in plant plastids could be tested on apicomplexans.

Electron transport chain? One line of evidence perhaps bears on the possible plastid function. The triazine class of herbicides, which target the D1 protein (PsbA) from the oxygen evolving complex of photosystem II, have anti-coccidial properties (Melhorn & al. 1984, Harder & Haberkorn 1989, Lindsay & al. 1995). A very preliminary report of a *psb*A gene from the coccidian *Sarcocystic* provides a possible rationale for the herbicide's anti-coccidial activity. It has been suggested that an electron-transport chain might exist and that could provide oxygen radicals facilitating entry of the parasites into host cells (Hackstein & al. 1995). As yet, there is no direct evidence for the plastid being the triazine target, and the observation that triazines also inhibit nematodes, microsporidians, myxozoans, and ciliates (Harder & Haberkorn 1989) – none of which are thought to harbour a plastid – leaves open the possibility that these compounds have a non-plastid target. Further characterisation of the putative *psbA* gene from *Sarcocystis* is required; particularly as the complete sequence of the *Plasmodium* 35 kb plastid genome did not reveal a *psbA* gene (Preiser & al. 1995).

The *P. falciparum* plastid genome (the best characterised) contains a total of 66 putative genes with 10 genes duplicated in the inverted repeat and one gene (*trn*M) existing in triplicate (Preiser & al. 1995; Wilson & al. 1996, accession Nos. X95275 and X95276). The majority of these genes (46) have known functions involved in the expression of the genome's information content (transcription and translation). Differential transcription, during the erythrocyte stages, of elements involved in both transcription and translation (*rpoB*, *rpoC*, *ssu*, and *Isu*,) suggests that this expression machinery is functional in the plastid (Feagin & Drew 1995) All but one (*clpC*) of the remaining nine genes have no ascribed function and represent open reading frames (ORFs). Initial efforts to establish a raison d'être for the parasite plastid will no doubt focus on these ORFs. However, the answer to the puzzle need not necessarily lie in the plastid genome. Based on our knowledge of the cooperation between nuclear and plastid genomes in plant and algal cells, it is reasonable to expect that the parasite nucleus will encode proteins that are synthesised in the cytoplasm and targeted into the apicomplexan plastid. Thus, one or more nuclear-encoded proteins may perform the essential function(s) that have ensured the plastids existence for hundreds of millions of years. The plastid may simply persist in order to provide a segregating, subcellular compartment in which a vital process occurs.

Though highly unlikely, it is remotely possible that the plastid has no vital function and is purely a vehicle for a selfish DNA. The plastid genome would then be a parasite of a parasite.

Origin of the plastid

The presence of a plastid implies that apicomplexans probably derive from photosynthetic ancestors, though other scenarios involving secondary acquisition of the plastid (cf. McFadden & Gilson 1995, McFadden 1993, Wilson & al. 1994, Palmer & Delwiche 1996) certainly cannot be excluded at present. Phylogenetic analysis of apicomplexan plastid genes yields conflicting results, with euglenoid plastids (Egea & Lang-Unnasch 1995; Gardner & al. 1994a,b; Howe 1992), red algal plastids (Williamson & al. 1994, Jeffries & Johnson 1996), and green algal plastids (Palmer & Delwiche 1996) being identified as the closest relatives of the apicomplexan plastid. The phylogenies are inferred from various genes but, with the exception of tufA trees (Palmer & Delwiche 1996), the taxon sampling is limited and the high AT content of the apicomplexan plastid DNA makes the analyses particularly difficult. The only firm conclusion that can be made at this point is that the apicomplexan genes are a part of the general plastid radiation.

The number of membranes surrounding plastids has been a useful indicator of plastid origin. Green algae, plants, red algae, and cyanophytes (which all have two bounding membranes) are suggested to have originated from a single primary endosymbiosis (Cavalier-Smith 1995b). A primary endosymbiosis involves the engulfment and retention of a prokaryote by a eukaryote. The four membrane plastids of cryptomonads, chlorarachniophytes, heterokont algae, and haptophyte algae arose through secondary endosymbioses, with at least two, and perhaps four, independent events implicated (cf. Cavalier-Smith & al. 1996 for a recent discussion). Secondary endosymbiosis occurs subsequent to primary endosymbiosis and involves a phagotrophic eukaryote engulfing and retaining a photosynthetic eukaryote (McFadden & Gilson 1995, Palmer & Delwiche 1996). Two groups of algae, euglenoids and dinoflagellates, have three plastid membranes. Their plastids could have either primary or secondary origins (Cavalier-Smith 1995b). Regrettably, the number of membranes surrounding the apicomplexan plastid is still equivocal. We observe only two membranes in many sections in which the membranes are cut perpendicularly. However, the inner membrane is particularly thick and dark (especially in the Hohlzylinder stage of *Toxoplasma*) and it sometimes appears to consist of two membranes where it diverges inwards from the outer membrane (e.g., Fig. 4b). The situation is further confounded by the occasional association of ER with the plastid. Appression of ER to the plastids of plants is common (Whatley 1977) and select images of *Toxoplasma* (e.g., D*ubremetz* 1995: Fig. 1) would appear to show the same phenomenon. Indeed, the almost total envelopment of the plastid by ER is sometimes observed to occur at the onset of division in the coccidian *Sarcocystis* (Radchenko 1987: Fig. 5a) and dividing plastids often appear to have four surrounding membranes (Radchenko 1987).

Ultrastructural chracterisation of plastids in other apicomplexans should clarify the number of membranes. If it is proven that only two membranes bound the plastid, this would suggest a primary origin, perhaps shared with red algae, green algae, plants, and cyanophytes. If more than two membranes are identified, a secondary endosymbiotic acquisition is possible. The lack of any gene sequences from plastids of dinoflagellates (the sister lineage to *Apicomplexa* on the basis of nuclear gene trees and ultrastructural grounds) obviates attempts to resolve whether or not the apicomplexans are ancestrally a plastid-carrying lineage.

Significance of the plastid

Previously, the only known plastid-containing organism responsible for human disease was *Prototheca* – a green alga associated with bursitis, lesions, and sprue (Kaplan 1978). Apicomplexans are responsible for numerous diseases in both humans and domesticated animals. The malarial agent, *Plasmodium*, is the most dire for humans. An estimated 500 million people are infected with *Plasmodium* annually, and each year the disease kills over 2 million people (The World Health Organisation Report 1996). Based on the high infant mortality, some demographers estimate that half the people who ever lived died of malaria. Parasites are transmitted by mosquito vectors of the genus *Anopheles. Plasmodium* cells ingested by the blood-sucking mosquito enter the insect gut, where they undergo the sexual part of the life cycle. Parasites are transferred to new human hosts when the mosquito takes subsequent blood meals. The parasite then invades and multiplies within the human host's liver and red blood cells. Proliferation of the parasites in the blood leads to cyclic lysis of the blood cells. Upon cell rupturing, toxins are released provoking cycles of fevers and chills, and the increase in levels of parasitized blood cells can ultimately result in fatal complications, particularly in the most virulent species, *P. falciparum*.

No feasible methods of treatment for malaria were available in the western world before the 1630s when Spanish missionaries discovered an extract from the bark of the Sourth American *Cinchona* tree. This extract, known as Jesuit's bark, contains the alkaloid quinine. Current treatment and prophylaxis centres on the drug chloroquine (a derivative of quinine), but parasites have developed resistance to chloroquine, particularly in Southeast Asia. Fears of similar development of resistance have been expressed concerning the recent worldwide distribution of two other quinine derivatives, halofantrine and mefloquine. Moreover the quinine derivatives are toxic to some patients (e.g., Cook 1995, Bradley & Warhurst 1995) further compromising their utility. Hopes of combating chloroquine resistance now rest with another drug group (Vanhensbroek & al. 1996, Hien & al. 1996), artemisin derivatives, which are sesquiterpine lactones from *Artemisia annual*. This herb has been used for centuries in traditional Chinese medicine to treat malaria and fever. A number of drugs (antibiotics) inhibiting either transcription or translation in prokaryote-like systems can block mitochondrial and/or plastid function and could potentially inhibit parasite growth. Interestingly, a number of these drugs are already prescribed against apicomplexan diseases. For instance, doxycycline (a tetracycline group antibiotic) is prescribed as an antimalarial (Bradley & Warhurst 1995), rifampicin (Pukrittayakamee & al. 1994) and thiostrepton (McConkey & al. 1997) inhibit malarial growth, spiramycin is prescribed for toxoplasmosis, and chloramphenicol is effective against cryptosporidiosis (Woods & al. 1996). Prior to the discovery of the plastid in these parasites, there was no rationale for the efficacy of some of these antibiotics. It was presumed that they acted against the parasite mitochondria, but it now seems likely that several of them (particularly thiostreptin, rifampicin and chloramphenicol) could target the plastid since they have no target in mitochondria (McConkey & al. 1997, Jeffries & Johnson 1996).

The need for more effective treatment of malaria is urgent. Tests of vaccines based on synthetic peptides and sporozoite antigens produced with recombinant DNA technology offer promise, but trials so far have not yet led to a practical vaccine (Brown 1994, Doolan & Good 1993). The development of any new avenues for treatment would be a welcome alternative. A plastid potentially offers an excellent target for chemotherapeutic agents. At least one herbicide (Harder & Haberkorn 1989, Melhorn & al. 1984, Lindsay & al. 1995) and several drugs inhibiting plastid transcription or translation have already been demonstrated to have anti-apicomplexan properties (Strath & al. 1993, Beckers & al. 1995, Pukrittayakamee & al. 1994, Woods & al. 1996). While it remains to be demonstrated that these or any other agents specifically inhibit processes in the parasite plastid, we are optimistic that useful anti-apicomplexan agents could exist among the armamentarium of herbicides and antimicrobials already in existence. Whether any of these can be adapted for use as treatments remains to be seen.

The tremendous success of antibiotics directed against bacteria stems from their specificity. Because they target prokaryote-specific functions, they generally have few contra-indications. Protozoan, fungal and viral diseases, on the other hand, have been far more difficult to treat since they share a great deal of the host's metabolic processes. The identification of a plastid in apicomplexans presents a unique, and potentially exploitable, difference between host and parasite.

References

Aikawa, M., 1966: The fine structure of the erythrocytic stage of three avian malarial parasites, *Plasmodium fallax, P. lophyrae* , and *P. cathemerium*. – Amer. J. Trop. Med. Hyg. **15**: 449–471.

– Hff, C.G., Strome, C. P. A., 1970: Morphological study of macrogametogenesis of *Leucocytozoon simondi*. – J. Ultrastruct. Res. **32**: 43–68.

Aldritt, S. M., Joseph, J. T., Wirth, D. F., 1989: Sequence identification of cytochrome b in *Plasmodium gallinaceum*. – Molec. Cell. Biol. **9**: 3614–3620.

Azevedo, C., 1989: Fine structure of *Perkinsus atlanticus*, new species (*Apicomplexa, Perkinsea*), Parasite of the clam *Ruditapes decussatus* from Portugal. – J. Parasitol. **75**: 627–635.

Baldauf, S. L., Palmer, J. D., 1993: Animals and fungi are each other's closest relatives: congruent evidence from multiple proteins. – Proc. Natl. Acad. Sci. USA **90**: 11558–11562.

Bardele, C. F., 1996: Elektronmikroskopische Untersuchungen an dem Sporozoon *Eucoccidium dinophili* Grell. – Z. Zellforsch. **74**: 559–595.

Barta, J. R., Jenkins, M. C., Danforth, H. D., 1991: Evolutionary relationships of avian *Eimeria* spp. among other apicomplexan protozoa: monophyly of the *Apicomplexa* is supported. – Molec. Biol. Evol. **8**: 345–355.

Beckers, C. J. M., Roos, D. S., Donald, R. G. K., Luft, B. J., Schwab, J. C., Cao, Y., Joiner, K. A., 1995: Inhibition of cytoplasmic and organellar protein synthesis in *Toxoplasma gondii*: implications for the target of macrolide antibiotics. – J. Clin. Invest. **95**: 367–376.

Bhattacharya, D., Helmchen, T., Melkonian, M., 1995: Molecular evolutionary analyses of nuclear-encoded small subunit ribosomal RNA identify an independent rhizopod lineage containing the *Euglyphidae* and the *Chlorarachniophyta*. – J. Euk. Microbiol. **42**: 65–69.

Borst, P., Overdlve, J. P., Weijers, P. J., Fase-Fowler, F., Berg, M. V. D., 1984: DNA circles with cruciforms from *Isospora* (*Toxoplasma*) *gondii*. – Biochim. Biophys. Acta **781**: 100–111.

Bradley, D. J., Warhurst, D. C., 1995: Malaria prophylaxis: guidelines for travellers from Britian. – Brit. Med. J. **310**: 709–714.

Brown, P., 1994: Guarded welcome for malaria vaccine. – New Scientist 1994: 14–15.

Brugerolle, G., Mignot, J. P., 1979: Observations sur le cycle i'ultrastructure et la position systematique de *Spiromonas perforans* (*Bodo perforans Hollande* 1938), flagelle parasite de *Chilomonas paramaecium*: ses relations avec les dinoflagelles et sporozoaires. – Protistologica **15**: 183–196.

Buttner, D. W., 1967: Elektronmicroscopische Studien der Vermehrung von *Theileria parva* im Rind. – Z. Tropenmed. Parasit. **18**: 245–268.

– 1968: Vergleichende Untersuchung der Feinstruktur von *Babesia gibsoni* und *Babesia canis*. – Z. Tropenmed. Parasit. **19**: 330–342.

Bzik, D. J., 1991: The structure and role of RNA polymerases in *Plasmodium*. – Parasitol. Today **7**: 211–214.

Cavalier-Smith, T., 1995a: Zooflagellate phylogeny and classification. – Cytology **37**: 1010–1029.

– 1995b: Membrane heredity, symbiogenesis, and the multiple origins of algae. – In Arai, R., Kato, M., Doi, Y., (Eds): Biodiversity and evolution, pp. 75–114. – Tokyo: National Science Museum.

– Couch, J., Thorsteinsen, K., Gilson, P., Deane, J., Hill, D., McFadden, G. I., 1996: Cryptomonad nuclear and nucleomorph 18S rRNA phylogeny. – Eur. J. Phycol. (in press).

Cermakian, N., Ikeda, T. M., Cedergren, R., Gray, M. W., 1996: Sequences homologous to yeast mitochondrial and bacteriophage T3 and T7 RNA polymerases are widespread throughout the eukaryotic lineage. – Nucl. Acids. Res. **24**: 648–654.

Colley, F. C., 1967: Fine structure of sporozoites of *Eimeria nieschulzi*. – J. Protozool. **14**: 217–220.

Cook, G. C., 1995: Mefloquine toxicity should limit its use to treatment alone. – Brit. Med. J. **311**: 190-191.

Creasey, A., Mendis, K., Carlton, J., Williamson, D., Wilson, I., Carter, R., 1994: Maternal inheritance of extrachromosomal DNA in malaria parasites. – Molec. Biochem. Parasitol. **65**: 95–98.

CyanoBase, 1996: The genome database for *Synechocystis* sp. strain PCC 6803 – http:// www. kazusa. or. jp/cyano/cyano. html

Dennis, D. T., 1989: Fatty acid biosynthesis in plastids. – In Boyer, C. D., Shannon, J. C., Hardison, R. C., (Eds): Physiology, biochemistry, and genetics of nongreen plastids, pp. 120–129 – Rockville: The American Society of Plant Physiologists.

– Miernyk, J. A., 1982: Compartmentation of nonphotosynthetic carbohydrate metabolism. – Annu Rev. Pl. Physiol. **33**: 27–50.

De Pamphilis, C. W., Palmer, J. D., 1989: Evolution and function of plastid DNA: a review with special reference to nonphotosynthetic plants. – In Boyer, C. D., Shannon, J. C.,

HARDISON, R. C., (Eds): Physiology, biochemistry, and genetics of nongreen plastids, pp. 182–202. – Rockville: The American Society of Plant Physiologists.

– 1990: Loss of photosynthetic and chlororespiratory genes from the plastid genome of a parasitic flowering plant. – Nature **348**: 337–339.

DOOLAN, D. L., GOOD, M. F., 1993: Developing a malaria sporozoite vaccine. – Today's Life Science **5**: 18–19.

DORE, E., FRONTALI, C., FORTE, T., FRATARCANGELI, S., 1983: Further studies and electron microscopic characterization of *Plasmodium berghei* DNA. – Molec. Biochem. Parasitol. **8**: 339–352.

DOUCE, R., ALBAN, C., JOURNET, E.-P., JOYARD, J., 1989: Biochemical properties of plastids isolated from cauliflower buds and sycamore cells and of their envelope membranes. – In BOYER, C. D., SHANNON, J. C., HARDISON, R. C., (Eds): Physiology, biochemistry, and genetics of nongreen plastids, pp. 99–119. Rockville: The American Society of Plant Physiologists.

DOUGLAS, S. E., 1994: Chloroplast origins and evolution. – In BRYANT, D. A., (Ed.): The molecular biology of *Cyanobacteria*, pp. 97–118. Dordrecht: Kluwer Academic Press.

DUBREMETZ, J. F., 1995: *Toxoplasma gondii*: cell biology update. – In BOOTHROYD, J. C., KOMUNIECKI, R. (Ed.): Molecular approaches to parasitology, pp. 345–358. – New York: Wiley-Liss.

EGEA, N., LANG-UNNASCH, N., 1995: Phylogeny of the large extrachromosomal DNA of organisms in the phylum *Apicomplexa*. – J. Euk. Microbiol. **42**: 679–684.

ESCALANTE, A. A., AYALA, F. J., 1995: Evolutionary origin of *Plasmodium* and other *Apicomplexa* based on rRNA genes. – Proc. Natl. Acad. Sci. USA **92**: 5793–5797.

FEAGIN, J. E., 1992: The 6 kb element of *Plasmodium falciparum* encodes mitochondrial cytochrome genes. – Molec. Biochem. Parasitol. **52**: 145–148.

– 1994: The extrachromosomal DNAs of apicomplexan parasites. – Annu. Rev. Microbiol. **48**: 81–104.

– 1995: Exploring the organelle genomes of malaria parasites – In BOOTHROYD, J. C., KOMUNIECKI, R. (Eds): Molecular approaches to parasitology, pp. 163–177. – New York: Wiley-Liss.

– DREW, M. E., 1995: *Plasmodium falciparum*: alterations in organelle transcript abundance during the erythrocytic cycle. – Exper. Parasitol. **80**: 430–440.

– GARDNER, M. J., WILLIAMSON, D. H., WILSON, R. J., 1991: The putative mitochondrial genome of *Plasmodium falciparum*. – J. Protozool. **38**: 243–245.

– WERNER, E., GARDNER, M. J., WILLIAMSON, D. H., WILSON, R. J., 1992: Homologies between the contiguous and fragmented rRNAs of the two *Plasmodium falciparum* extrachromosomal DNAs are limited to core sequences. – Nucl. Acids Res. **20**: 879–87.

FRASER, J. D., GOCAYNE, O., WHITE, M. D., ADAMS, R. A., CLAYTON, R. D., VENTER, C. J., 1995: The minimal gene complement of *Mycoplasma genitalium*. – Science **270**: 397–403.

FLEISCHMANN, R. D., ADAMS, M. D., WHITE, O., CLAYTON, R. A., KIRKNESS, E. F., KERLAVAGE, A. R., BULT, C. J., TOMB, J. F., DOUGHERTY, B. A., MERRICK, J. M., VENTER, C. J., 1995: Whole-genome random sequencing and assembly of *Haemophilus influenzae* Rd. – Science **269**: 496–512.

FOISSNER, W., FOISSNER, I., 1984: First record of an ectoparasitic flagellate on ciliates. – Protistologica **20**: 635–648.

GAGNON, S., LEVESQUE, R. C., SOGIN, M. L., GAJADHAR, A. A., 1993: Molecular cloning, complete sequence of the small subunit ribosomal RNA coding region and phylogeny of *Toxoplasma gondii*. – Molec. Biochem. Parasitol. **60**: 145–148.

GAJADHAR, A. A., MARQUARDT, W. C., HALL, R., GUNDERSON, J., ARIZTIA, C. E. V., SOGIN, M. L., 1991: Ribosomal RNA sequences of *Sarcocystis muris, Theileria annulata* and

Crypthecodinium cohnii reveal evolutionary relationships among apicomplexans, dinoflagellates, and ciliates. – Molec. Biochem. Parasitol. **45**: 147–154.

Gardner, M. J., Bates, P. A., Ling, I. T., Moore, D. J., McCready, S., Gunasekera, M. B. R., Wilson, R. J. M., Williamson, D. H., 1988: Mitochondrial DNA of the human malarial parasite *Plasmodium falciparum*. – Molec. Biochem. Parasitol. **31**: 11–18.

– Feagin, J. E., Moore, D. J., Spencer, D. F., Gray, M. W., Williamson, D. H., Wilson, R. J., 1991a: Organization and expression of small subunit ribosomal RNA genes encoded by a 35-kilobase circular DNA in *Plasmodium falciparum*. – Molec. Biochem. Parasitol. **48**: 77–88.

– Williamson, D. H., Wilson, R. J. M., 1991b: A circular DNA in malaria parasites encodes an RNA polymerase like that of prokaryotes and chloroplasts. – Molec. Biochem. Parasitol. **44**: 115–124.

– Feagin, J. E., Moore, D. J., Rangachari, K., Williamson, D. H., Wilson, R. J., 1993: Sequence and organization of large subunit rRNA genes from the extrachromosomal 35 kb circular DNA of the malaria parasite *Plasmodium falciparum*. – Nucl. Acids Res. **21**: 1067–1071.

– Preiser, P., Rangachari, K., Moore, D., Feagin, J. E., Williamson, D. H., Wilson, R. J., 1994a: Nine duplicated tRNA genes on the plastid-like DNA of the malaria parasite *Plasmodium falciparum*. – Gene **144**: 307–8.

– Goldman, N., Barnett, P., Moore, P. W., Rangachari, K., Strath, M., Whyte, A., Williamson, D. H., Wilson, R. J., 1994b: Phylogenetic analysis of the rpoB gene from the plastid-like DNA of *Plasmodium falciparum*. – Molec. Biochem. Parasitol. **66**: 221–231.

Goff, L. J., 1982: The biology of parasitic red algae. – Prog. Phycol. Res. **1**: 289–369.

Gozar, M.M.G., Bagnara, A. S., 1993: Identification of a *Babesia bovis* gene with homology to the small subunit ribosomal RNA gene from the 35-kilobase circular DNA of *Plasmodium falciparum*. – Int. J. parasitol. **23**: 145–148.

– – 1995: An organelle-like small subunit ribosomal RNA gene from *Babesia bovis*: nucleotide sequence, secondary structure of the transcript and preliminary phylogenetic analysis. Int. J. Parasitol. **25**: 929–938.

Gray, M., 1992: Origin and evolution of organelle genomes. – Curr. Opinion Genet. Developm. **3**: 884–890.

Grimm, B., 1990: Primary structure of a key enzyme in plant tetrapyrrole synthesis: glutamate 1-semialdehyde aminotransferase. – Proc. Natl. Acad. Sci. USA **87**: 4169–4173.

Hackstein, J. H., Mackenstedt, U., Mehlhorn, H., Meijerine, J. P., Schubert, H., Leunissen, J. A., 1995: Parasitic apicomplexans harbor a chlorophyll a-D1 complex, the potential target for therapeutic triazines. Parasitol. Res. **81**: 207–16.

Harada, T., Ishikawa, R., Nizeki, M., Saito, K.-I., 1992: Pollen-derived rice calli that have larger deletions in plastid DNA do not require protein synthesis in plastids for growth. – Molec. Gen. Genet. **233**: 145–150.

Harder, A., Haberkorn, A., 1989: Possible mode of action of toltrazuril: studies on two *Eimeria* species and mammalian and *Ascaris suum* enzymes. – Parasitol. Res. **76**: 8–12.

Harris, E. H., Boynton, J. E., Gillham, N. W., 1994: Chloroplast ribosomes and protein synthesis. – Microbiol. Rev. **58**: 700–754.

Hart, A. S., Ridigner, M. T., Soundarajan, R., Peters, C. S., Swiatlo, A. L., Kocka, F. E., 1990: Novel organism associated with chronic diarrhoea in AIDS. – Lancet **335**: 169–170.

Hien, T. T., Day, N. P. J., Phu, N. H., Mai, N. T. H., Chau, T. T. H., Loc, P. P., Sinh, D. X., Chuong, L.V., Vinh, H., Waller, D., Peto, T. E. A., White, N. J., 1996: A controlled trial of artemether or quinine in vietnamese adults with sever falciparum malaria. – New Engl. J. Med. **335**: 76–83.

HOGE, C. W., SHLIM, D. R., RAJAH, R., TRIPLETT, J., SHEAR, M., RABOLD, J. G., ECHEVERRIA, P., 1993: Epidemiology of diarrhoeal illness associated with coccidian-like organism among travellers and foreign residents in Nepal. – Lancet **341**: 1175–1179.

HOWE, C. J., 1992: Plastid origin of an extrachromosomal DNA molecule from *Plasmodium*, the causative agent of malaria. – J. Theor. Biol. **158**: 199–205.

– SMITH, A. G., 1991: Plants without chlorophyll. – Nature **349**: 109.

JEFFRIES, A. C., JOHNSON, A. M., 1996: The growing importance of the plastid-like DNAs of the *Apicomplexa*. – Int. J. Parasitol. **26**: 1139–1150.

JOSEPH, J. T., ALDRITT, S. M., UNNASCH, T., PUIJALON, O., WIRTH, D. F., 1989: Characterization of a conserved extrachromosomal element isolated from the avian malarial parasite *Plasmodium gallinaceum*. – Molec. Cell. Biol. **9**: 3621–3629.

KANNANGARA, C. G., GOUGH, S. P., BRUYANT, P., HOOBER, J. K., KAHN, A., WETTSTEIN, D. V., 1988: $tRNA^{Glu}$ as a cofactor in δ-aminolevulinate biosynthesis: steps that regulate chlorophyll synthesis. – Trends Biochem. Sci. **13**: 139–143.

KAPLAN, W., 1978: Protothecosis and infections caused by morphologically similar green algae. – Pan American Health Organization Scientific Publication **356**: 218–232.

KEELING, P. J., R. L., DOOLITTLE, W. F., 1996: Alpha-tubulin from early diverging eukaryotic lineages and the evolution of the tubulin family. – Molec. Biol. Evol. **13**: 1297–1305.

– CHARLEBOIS, R. L., DOOLITTLE, W. F., 1994: Archaebacterial genomes: eubacterial form and eukaryotic content. – Curr. Biol. **4**: 816–822.

KEPKA, O., SCHOLTYSECK, E., 1970: Weitere Untersuchungen der Feinstruktur von *Frenkelia* spec. (=M-Organismus, *Sporozoa*). – Protistologica **6**: 249–266.

KILEJIAN, A., 1975: Circular mitochondrial DNA from the avian malarial parasite *Plasmodium lophurae*. – Biochim. Biophys. Acta **390**: 276–284.

KLEINSCHUSTER, S. J., PERKINS, F. O., DYKSTRA, M. J., SWINK, S. L., 1994: The in vitro life cycle of a *Perkinsus* species (*Apicomplexa*, *Perkinsidae*) isolated from *Macoma balthica* (*Linneaus* 1758). – J. Shellfish Res. **13**: 461–465.

KONISHI, T., SASAKI, Y., 1994: Compartmentalization of two forms of acetyl-CoA carboxylase in plants and the origin of their tolerance towards herbicides. – Proc. Natl. Acad. Sci. USA **91**: 3598–3601.

KOWALLIK, K. V., STOEBE, B., SCHAFFRAN, I., KROTH-PANCIC, P., FREIER, U., 1995: The chloroplast genome of a chlorophyll a + c containing alga, *Odontella sinensis*. – Pl. Molec. Biol. Reporter **13**: 336–342.

KUIJT, J., 1996: The biology of parasitic flowering plants. Berkeley: University of California Press.

KUMAR, S., RZHETSKY, A., 1996: Evolutionary relationships of eukaryotic kingdoms. – J. Molec. Evol. **42**: 183–193.

LEA, P. J., MILLS, W. R., WALLSGROVE, R. M., MIFLIN, B. J., 1989: Assimilation of nitrogen and synthesis of amino acids in chloroplasts and cyanobacteria (blue-green algae). – In SCHIFF, J. A., (Ed.): On the origins of chloroplasts, pp. 149–178. – Amsterdam: Elsevier North Holland.

LEVINE, N. P., 1988: The protozoan phylum *Apicomplexa*. – Boca Raton: CRC Press.

LIM, S. H., WITTY, M., WALLACE, C. A. D. M., ILAG, L. I., SMITH, A. G., 1994: Porphobilinogen deaminase is encoded by a single gene in *Arbidopsis thaliana* and is targeted to the chloroplasts. – Pl. Molec. Biol. **26**: 863–872.

LINDSAY, D. S., RIPPEY, N. S., TOIVIO-KINNUCAN, M. A., BLAGBURN, B. L., 1995: Ultrastructural effects of diclazuril against *Toxoplasma gondii* and investigations of a diclazuril-resistant mutant. – J. Parasitol. **81**: 459–466.

Long, E. G., White, E. H., Carmichael, W. W., Quinslisk, P. M., Raja, R., Swisher, B. L., Daugharty, H., Cohen, M. T., 1991: Morphologic and staining characteristics of a cyanobacterium-like organism associated with diarrhea. – J. Infect. Dis. **164**: 199–202.
– Ebrahimzadeh, A., White, E. H., Swisher, B. L., Callaway, C. S., 1990: Alga associated with diarrhea in patients with acquired immunodeficiency syndrome and in travelers. – J. Clin. Microbiol. **28**: 1101–1104.
MacDonald, C. M., Darbyshire, J. F., 1977: The morphology of a soil flagellate, *Spiromonas angsta* (Dij.) Alexeieff (*Mastigophorea*: *Protozoa*). – Protistologica **13**: 441–450.
McConkey, G. A., Rogers, M. J., McCutchan, T. F., 1997: Inhibition of *Plasmodium falciparum* protein synthesis: targeting the plastid-like organelle with thiostreptin. – J. Biol. Chem. (in press).
McFadden, G., 1991: Molecular cytology goes ultrastructural. – In Hall, J., Hawes, C., (Eds): Electron microscopy of plant cells, pp. 219–255. – London: Academic Press.
– 1993: Second-hand chloroplasts: evolution of cryptomonad algae. – Adv. Bot. Res. **19**: 189–230.
– Gilson, P., 1995: Something borrowed, something green: lateral transfer of chloroplasts by secondary endosymbiosis. – Trends Ecol. Evol. **10**: 12–17.
– – Hill, D., 1994: *Goniomonas*: rRNA sequences indicate that this phagotrophic flagellate is a close relative of the host component of cryptomonads. – Eur. J. Phycol. **29**: 29–32.
– Reith, M. E., Mulholland, J., Lang-Unnasch, N., 1996: Plastid in human parasites. – Nature **381**: 482.
Melhorn, H., Schein, E., 1984: The Piroplasms: life cycle and sexual stages. – Adv. Parasitol. **23**: 37–103.
– Ortmann-Flakenstein, G., Haberkorn, A., 1984: The effects of sym. triazines on development of *Eimeria tenella, E. maxima* and *E. acervulina*: a light and electron microscopical study. – Z. Parasitenk. **70**: 173–182.
Mock, H. P., Trainotti, L., Kruse, E., Grimm, B., 1995: Isolation, sequencing and expression of cDNA sequences encoding uroporphyrinogen decarboxylase from tobacco and barley. – Pl. Molec. Biol. **28**: 245–256.
Myl'nikov, A. P., 1991: Ultrastructure and biology of certain representatives of the order *Spiromonadida* (*Protozoa*). – Zool. Zhurn. **7**: 5–15.
Nelissen, J., van de Peer, Y., Wilmotte, A., Wachter, R. D., 1995: An early origin of plastids within the cyanobacterial divergence is suggested by evolutionary trees based on complete 16S rRNA sequences. – Molec. Biol. Evol. **12**: 1166–1173.
Ogino, N., Yoneda, C., 1996: The fine structure and mode of division of *Toxoplasma gondii*. – Arch. Ophthal. **75**: 218–227.
Olliaro, P. L., Goldberg, D. E., 1995: The *Plasmodium* digestive vacuole: metabolic headquarters and choice drug target. – Parasitol. Today **11**: 294–297.
Ortega, Y. R., Sterling, C. R., Gilman, R. H., Cama, V. A., Diaz, F., 1993: *Cyclospora* species-new protozoan pathogen of humans. – New Engl. J. Med. **328**: 1308–1312.
Palmer, J. D., 1992a: Green ancestry of malarial parasites? – Curr. Biol. **2**: 318–320.
– 1992b: Comparison of chloroplasts and mitochondrial genome evolution in plants. – In Hermann, H., (Ed.): Plant gene research. Cell organelles, pp. 137–163. – Wien: Springer.
– Delwiche, C. F., 1996: Second-hand chloroplasts and the case of the disappearing nucleus. – Proc. Natl. Acad. Sci. USA **93**: 7432–7435.
Pawlovskki, J., Bolivar, I., Fahrni, J. F., Cavalier-Smith, T., Gouy, M., 1996: Early origin of foraminifera suggested by SSU rRNA gene sequences. – Molec. Biol. Evol. **13**: 445–450.

PERKINS, F. O., 1969: Ultrastructure of vegetative stages in *Labyrinthomyxa marina* (=*Dermocystidium marinum*), a commercially significant oyster pathogen. – J. Invert. Pathol. **13**: 199–222.
– 1976: Zoospores of the oyster pathogen, *Dermocystidium marinum*. I. Fine structure of the conoid and other sporozoan-like organelles – J. Parasitol. **62**: 959–974.
– MENZEL, R.W., 1967: Ultrastructure of sporulation in the oyster pathogen *Dermocystidium marinum*. – J. Invert. Pathol. **9**: 205–229.
PIETROKOVSKI, S., 1994: Conserved sequence feature of inteins (protein introns) and their use in identifying new inteins and related proteins. – Protein Sci. **3**: 2340–2350.
PORCHET-HENNERE, E., 1972: Observations en mictoscopie photonique et electronique sur la sporogenese de *Dehornia* (1) *sthenelais* (n. gen., sp. n.), sporozoaire parasite de I'annelide polychete *Sthenelais boa* (*Aphroditides*). – Protistologica **8**: 245–255.
– RICHARD, A., 1970: Ultrastructure des stades végétatifs d' *Aggregata eberthi Labbé*: le trophozoite et le schizonte. – Z. Zellforsch. **103**: 179–191.
PREISER, P., WILLIAMSON, D. H., WILSON, R. J., 1995: tRNA genes transcribed from the plastid-like DNA of *Plasmodium falciparum*. – Nucl. Acids Res. **23**: 4329–4336.
PUKRITTAYAKAMEE, S., VIRAVAN, C., CHAROENLARP, P., YEAMPUT, C., WILSON, R. J., WHITE, N. J., 1994: Antimalarial effects of rifampin in *Plasmodium vivax* malaria. – Antimicrob. Agents Chemother. **38**: 511–514.
RADCHENKO, A. I., 1987: *Sarcocystis muris* (*Sporozoa, Apicomplexa*): the mode of division of the intermediate cell in the cyst as revealed by electron microscope. – Tsitologiia **29**: 404–409.
RAVENTOS-SUAREZ, C., POLLACK, S., NAGEL, R., 1982: *Plasmodium falciparum*: inhibition of in vitro growth by desferrioxamine. – Amer. J. Trop. Med. Hyg. **31**: 919–922.
REARDON, E. M., PRICE, C. A., 1995: Plastid genomes of three non-green algae are sequenced. – Pl. Molec. Biol. Reporter **13**: 320–326.
– TURANO, F. J., WEISEMANN, J. M., WILSON, B. J., MATTHEWS, B. F., 1989: Amino acid biosynthesis and nitrogen assimilation in higher plants. In: BOYER, C. D., SHANNON, J. C., HARDISON, R. C., (Eds.): Physiology, biochemistry, and genetics of nongreen plastids, pp. 130–140. – Rockville: The American Society of Plant Physiologists.
REITH, M. E., MUNHOLLAND, J., 1993: A high resolution gene map of the chloroplast genome of the red alga *Porphyra purpurea*. – Pl. Cell **5**: 465–475.
RELMAN, D. A., SCHMIDT, T. M., GARJADAR, A., SOGIN, M., CROSS, J., YODER, K., SEHTABUTR, O., ECHEVERRIA, P., 1996: Molecular phylogenetic analysis of *Cyclospora*, the human intestinal pathogen, suggests that it is closely related to *Eimeria* species. – J. Infect. Dis. **173**: 440–445.
ROBERTS, W. L., HAMMOND, D. M., ANDERSON, L. C., SPEER, C. A., 1970: Ultrastructural study of schizogony in *Eimeria callospermophili* and *E. Iarmerensis*. – J. Protozool. **17**: 584–592.
RUDZINSKA, M. A., TRAGER, W., 1962: Intracellular phagotrophy in *Babesia rodhaini* as revealed by electron microscopy. – J. Protozool. **9**: 279–288.
– VICKERMAN, K., 1968: The fine structure – In WEINMAN, D., RISTIC, M., (Eds): Infectious blood diseases of man and animals: Diseases caused by *Protista*. I, pp. 217–306. – New York: Academic Press.
RUSH, M. G., MISRA, R., 1985: Extrachromosomal DNA in eukaryotes. – Plasmid **14**: 177–191.
SADLER, L. A., MCNALLY, K. L., GOVIND, N. S., BRUNK, C. F., TRENCH, R. K., 1992: The nucleotide sequence of the small subunit ribosomal RNA gene from *Symbiodinium pilosum*, a symbiotic dinoflagellate. – Curr. Genet. **21**: 409–416.
SCHOLTYSECK, E., 1979: Fine structure of the parasitic protozoa: an atlas of micrographs, drawings and diagrams. – Berlin: Springer.

– Piekarski, G. P., 1965: Electronenmikroscopische Untersuchungen an Merozoiten von Eimerien (*Eimeria perforans* und *E. stiedae*) und *Toxoplama gondii* zur systematischen Stellung von *T. gondii*. – Z. Parasiten. **26**: 91–115.

Schrevel, J., 1971: Contribution a l'étude des *Selenidiidae* parasites d'annélides polychétes II. ultrastructure de quelques trophozoïtes. Protistologica **7**: 101–130.

Sheffield, H. G., 1966: Electron microscope study of the proliferative form of *Besnoitia jellisoni*. – J. Parasitol. **52**: 583–594.

– Melton, M. L., 1968: The fine structure and reproduction of *Toxoplasma gondii*. – J. Parasitol. **54**: 209–226.

Siddall, M. E., 1992: Hohlzylinders. – Parasitol. Today **8**: 90–91.

– Stokes, N. A., Burreson, E. M., 1995: Molecular phylogenetic evidence that the phylum *Haplosporida* has an alveolate ancestry. – Molec. Biol. Evol. **12**: 573–581.

Silberman, J. D., Sogin, M. L., Leipe, D. D., Clark, C. G., 1996: Human parasite finds taxonomic home. – Nature **380**: 398.

Simpson, A. F. B., Patterson, D. J., 1996: Ultrastructure and identification of the predatory flagellate *Colpodella pugnax Cienkowski* (*Apicomplexa*) with a description of *Colpodella turpis* n. sp. and a review of the genus. – Syst. Parasitol. **33**: 187–198.

Slater, A. F. G., Cerami, A., 1992: Inhibition by chloroquine of a novel haem polymerase enzyme activity in malaria trophozoites. – Nature **355**: 167–169.

Small, I., Suffolk, R., Leaver, C. J., 1989: Evolution of plant mitochondrial genomes via substoichiometric intermediates. – Cell **58**: 69–76.

Sterling, C., DeGiusti, D., 1972: Ultrastructural aspects of schizogony, mature schizonts, and merozoites of *Haemoproteus metchnikovi*. – J. Parasitol. **58**: 641–652.

Strath, M., Scott, F. T., Gardner, M., Williamson, D., Wilson, I., 1993: Antimalarial activity of rifampicin in vitro and in rodent models. – Trans Roy. Soc. Trop. Med. Hyg. **87**: 211–216.

Surolia, N., Padmanaban, G., 1992: De novo biosynthesis of heme offers a new chemotherapeutic target in the human malarial parasite. – Biochem. Biophys. Res Com. **187**: 744–750.

Taylor, F. J. R., 1990: Phylum *Dinoflagellata*. – In Margulis, L., Corliss, J. O., Melkonian, M., Chapman, D. J., (Eds): Handbook of *Protoctista*, pp. 549–573. –Boston: Jones & Bartlett.

Vaidya, A. B., Akella, R., Suplick, K., 1989: Sequences similar to genes for two mitochondrial proteins and portions of ribosomal RNA in tandemly arrayed 6-kilobase-pair DNA of a malarial parasite. – Molec. Biochem. Parasitol. **35**: 97–108.

Vanhensbroek, M. B., Onyiorah, E., Jaffar, S., Schneider, G., Palmer, A., Frenkel, J., Enwere, G., Forck, S., Nusmeijer, A., Bennett, S., Greenwood, B., Kwiatkowski, D., 1996: A trial of arthemeeter or quinine in children with cerebral malaria. – New Engl. J. Med. **335**: 69–75.

Van de Peer, Y., Rensing, S., Maier, U.-G., Wachter, R. D., 1996: Substitution rate calibration of small subunit rRNA identifies chlorarachniophyte endosymbionts as remnants of green algae. – Proc. Natl. Acad. Sci. USA **93**: 7732–7736.

Van der Zypen, E., Piekarski, G., 1967: Ultrastrukturelle Unterschiede zwischen der sog. Proliferationsform (RH-Stamm, BK-Stamm) und dem sog. Cysten-Stadium (DX-Stamm) von *Toxoplasma gondii*. Zentralbl. Bakteriol. Parasitenk. **203**: 495–517.

Vivier, E., Desportes, I., 1990: *Apicomplexa*. – In Margulis, L., Corliss, J. O., Melkonian, M., Chapman, D. J., (Eds): Handbook of *Protoctista*, pp. 549–573. – Boston: Jones & Bartlett.

– Hennere, E., 1965: Ultrastructure des stades végétatifs de la coccidie *Coelotropha durchoni*. – Protistologica **1**: 89–104.

– PETITPREZ, A., 1969: Observations ultrastructurales sur l'hematozoaire *Anthesoma garnhami* et examen de critéres morphologiques utilisables pour la taxonomie chez les sporozoaires. – Protistologica **5**: 363–379.

– PETITPREZ, A., LANDAU, I., 1972: Observations ultrastructurales sur la sporoblastogenése de l'hémogregarine, *Hepatozoon domerguei*, Coccide *Adeleidea*. – Protistologica **8**: 315–334.

WALSH, M. A., RECHEL, E. A., POPOVICH, T. M., 1980: Observations of plastid fine structure in the holoparasitic angiosperm *Epifagus virginiana*. – Amer. J. Bot. **67**: 833–837.

WHATLEY, J. M., 1977: Variation in the basic pathway of chloroplast development. – New Phytol. **78**: 407–420.

WILDFÜHR, W., 1966: Elektronenmikroskopische Untersuchungen zur Morphologie and Reproduktion von *Toxoplama gondii*. – Zentralbl Bakteriol. I. Abt. orig. **201**: 110–130.

WILLIAMSON, D. H., WILSON, R. J. M., BATES, R. A., MCCREADY, S., PERLER, R., QIANG, B.-U., 1985: Nuclear and mitochondrial DNA of the primate parasite *Plasmodium knowlesi*. – Molec. Biochem. Parasitol. **14**: 199–209.

– GARDNER, M. J., PREISHER, P., MOORE, D. J., RANGACHARI, K., WILSON, R. J. M., 1994: The evolutionary origin of the 35 kb circular DNA of *Plasmodium falciparum*: new evidence supports a possible rhodophyte ancestry. – Molec. Gen. Genet. **243**: 249–252.

– PREISER, P., WILSON, R. J. M., 1996: Organelle DNAs: the bit players in malaria parasite DNA replication. – Parasitol. Today **12**: 357–362.

WILSON, C. M., SMITH, A.B., BAYLON, R. V., 1996: Characterization of the delta-aminolevulinate synthase gene homologue in *Plasmodium falciparum*. – Molec. Biochem. Parasitol. **75**: 271–275.

WILSON, I., 1993: Plastids better red than dead. – Nature **366**: 638.

WILSON, R. J. M., GARDNER, M. J., FEAGIN, J. E., WILLIAMSON, D. H., 1991: Have malaria parasites three genomes? – Parasitol. Today **7**: 134–136.

– FRY, M., GARDNER, M. J., FEAGIN, J. E., WILLIAMSON, D. H., 1992: Subcellular fractionation of the two organelle DNAs of malaria parasites – Curr. Genet. **21**: 405–408.

– GARDNER, M. J., RANGACHARI, K., WILLIAMSON, D. H., 1993: Extrachromosomal DNAs in the *Apicomplexa*. – In SMITH, J. E., (Ed.): Toxoplasmosis, NATO ASI Series H**78**: 51–60. – Berlin: Springer.

– DENNY, P. W., PREISSER, P. R., RANGACHARI, K., ROBERTS, K., ROY, A., WHYTE, A., STRATH, M., MOORE, D. J., MOORE, P. W., WILLIAMSON, D. H., 1996: Complete gene map of the plastid-like DNA of the malaria parasite *Plasmodium falciparum*. – J. Molec. Biol. **261**: 155–172.

– WILLIAMSON, D. H., PREISER, P., 1994: Malaria and other Apicomplexans: the "plant" connection. – Infect. Agents Disease **3**: 29–37.

WOLFE, K. H., MORDEN, C. W., PALMER, J., 1992: Function and evolution of a minimal plastid genome from a nonphotosynthetic parasitic plant. Proc. Natl. Acad. Sci. USA **89**: 10648–10652.

WOLTERS, J., 1991: The troublesome parasites: Molecular and morphological evidence that *Apicomplexa* belong to the dinoflagellate-ciliate clade. – BioSystems **25**: 75–84.

WOODS, K. M., NESTERENKO, M. V., UPTON, S. J., 1996: Efficacy of 101 antimicrobials and other agents on the development of *Cryptosporidium parvum* in vitro. – Ann. Trop. Med. Parasitol. **90**: 603–615.

Springer-Verlag
and the Environment